Triumph of Love

A Mathematical Exploration of Being, Becoming, Life, and
Transhumanism in a Cosmology of Light

Pravir Malik, Ph.D.

ISBN-13: 978-1-7342743-1-8

"The Spirit shall look out through Matter's gaze
And Matter shall reveal the Spirit's face."

(Sri Aurobindo, *Savitri*)

Author's Introduction

Broadly speaking transhumanism refers to a philosophy whose focus is to move beyond current human limits. This book suggests an approach to transhumanism based on a perception of Light elaborated in the previous nine books in the Cosmology of Light book series.

In this perception light is envisioned as existing at different constant speeds. We know that in our physical universe light travels at a constant speed referred to as 'c', of 186,000 miles per second. This has a concrete bearing on how we experience time and space, and on how matter arises, and in a Cosmology of Light it can be said that light at c is an intentionality to filter infinite potentiality from Light's native state of traveling infinitely fast, so life can arise in a particular way.

So, imagine light traveling infinitely fast. This would suggest another reality packed with infinite possibility due to Light's omnipresence, omnipotence, omniscience, and omninurturing fourfoldness. Omnipresence, because Light would be instantaneously present in whatever volume considered. Omnipotence, because anything not of the nature of light will be overcome by it. Omniscience, because light being everywhere will know what arises or disappears in it. Omninurturing, because all would exist in one nature and be bound by it.

As the previous books have elaborated, and as this book will adequately summarize, a cosmology arises when we consider the simultaneous and interpenetrating realities created when Light exists at multiple constant speeds simultaneously. The dynamics and information resident in each of these simultaneous realities is arbitrated into material reality through a constant and persistent quantum-level computation that generates genetic-type information that effectively becomes "law". Quanta in such a cosmology is perceived as being the mechanism by which information in faster-moving layers of light is materialized in a slower-moving layer of light. "Law" manifests as quadrumvirate mechanisms such as space-time-energy-gravity, the electro-magnetic-wavearchetype-masspotential spectrum, quark-lepton-boson-Higgsboson particles, 's'shell-'p'shell'-'d'shell-'f'shell atoms, nucleicacid-lipid-protein-polysaccharide cells, amongst other mechanisms, that derive their ability to "become" based on the native "being" of Light's omnipresent-omnipotent-omniscient-omninurturing fourfoldness. It may even be said that such "becomings" resulting in quadrumvirate-based law, reveal essential "beings". Hence a

11

being or a particular kind of taxonomic classification akin to the species-genus-family-order-class-phylum-kingdom-domain as in the hierarchy of life, is revealed with successful becoming. Plate 1, that follows shortly, summarizes such a classification by way of a contemplative map of being. Hence Light in its native state, traveling with infinite speed, may be thought of as sitting at the top of such a taxonomic hierarchy generating light-based domains, kingdoms, phyla, classes, and so on.

But, such fourfoldness that manifests as effective law, or through the becoming involving persistent quantum-level computation that reveals a being, is none other than a triumph of love. For it is only the power of love, of that innate need to maintain the integrality of light's fourfoldness even as it continues to materialize, that can be the foundation of a sustainable becoming. The more powerful such love, so that in any materialization there is integration not only of the fourfoldness of Light's implicit properties, but also of an integration of the many layers of light existing at different speeds, the more fully will a becoming be founded on completeness of potentiality in light to itself become a being capable of engendering light-based life. Hence, as will be suggested in this book, not only does a particular 'type' of being engender vast variation within that type, but further, beings can combine with beings, which is essentially an act of love, to create more comprehensive beings resulting in all the complexity of life. Life itself will be seen to originate in the native state of Light, adding function and variation as light precipitates into more and more material reality emerging as a fullness of life.

At the base of all possible variation and advancement in being, becoming, and life, is the ability to influence the process of the persistent quantum-level computation so that there is a more complete horizontal and vertical integration of Light - a more complete light-based-singularity as it were - and therefore of the output of genetic-type information and consequently of laws that are active in a being's becoming. This is what is considered to be the basis of a system or framework of transhumanism in a cosmology of light.

Pravir Malik,
San Francisco

While this book builds on the previous nine books in the Cosmology of Light book series, it is complete in itself, and therefore necessarily denser and more compact. The mathematics in the Cosmology of Light (COL) book series is fractal and holographic, as any universal mathematics should be, and so the same fundamental formulations are leveraged to model different domains. Readers of the previous COL books will notice familiar mathematical themes named differently so as to apply to the current domain of being, becoming, life, and transhumanism.

The Contemplative Map and the Section Overview that follow, prior to the main materials of this book, should be referred to often as you get further in the book. These will serve as a reality check and help to summarize concepts and the gist of the overall flow of the book.

Section 1 is the only non-mathematical section of the book and serves to conceptualize key ideas that will be elaborated mathematically through the rest of the book.

The mathematics in this book serves as an artistic expression to attempt to grasp in more compact form the wonder that exists in light. Hence, while such traditional mathematical structures as sets, linear algebra, calculus, and imaginary numbers are used, the alert reader will also notice that the mathematical formulation sometimes, in keeping with being used as a medium for artistic expression, deviates from, or suggests something that may be considered new or unusual.

For that matter, many suggestions and modeling that are traditionally considered to be part of physics, chemistry, biology, astrophysics, sciences in general, mythology, sociology, anthropology, and history as well, amongst other areas, have been combined or perceived differently in this book, following the primary inspiration of the possible multi-dimensional and information-dense reality of light.

CONTEMPLATIVE MAP OF BEING & SINGULARITIES IN A COSMOLOGY OF LIGHT

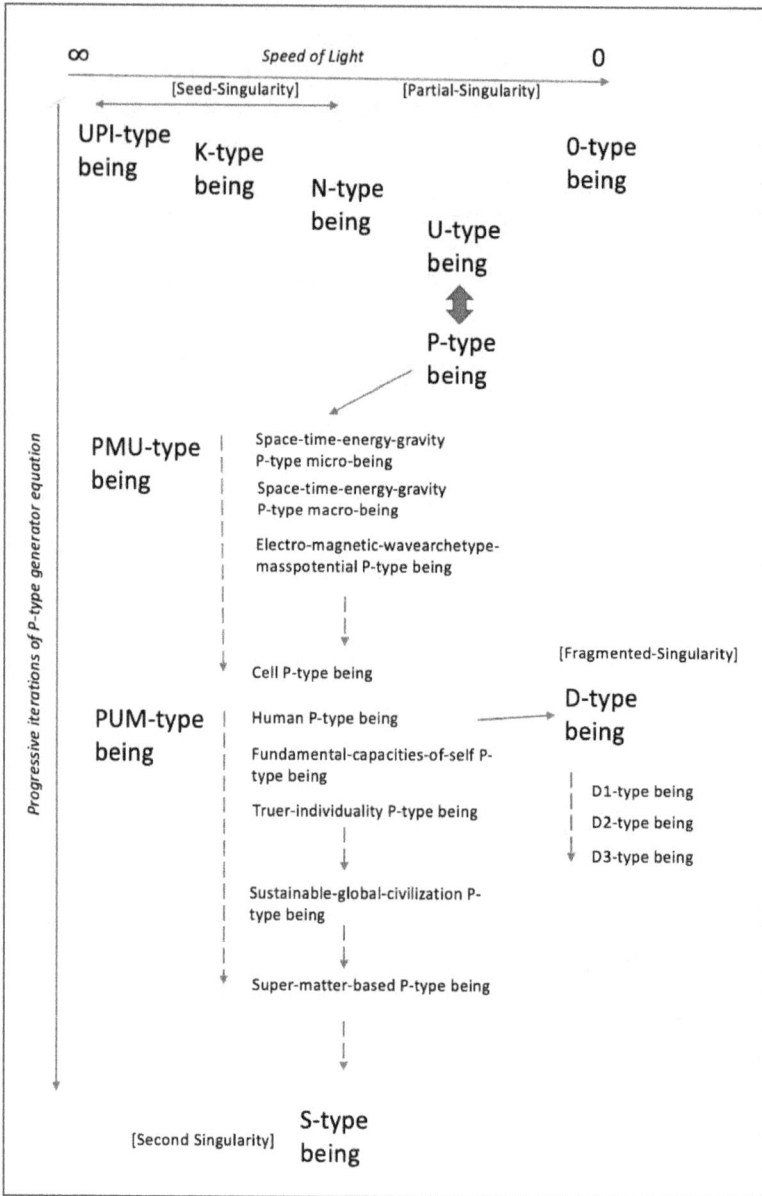

Plate 1: Contemplative Map of Being & Singularities in a Cosmology of Light

SECTION OVERVIEW

Section 1, Fundamental Concepts, explores fundamental concepts necessary to understanding light-based singularities. These concepts are initially presented in non-mathematical form. In following Sections these concepts will be elaborated with mathematical notation and be weaved together into a comprehensive mathematical framework.

Hence, Chapter 1.1 explores the impact that light has on the nature of reality. In particular the impact of light traveling at potentially different speeds will be examined for its ability to create realities based on its speed. Quanta are positioned as being the bridge between one layer of light and another slower moving layer of light. Further the Big Bang and a pervasive light-based-singularity superstructure are posited as existing as a result of the slowing down of Light. The information codified in Light will be positioned as being the origin of genetics. Quantum-level computations, generation of genetic-type information, entanglement, and superposition are positioned as being the base dynamics of any light-based-singularity.

Chapter 1.2 explores a theory for the creation of quanta related to the speed of light. Quanta are intimately associated with the light-based-singularity superstructure and acts as a doorway, as it were, into the multi-layered structure. Further, time, space, matter, and gravity being emergent from Light are also positioned as requiring quantization to be expressed.

Chapter 1.3 explores the relationship between Light and Love. Essentially the oneness of light seeks to maintain that oneness even as it materializes. This maintenance of the nature of oneness is nothing other than Love. The chapter begins to trace concrete examples where materialized quadrumvirates starting with past-present-future-matter and space-time-energy-gravity and running through lipids-polysaccharides-proteins-nucleic_acids are each examples of actions of Love. Such horizontal integrity creates stability of becoming, resulting in a new being, while also enabling a vertical integrity that results in the enrichment of Life.

15

Chapter 1.4 summarizes the high-level aspects of the composite light-based-singularity superstructure in which the seed-singularity, partial-singularities, and the Second Singularity occur and positions this as being the matrix that gives rise to the process of becoming, the different types of beings, and the phenomenon of life, in a Cosmology of Light.

Section 2, Modeling the Mathematical Structure of Being & The Becoming of the P-Type Being, explores the mathematical structure of being, and the becoming of the P-type being in a cosmology of light.

There are five representative levels of light that are illustrated in this section, that are themselves determined by the play of four ubiquitous bases or properties of light. These levels spawn five types of beings corresponding to each layer of light. While the number of levels or types of beings may potentially be infinite, the five levels seem to represent important and perhaps even fundamental quantum-based shifts or "quantizations", as light projects itself at slower and slower speeds to progressively reveal materially more of the vast information implicit in it. The quantized appearance of a level relates to a fundamental genre of emergence or being replete with different kinds of dynamics. Note that "information" relates to the four fundamental properties intuited to occur in Light.

Chapter 2.1 explores the Light-Matrix or the Light-Based-Singularity Superstructure, comprising of the five representative levels of light, that is also the source of the types of beings that are envisioned to exist in our universe. The light-based-singularity superstructure as suggested in the previous section, is such because the seed-singularity, partial-singularities, and the Second Singularity occur in it. The light-based-singularity superstructure gives rise to the first apparent layer or type of being of the "ubiquitous-point-instant" and to subsequent layers culminating in the layer that results in the vast diversity of life. Each layer, and the subtle-libraries of information they automatically spawn, thereby materializing a set of laws that the corresponding being will generally be subject to, is then explored in more detail in subsequent chapters through the ubiquitous-point-instant or UPI-type being (Chapter 2.2), the architectural forces or the K-type being (Chapter 2.3), and uniqueness of organization or the N-type being (Chapter 2.4).

The fourth layer in the Light-Matrix is created by light traveling at speed c and is the foundation for the U-type being, where 'U' suggests 'untransformed'. This layer, hence, is where material reality emerges. But the nature of the emergences will themselves be influenced by conditions that are prevalent in space and time. It is the combined dynamics of the seed-singularity, comprised of the three antecedent layers of light, with the play of the emergences in space and time arbitrated through a persistent quantum-level computation that outputs genetic-type information, that constitutes the process of becoming, also detailing the dynamics of any partial-singularity and emergence of the P-type being, where 'P' stands for 'partial-singularity'. The P-type being is hence an instance of the U-type being that has iterated through a process of becoming to reveal a stability of some aspect of Light that originated in pure potentiality in Light at its native state.

Chapter 2.5 suggests a mathematical basis for a process of becoming involving mutations for the P-type being. To be clear, the very basis of mutation derives from the seed-singularity itself, as will be elaborated. This mutational sequence thereby also suggests the conditions by which a partial-singularity or the P-type being could eventually evolve into the Second Singularity or the S-type being, where 'S' stands for 'Second Singularity'. Chapter 2.6 attempts a broader systematization of the process of becoming of the P-type being.

The mathematical foundations explored in Section 2 will set up the basis for the creation and dynamics of chains of four-base ecosystems that encode logic and provide insight into the biography of becoming of partial-singularities or the P-type being, leading also to the culmination in the S-type being, as explored in Section 3.

The UPI-type, K-type, and N-type beings exist in the seed-singularity portion of the downward-strand of subtle-DNA. Each of these beings is related to layers of light traveling faster than c. As light slows down to c the material layer U comes into existence. Dynamics at U are the start of the becoming of the vast range of the P-type being. The downward-strand is caused by light slowing down in quantized-decelerations, as it were, progressively concretizing more of the information in light. The becoming of beings in this downward-strand are therefore envisioned as precise "quantized" accumulations architected to bring about unfolding possibilities in our material universe. Other sets of decelerated-quantizations may serve other purposes and create other kinds of universes that would in effect exist simultaneously with ours. As such the UPI-type, K-type, and N-type beings are envisioned to play a significant role in any effort at transhumanism.

The upward-strand is envisioned as prescribing a time-variable sequence wholly determined by the nature of the beings in the downward-strand. The time-variability is due to the result of the interplay of the active influences of the UPI-type, K-type, and N-type beings in the downward-strand. But further, as already alluded to, each of these beings spawn subtle-libraries of pre-genetic information as part of their dynamics. These subtle-libraries are effectively infinitely large, encoding many different possibilities. Such genetic-type information is the language in which biographies of light-based singularities and all light-based beings, including the P-type being, are written.

Section 3, Further Elaboration of the Mathematics of the Becoming of the P-Type Being, will illustrate some of the interplay involving the subtle-libraries and a first projection of their possibilities in a space-time-energy-gravity quadrumvirate construct. The space-time-energy-gravity construct, itself one of the first instances of the P-type being, is positioned as being fundamental in altering any genetic information at the material level. Further, numerous other constructs that have their logic determined in 'four-base logic-encoding ecosystems' (FBLEE), envisioned as existing in the quantum layer antecedent to the material layer, will also be explored in subsequent sections. The process of becoming of the P-type being, hence, involves encoding at the quantum level as a prerequisite to further materialization of becoming. It is such ecosystems that can change due to interplay with the material layer. In the interplay or call from 'below' in which emergences find that their forms are inadequate in expressing the fullness of Light, as it were, there can be thought to be a response from 'above', from more of the fullness that Light is, that may precipitate one of the infinite functions already created by the subtle-libraries that are part of the UPI-type, K-type, and N-type beings. Such a call and response system can be thought of as operating like a lock-and-key mechanism.

As such, this section will reinterpret some basic equations and existing quantum mechanics fundamentals based on the Cosmology of Light view adopted in this treatise. Further, it will focus on the derivation of the Light-Space-Time Emergence or P-Type Being Generator equation

18

leveraged in subsequent quantization analyses. The Light-Space-Time Emergence equation models the basis for quantization suggesting emergent reality for all phenomena from Light. Schrodinger's wave equation and Heisenberg's uncertainty principle are also interpreted from the point of view of the Light-based Interpretation of quantum mechanics central to this treatise, to reinforce the notion that even when considered from these points of view the existence of multiple layers of light is feasible. The chapter on quantization of space, time, matter and gravity, models how these phenomena are related to Light and subsequently also models how these fundamentals work together to potentially impact any genetic-type information.

This section therefore elaborates dynamics central to the becoming of the P-type being.

Section 4, Overview of Generation of Light-Based-Singularity Beings, briefly summarizes the generation of different types of beings and some of the associated genetic-type code resulting from the persistent Light-Space-Time Matrix based computations ranging from an era pre-dating the Big Bang to modern day Global Civilization, and beyond.

The implications of this are that all light-based beings are imbued with code. It is just the way the code is housed that changes. This also implies that there is likely code within all living cells, as an example, that has to do with space, time, energy, gravity, the functioning of the electromagnetic spectrum, and all the stages of material elaboration that precede the emergence of a living cell. But this has to be, in a process of love, in which emergent P-type beings give themselves or are the foundation for the subsequent emergences of P-type beings with greater spheres of influence.

It is the sharing of such large swathes of code that precede an emergent form, with the emergent form, that typifies the notion of a light-based-singularity. In fact, it is a "singularity" because the most current emergent form has all previous code embedded or available to it thereby inherently abiding with all "laws" that have thus far emerged in previous singularities. This means that so long as an emergence is a light-based-singularity it inherently follows the law of love in that more and more of the expressed nature of light is integrated into a single, more-material whole. This also implies the notion that the seed-singularity itself will

19

have gone through a process of becoming articulated by a progressive biography and resulting in partial-seed-singularities along the way, until the seed-singularity has itself emerged.

Section 5, Generation of the Electro-Magnetic-Wavearchetype-Masspotential P-Type Being, will explore the computation involved in the creation of the electro-magnetic-wavearchetype-masspotential P-type being and its associated pre-genetic, material-fabric code.

The electromagnetic spectrum is a technical and contemporary way to refer to light. So, as light becomes more concrete to us or as the possibilities within it begin to emerge, one of the first forms it takes is as the electromagnetic spectrum. Note that while the generation of photons, the carrier particle for the electromagnetic spectrum, is usually associated with excitation of electrons in an atom, in this treatise light itself has a more fundamental place, being the very matrix from which everything else emerges. As explored in sections 1 and 2, the Matrix of Being is constituted by light traveling at different speeds. The fact that these speeds are constant is significant and implies an intentionality, that is itself being comprehensively explored in the Cosmology of Light book series. That said, it must be the case that the previously surfaced properties of light – Presence, Power, Knowledge, and Harmony – emerge so as to define the very architecture of the electromagnetic spectrum.

Chapter 5.1, Generation of Electro-Magnetic-Wavearchetype-Masspotential P-Type Being, summarizes the emergence of the electromagnetic spectrum in terms of the underlying Light-Space-Time Emergence equation and the process of quantization that must occur to create the logic of the electromagnetic spectrum ecosystem that precipitates into the material-fabric. So we find that the four underlying properties of Light that we call Harmony, Knowledge, Power, and Presence are of the essence of the speed with which the electromagnetic spectrum moves, the wave-range within the electromagnetic spectrum, the energy-gradient within the electromagnetic spectrum, and the mass-possibilities due to the electromagnetic spectrum, respectively. Further the description – electromagnetic – seems to have captured the Power-Harmony aspects implicit in light. In reality the electro-magnetic spectrum can likely be more completely described as electro-magnetic-wavearchetype-masspotential spectrum.

In **Section 6, Generation of P-Type Beings in the Surfacing of Matter,** we will find that layers of matter – quantum particles, which include bosons, and atoms – are also structured or emerge along the same property-lines or property-families of Light. There is a continuous process of computation that involves the quantum-realms and quantization to create the realities of quantum particles, including bosons, and atoms, to generate matter-based P-type beings with its associated material-fabric, pre-genetic code.

However, there is an apparent difference in the way matter materializes, in contrast to the electro-magnetic-wavearchetype-masspotential P-type being. Chapter 6.1 explores this in greater detail. Chapters, 6.2, 6.3, and 6.4, then, describe a process of computation by which the quantum particle, boson, and atom P-type beings and their associated material-fabric, pre-genetic code are generated respectively.

Section 7, Generation of P-Type Beings in the Surfacing of Life, explores the quantum-level computation and the generation of genetic code that is associated with the emergence of life through the cell, complex human attributes such as thoughts and feelings, and uniqueness of individuality. As suggested genetic code articulates a biography, and in this case, the emerging biography depicting an increase in the P-type being's sphere of influence through the emergence of additional light-generated "laws".

So far, the action of the Light-Space-Time Emergence has generated pre-genetic code. After the Big Bang this pre-genetic code is housed in the material-fabric, which has a universal action on all constructs arising in the universe. The emergence of matter precipitates a phenomenon of material containerization. This is hypothesized as being due to an essential action of the space-time-energy-gravity quadrumvirate P-type macro-being contained within a smaller "space" as a result of the number of driver seeds.

It is assumed that just as the space-time-energy-gravity quadrumvirate P-type micro and macro beings have an impact on every material emergence, so too will the action of the electro-magnetic-wavearchetype-masspotential P-type being have an impact on every emergence of life. Some dynamics of this possibility are explored in Chapter 7.1.

Chapter 7.2 will explore the generation of the genetic code responsible for the emergence and complexification of four-foldness through the primary molecular plans of nucleic acids, proteins, lipids, and polysaccharides as manifest in the body of the cell P-type being.

Chapter 7.3 will relate key human attributes of sensations, urges, desires, wills, feelings, emotions, and thought to the continuing journey of fourfold complexification and suggest that genetic code is also generated to support these attributes as manifest in the body of the fundamental-capacities-of-self P-type being.

Chapter 7.4 will relate truer individuality to the fourfold properties implicit in Light and suggest the generation of the genetic code tied to this as manifest in the body of the truer-individuality P-type being.

Note that the dynamics of the generation of the fundamental-capacities-of-self P-type being as covered in Chapter 7.3, and the generation of the truer-individuality P-type being as covered in Chapter 7.4, is central to envisioning any path toward transhumanism. This is because it is with the integration of any light-based-singularity – be it a seed-singularity or partial-singularity based being – as opposed to a fragmented-singularity, which appears to be the current trend in thinking, that transhumanism will have its most fruitful foundation. This has to be since a light-based singularity culminating in the emergence of the S-type being is nothing other than the triumphant expression of the infinite formative forces of light and love.

In **Section 8, Generation of P-Type Beings in the Surfacing of Complex Organization,** we turn our attention to study the computation resulting in the generation of P-type beings related to complex organizations.

In the previous sections we have seen that there is a gradual addition to P-type being biographies as pre-genetic and then genetic code capturing more complex fourfold-functionality is generated. In these biographies the condition for constructive mutation is that some meta-level is primarily active. When this happens, there is a possibility that the existing four-base logic-encoding ecosystems (FBLEE) are enhanced or that new ones are created. Note that it is such FBLEE action that is envisioned to alter the quantum-level interface between the material and antecedent layers of light in effect changing the basis by which matter materializes. In other words, FBLEE changes what matter can be.

But at its heart it is good to get clear that it is the complexification of four-foldness that is the driver of changes to FBLEE, and therefore of possible changes to the material-fabric, the genetic substance of living cells, and of any post-genetic substance that were to arise. This is so because such complexification essentially necessitates that existing habits, of any kind, are continually broken and in the process can admit of greater and more sweeping actions of light from the meta-levels. All of light is ever-present but requires the material receptacle to alter in order that it may precipitate more of its infinite possibility. It is such action that makes clear how the emergence of complex organization actually links to changes such as to matter and to the make-up of genetic content.

Such a line of development also makes clear how being, becoming, life itself, and therefore possible transhumanism is intimately related to larger forms that we create. The path to transhumanism is clear and apparent from this point of view. There are no additional technologies that need to be developed. There are no minds that need to be uploaded into a silicon-based computing infrastructure. Instead we have to become more present to what is already before us. We need to change our habitual reactions to things and master ourselves and our relationship with things, so that in this process more of the infinite possibility in light can effectively materialize to even change the very bases of matter through the action of FBLEE. The path to any transhumanism is therefore already staring at us in the face. We have only to recognize this, grasp it, and let Light play more fully, materially, so that the infinity in it can firmly express itself through us.

Chapter 8.1, therefore, examines the generation of the stable-mega-organization FBLEE-based P-type being. Similarly, Chapter 8.2 examines the generation of the FBLEE-based sustainable-global-civilization P-type being. In both these cases the nature of the P-type being is FBLEE-based, as opposed to being both FBLEE and MF-based. This is so because both the stable mega-organization and the sustainable global civilization are in process of being worked out. Their becoming realities means that there is a clear material and therefore genetic-type foundation within the being, so that in effect these forms of collectivity are nothing other than extensions of being. Hence there is a post-human structure and possibly a post-cell genetic-type structure to house their respective ecosystem logic materially that will need to be worked out. FBLEE is the realm where human aspiration meets with a response from functions and libraries of possibility in meta-levels of light, and gets housed in logic, that then becomes available to the human being. The precipitation of such logic alters matter, and it becomes clearer through the generation of such P-type beings that house such enhanced genetic-type information that in its reality matter is nothing other than a crystallization of possibilities of light, and will one day house light so much more completely that it will be entirely different from what we perceive it as being at present time.

Such post-cell genetic structure, as will be further explored in Section 9 on super-matter-based P-type beings, requires a much higher threshold of functional richness to come into material reality.

Also, to be noted is that the generation of P-type beings is a variable process, meaning it is not inevitable that particular kinds of P-type beings be developed. This variability implies that in a Cosmology of Light in which light-based-singularities are the foundation for materialization of infinite possibility in matter, there is a variable path to transhumanism. The important thing is the increase in functional richness dues to a deep urge or aspiration or willing as the case may be.

Section 9, Extrapolation to Super-Matter-Based P-Type Beings, examines micro-level dynamics due to the action of the space-time-energy-gravity micro-being leading to meta-level induced change in specific biographies, and macro-implications that will result in possible creation of super-matter.

To summarize, sections 2 and 3 proposed a mathematical framework for being and becoming. Sections 4 through 8 leveraged the mathematical framework to explore the biographies of seed-singularity based and P-type beings. Such biographies are entirely generated by key aspects of the light-space-time emergence equation. Components of such biographies include the essential architecture of the P-type being generated from the four principal characteristics of light - which is an action of the UPI-type being, the four sets of properties related to the principal characteristics of light - which is an action of the K-type being, unique seeds created from combinations of elements derived from the four sets - which is an action of the N-type being, and the essential dynamics of maturation, or becoming, culminating in possible quantization. Hence, in the biography of any P-type being there is active participation by all seed-singularity based beings – the UPI-type being, the K-type being, and the N-type being - as summarized by different levels of light in the Light-Matrix, and by the Space-Matrix and Time-Matrix that detail the conditions of becoming.

While the previous sections elaborated the broad lines of emergent biographies of the P-type being, this section will go a step further and examine micro-level dynamics due to the action of the space-time-energy-gravity micro-being, and Chapter 9.1 will look at insect adaptation leveraging a discussion on cells in Chapter 7.2, and Chapter 9.2 will look at sustainable global civilization adaptation leveraging a discussion in Chapter 8.2, leading to meta-level induced change in specific biographies, and macro-implications that will result in possible creation of super-matter in Chapter 9.3.

Super-matter is suggested to be that type of matter created through the intervention of conscious will or cohesive want. While complexification of matter as summarized in previous sections is often an "automatic" process of Nature, super-matter is a foundation based on will or cohesive want and sets the stage for a potentially unending willed development in which functional-richness existing in Light in its native state traveling at infinite speed, can manifest in this material universe. From a macro-level, meta-level bases or more conscious integration with seed-singularity beings in an increasing triumph of love, is what will ensure the development of super-matter.

Creation of such super-matter is tied to the meaning of space, and to cosmic observations of the expansion of the universe (Chapter 9.4), hence involving the space-time-energy-gravity macro-being. Chapter 9.4 hence focuses on the nature of matter by drawing on observations from the field of astrophysics.

The gestalt of such an analysis may engender a visceral sense of the intimacy and oneness of the light-based adventure that results in the vastness of the cosmos tied to human-precipitated determinable dynamics at the quantum-level. As such the persistent quantum-level computation resulting in a constant stream of genetic-type information alters the code and the very light-based computational machinery emergent as more and more sophisticated pre-matter, matter, and post or super-matter structures. Genetics then has to be viewed more broadly as also integrating instructions or insights from our larger and larger collective endeavors to hook or connect us more intimately with the world around us. It is foreseeable that such genetic-type information expands the action of nucleic acids, proteins, lipids, and polysaccharides to even allow the cell and therefore a body to act or morph in unforeseen ways. The boundaries between the world outside and the world inside may thin and dissolve, and all life progressively become one unbroken material though intensely diverse totality precipitating the oneness and infinity of light. Matter then, or super-matter, becomes the medium in which this grand adventure of light is experienced. Any transhumanism traces the steps by which the adventure is pursued.

Section 10, Triumph of Love and the Emergence of the S-Type Being, summarizes the journey of light-based-singularities through to the generation of the S-type being. The S-type being is envisioned as being the result of an increasing and more comprehensive unity of light along both the vertical and horizontal dimensions.

In its automaticity, starting with the emergence of the space-time-energy-gravity beings through to the cell P-type being, horizontal unity is ensured, also revealing that the deep essence of light is nothing other than love. But as emergence approaches the human P-type being, conscious choice and will become more important and emergence of the truer-individuality P-type being and beyond requires a more conscious opening to meta-layers of light. In such an opening the deeper essence of love become more active and gradually as this becomes stronger there is a greater and deeper yearning or even a surrendering to the possibilities embodied in the meta-layers of light, that allows a more and more profound unity to take place along the vertical dimension.

The increasing dynamism of such a comprehensive unity, a triumph of love, culminates in a foundation for the emergence of the S-type being.

Chapter 10.1 summarizes the unbroken continuity that exists in the P-type being starting from the space-time-energy-gravity quadrumvirate beings, to the emergence of super-matter-based P-type beings. The latter is seen as a prerequisite for the emergence of the S-type being.

Chapter 10.2 leverages the method of contrasts by focusing on fragmented-singularities, exploring a hypothetical creation generated from space, time, energy, and gravity alone, and the resulting creation and nature of D-type or derivative beings.

Chapter 10.3 explores the use of a set of mathematical operators in expediting the arrival of the S-type being. These operators are non-exhaustive, but rather are indicative of the nature of dynamics that will increasingly arise in the extrapolation of P-type beings to super-matter-based P-type beings and beyond. The use of such operators assists in perceiving reality as derivate from Light, and hence their active adoption will accelerate the overcoming of limiting patterns and subsequently the integration of the multiple layers of light in any light-based-singularity.

Chapter 10.4, which focuses on altering the microcosmic-macrocosmic balance, examines on-going conditions for the generation of the S-type being. A key condition is the audacity of the human-species made possible because the human is the result of 30 orders of microcosmic magnitude implicit to it. Fields, quanta, atoms, cells, the incredible four-fold emergences of properties codified in Light have all contributed to the emergence of the human. Now in a possible reversal of causality the human appears to be positioned to similarly influence 30 orders of magnitude explicit to it and even structure the further development of the macrocosm by mastery of the microcosm.

Chapter 10.5 reviews key insights that have surfaced through tracing the emergence of the series of light-based beings culminating in the emergence of the S-type being. driven by enlightened heart, will, and mind dynamics.

Chapter 10.6 summarizes the mythology in the cosmology of light, highlighting themes of darkness, light, sacrifice, priests, gods, that culminate in the S-type being and an increasing triumph of love.

SECTION 1: FUNDAMENTAL CONCEPTS

Section 1 explores fundamental concepts necessary to understanding light-based singularities. These concepts are initially presented in non-mathematical form. In following Sections these concepts will be elaborated with mathematical notation and be weaved together into a comprehensive mathematical framework.

Hence, Chapter 1.1 explores the impact that light has on the nature of reality. In particular the impact of light traveling at potentially different speeds will be examined for its ability to create realities based on its speed. Quanta are positioned as being the bridge between one layer of light and another slower moving layer of light. Further the Big Bang and a pervasive light-based-singularity superstructure are posited as existing as a result of the slowing down of Light. The information codified in Light will be positioned as being the origin of genetics. Quantum-level computations, generation of genetic-type information, entanglement, and superposition are positioned as being the base dynamics of any light-based-singularity.

Chapter 1.2 explores a theory for the creation of quanta related to the speed of light. Quanta are intimately associated with the light-based-singularity superstructure and acts as a doorway, as it were, into the multi-layered structure. Further, time, space, matter, and gravity being emergent from Light are also positioned as requiring quantization to be expressed.

Chapter 1.3 explores the relationship between Light and Love. Essentially the oneness of light seeks to maintain that oneness even as it materializes. This maintenance of the nature of oneness is nothing other than Love. The chapter begins to trace concrete examples where materialized quadrumvirates starting with past-present-future-matter and space-time-energy-gravity and running through lipids-polysaccharides-proteins-nucleic_acids are each examples of actions of Love. Such horizontal integrity creates stability of becoming, resulting in a new being, while also enabling a vertical integrity that results in the enrichment of Life.

Chapter 1.4 summarizes the high-level aspects of the composite light-based-singularity superstructure in which the seed-singularity, partial-singularities, and the Second Singularity occur and positions this as being the matrix that gives rise to the process of becoming, the different types of beings, and the phenomenon of life, in a Cosmology of Light.

Chapter 1.1: Light and the Realities it Creates

This chapter will look at the impact light has on creating practical realities.

Light @Infinity and Quanta

Let us imagine that light can travel infinitely fast. Think about a big area or volume with a light source at the center. Now since the light travels infinitely fast it will fill up the entire volume instantly. This will be true no matter how large the volume is – it could be the entire solar system, an entire galaxy, or an entire universe. So that light is going to be instantaneously present everywhere – it is going to be omnipresent.

A further thought experiment may give insight into such omnipresence by considering the night sky. It is because of the finite speed of light that the night sky has only spots of light, however many, across it. But if light were traveling infinitely fast then the night sky would be a canvas or a sea of brilliant white since the light from every corner of the universe, wherever a light source or star existed, would immediately be present. It would appear, in such a thought experiment, that we existed in a sea of light.

Now, since the light is already present everywhere in whatever volume, that is, since that light has already filled up the entire volume, there is nothing else that can similarly arise there that is not of the nature of light. Even if something else were to arise, being surrounded by light it would eventually succumb to that light. So, the light is all-powerful or omnipotent within that volume.

Further, since the light exists simultaneously everywhere in that space and has a complete knowledge of itself, it therefore has a complete knowledge of that space or of anything that can arise in that space. So, it is all-knowing or omniscient in that space.

Finally, the light connects everything together instantaneously and holds these connections and the things connected in its nature, so it is all-nurturing or omninurturing within that space.

So, as a result of the infinite speed of light, it appears that light will have properties of omnipresence, omnipotence, omniscience, and omninurturance in such a space. Such properties also imply that light has infinite information or potentiality within it.

This book will take the position that Light's native state is in fact such a reality where it travels infinitely fast. This native state is the foundation of existence and in order to materialize the infinite potentiality within it light projects itself at slower speeds. In the next Section we will construct a minimalistic mathematical model that will allow the unity of light traveling at infinite speed to express itself as infinite material diversity. The key mechanism that allows infinite possibility to express itself materially is 'quanta'.

Quanta have to be understood as the bridge or device that tie different layers of light together. Quanta will show up only in slower-moving layers of light. This is because they serve the purpose of gathering possibility from a faster moving layer of light in order to materialize it. Quanta can be thought of as light itself, but light packetizing itself in order to parcel out some aspects of the infinite potentiality contained in Light at its native state.

Further, the slower speed of light implies that light will need to travel some unit distance in order to be expressed. This is so because light at a slower speed is slower because it is materializing something. There is no other reason for it to be slower. We can assume that light traveling slower than at an infinite speed is in fact laying the foundation for reality that emerges at that speed of light.

This unit distance that light must travel in order to express itself is related to such quanta, and to distance called the 'Planck-distance' reinforcing an inverse relationship between the speed of light and this unit of quantum. A universe, therefore, will arise with the speed of light and its related quantum expressing some fundamental upper and lower bound in that existence respectively. This notion will be further explored in more detail subsequently.

Light@c and the Reality it Creates

'c' is the notation for light traveling at a constant speed of approximately 186,000 miles per second in vacuum. This is the speed of light in our universe. The fact that light travels with speed 'c' implies that quanta is

30

fundamental to this universe, likely related to there being a particular kind of materialization that must occur in this universe.

Such materialization expresses itself as the fundamental building block in quantum particles, and subsequently in atoms. Another way of saying this is that all matter, hence, is the result of light traveling at c.

But what else is implicit or made possible because of the finite speed of light? Imagine traveling on a ray of light from the sun to the earth. Imagine that you are in minute 4 of the approximately 8-minute journey. As you look back you will see that 4 minutes in the past you were at the sun. 4 minutes in the future you will be at the earth. And in the present moment you are somewhere between the sun and the earth. So, this limited speed of light already creates the concept of time and specifically of the past, the present, and the future.

So, four incredibly fundamental things are created because of the finite speed of light: matter, the past, the present, and the future. It could be said, therefore, that this matter-based time-experienced universe is a result of the finite speed of light at c.

Looking at this in another way, it is known too that in a denser medium the speed of light further slows down. Hence, light travels at some fraction of c when moving through water for instance. So by reversing the logic of such a process this may show that if light slows down from an infinite speed to some other lesser speed whether a multiple or a fraction of c, then the material reality will have to alter, perhaps in a similar way as the material reality between vacuum and water is different.

The Big Bang, Being, and Becoming

We hear about the Big Bang as the start of the universe. In this Big Bang matter is created. But in the examinations just presented the creation of matter is nothing other than the result of light traveling at the finite speed, and in this universe at 186,000 miles per second in vacuum. So, we can say that the Big Bang, the apparent start of the universe, is the result of a slowing down of light from an infinite speed to some finite speed.

When light slows down then some energy or information existing in Light traveling at a speed greater than c accumulates in packets or quanta and this results in what was inexpressible being able to express itself in more material form.

As an analogy, this can also be thought of in terms of an incredibly rapidly moving stream of water. If the water is traveling so fast, then no boundary will be able to contain it and the energy will be continuous over the length of the stream. If it is traveling slower though, then the water will be able to be held by boundaries along the length of the stream. The energy in this case will be discontinuous and will appear in "packets". In other words, there is a process by which information implicit in light traveling infinitely fast is rearranged into a more and more material basis as light travels slower. This rearrangement or elaboration of implicit information is also the origin of genetics as explored in a previous book on genetics (Malik, 2019a).

In this process of slowing down, the implicit omnipresent-omnipotent-omniscient-omninurturing nature of light becomes or transforms into an explicit or emergent matter-past-present-future nature of light. So, there is implicit in this transformation also a high degree of 'entanglement' as it were. That is information exists in a different and highly connected manner. Ab initio, everything that appears in this universe is highly entangled. In other words, every fundamental particle, whether it be a boson or lepton, or any subsequent emergence of matter is already deeply entangled by virtue of having emanated from the same single starting-point conceptualized as the Big Bang.

This process of transformation creates a light-based-singularity superstructure. This light-based-singularity superstructure has implicit in it, what will be referred to and explained in greater detail in Section 2: the seed-singularity. Such light-based-singularities, which includes the seed-singularity, and what will subsequently be introduced as a partial-singularity, is also the basis of "being". Such beings can be thought of as having universality and presence in a range of operations that will be described subsequently. In mythology such beings may be termed as 'gods' and 'goddesses'. In a cosmology of light such beings are not mythological though but are the result of the play and possibilities of Light as will be discovered in subsequent sections.

The dynamics of a resulting universe are contained in the dynamics of such a light-based-singularity superstructure. The light-based-singularity superstructure provides insight into the subtle-structure which houses genetic-type information, and the dynamic of entanglement, that provides insight into the universal availability of such antecedent or subtle genetic-type information. The light-based-singularity superstructure is also the basis of an ongoing quantum-level computation. That is, constant or persistent quantum-level computation within the light-based-singularity superstructure is what generates genetic-type information and propels any subsequent materialization. This ongoing quantum-level computation is therefore the basis of "becoming". The process of becoming also arrives at a partial-singularity, and any partial-singularity in turn reveals a being. The notion of partial-singularities is perhaps similar to Gebser's notion of the various mutations that are considered to be partial manifestations of a single 'ever-present origin' (Gebser, 1986). Such a cycle of becoming and being must also be the basis of any transhumanism in a cosmology of light.

We have arisen in the field of light that has a finite speed. It is difficult therefore to feel the reality of the omnipresent-omnipotent-omniscient-omninurturing light. We are attuned to perceiving in the material world that has resulted due to the finite speed of light. But if we could step back into the fullness of light in its pure state at infinity, from there we would likely see that there can be different universes created as a result of light selectively slowing down to different levels.

Note that there have been experiments to slow down light so that it practically moves at a snail's pace. This slowing down has to be put into context. Even if light were to slow down to 1 mile per second, say, from 186,000 miles per second, keep in mind that it is possible that this change in speed is likely only a miniscule fraction of the change in speed from light traveling infinitely fast to light traveling at c, that is, 186,000 miles per second. So, in effect what could be said is that when light is made to slow down in experiments, the range or band that it slows down to deviates only slightly from its relative-to-infinity slower speed in vacuum.

Necessity of Constant Speed of Light and Variable Time and Space

The constant speed of light, at a speed less than an infinite speed, allows a buildup of energy at quantum levels, that in turn allows any properties or possibility in light to express itself materially. When light travels at the speed of c – 186,000 miles per second – then the result of that is the material universe as we see and experience it, also with its division of time into a past, a present, and a future. In fact, we could say that for the known

33

universe, as we experience it, light had to travel at c for it to arise. The speed of light had to be constant else there would be a variable, fluid reality to matter, likely displaying barely forming islands of matter subject to sudden disappearance, reappearance, continuing ad infinitum.

But also, since the distance traveled by an object or a ray is the result of the speed it is traveling at for the time involved, this gives us an insight that Einstein based his Theory of Relativity on (Einstein, 1995). Since the speed of light is constant and has to be for the universe to be in its observable stable condition, this means time and distance (or space) potentially can vary. So, if an object is traveling very fast, then time and space are going to be experienced differently by it, as compared with an object that is traveling considerably slower: as the speed of an object approaches c, time slows down and distance contracts.

So, an object that manages to travel at a speed of c will in some sense partake of the reality as experienced by c. It will transcend the conventions of time and space that are set up because of c and experience these differently.

Light at Speed Close to Zero

On the flipside imagine light traveling at a speed close to or approaching zero. In this case the experienced reality is going to be the opposite of reality as experienced when light is traveling at an infinite speed.

First of all, light emanating from any point will remain fundamentally isolated and become the basis of extreme fragmentation. This has to be since regardless of the amount of elapsed time the light will still be only at its point of origin or source. So instead of a Presence of Light in any considered volume, there will be an Absence of Light.

Further, in imagining an area or volume, such light will have no knowledge of what is going on in the volume, and hence be completely ignorant, will further, have no power to effect any circumstances and hence be completely powerless, and finally, will not be able to coordinate or be present in any relationships. Hence it will be detrimental to or agnostic of any relationships instead of harmonious or nurturing to relationships.

34

Hence, reality in such a scenario would be stark, dark, desolate, fruitless, completely fragmented, and even perhaps pointless.

Such a reality may represent a negative-infinity and may be thought of as an extreme limit of what may be possible in worlds in which Light projects itself.

Seen from this point of view a Big Bang would be a fortuitous event since it will allow a fruitful reintegration of different worlds, each created by light traveling or projected at a different speed, to perhaps begin to take place as will be discussed further.

Impact of Superposition in a Light-Based Model

Light traveling at different speeds can, as illustrated by the cases of an infinite speed, c, and 0, create different realities. But if the infinite speed case is the native state, and if every other case is a projected state, then it is possible that these realities all exist simultaneously. In other words, phenomena resident in the reality created by the native state and by multiple projected states can be superposed. This also implies that categorization of genetic-type information and interrelation of genetic-type information will be superposed.

Further, as will be explored in detail in subsequent chapters such states of superposition must be considered as superposition because they are inherently related to each other. A phenomenon that occurs in the reality where light travels at c, has its origin in phenomena or functionality existing in reality where light travels at a speed greater than c, and is influenced by phenomena existing in a reality where light exists at 0 miles per second.

Chapter 1.2: Quanta

This chapter explores quanta, its relation to light in more detail, and other emergent properties that must also be quantized because of multiple speeds of light.

Quanta & Other Basic Phenomenon Related to Singularities

When we think of light traveling at the speed of c, there are properties in the reality that emerge as a result of this. These properties as we have examined are of the nature of matter, the past, the present, and the future. These properties are the result of light traveling at c and so can be thought of as emerging from light.

But what do these properties actually mean? And further, how do these properties relate to the apparently quite different properties of omnipresence, omnipotent, omniscience, and omninurturance that can appear in a reality where the speed of light is infinite?

Let us consider first the properties of light that emerge when light is traveling at c: the past, the present, the future, and matter.

What is the past? It is the perceivable result of all the work and effort that has taken place so far. It is the foundation upon which the present and future will be built. It represents a status quo, a stability, and even a rigidity, and given that it is the result of the long play of time, it will not easily be persuaded to become another thing. It can be thought of as that which the eye can see when it looks around it. There is "physicality" to what the eye can see and so the essence of the past is a kind of physical-ness. So, ingrained in light, is this ability to project or create physical-ness.

What is the present? It is the tremendous play of forces of all kinds to express themselves here and now. There is "vitality" that is present in this play and often it is the most energetic or forceful of the forces that will win out, as opposed to the most insightful or thoughtful. All the tremendous possibility of the future is seeking for expression now and so this essential vitality can also be thought of as a projection or possibility implicit in light.

What is the future? It is the inevitability of what will manifest. The great thoughts, the great ideas, the purpose, the possibilities will sooner or later express themselves in what we call the future. And the essence of this is

36

thoughtfulness or a curiosity or a purpose that we can summarize as an essential "mentality". So embedded in light is this ability to project mentality.

And what is matter? It is the myriad crystallization into apparent diversity of the one essential reality of light, to allow for a play between these different sides or possibilities in an increasingly harmonious interaction. So, its essence is "harmony", and this too can be thought of as an essential property projected from or implicit in light.

But what about when light is considered to move at an infinite speed? Then the properties that become apparent, as already explored, are those of omnipresence, omnipotence, omniscience, and omninurturance. But omnipresence has physical-ness to it, omnipotence has vitality to it, omniscience a mentality to it, and omninurturance a harmony to it. It may be said that physical-ness comes from omnipresence, vitality comes from omnipotence, mentality comes from omniscience, and matter comes from omninurturance.

So, whether light is traveling at an infinite speed or the speed we know as c, there is something about the properties it projects, that in essence is the same. So, let us refer to these essential properties made apparent through the worlds that are created, as Presence (from omnipresence), Power (from omnipotence), Knowledge (from omniscience), and Harmony (from omninurturance). These properties can be thought of as essential to the functioning of any light-based-singularity or being. In a partial-singularity, as we explore further, materialization progressively culminates to a fourfold reality of physical-ness, vitality, mentality, and matter.

Further, quanta can be thought of as existing on that very border of the world or reality created due to Light traveling at c, and realities created as Light travels faster, which at its limit is an infinite speed. So, quanta are a doorway or an interface into worlds of Light, and a doorway by which possibilities in deeper worlds of Light can express themselves here in this material realm. Being so, phenomenon such as superposition and entanglement are natural to quanta: superposition, because different possibilities created by light traveling at different speeds can be thought of as being present simultaneously, and entanglement, because possibilities created by light traveling at different speeds operate in a different time and space consistent with the reality created by light

37

traveling at that said speed. But further, any process of genetics or the creation of genetic-type information, which by definition is associated with dynamics of superposition and entanglement, will also intimately be associated with quanta. This has to be since quanta are the interface between different layers of light. In fact, a process of quantum computation can be thought of as the means by which genetic mutation or becoming takes place and any future genetic-type possibilities unveiled, as will be discussed further. The phenomenon and mechanics of quanta, entanglement, superposition, genetics, quantum-level computation are natural to any being and becoming. This will be taken up in later sections of this book.

Structured Time, Structured Space, and Their Relation to Quanta

In considering the worlds that are created due to the way light moves, we see properties that are projected because of it. In the finite world, the world that results from light traveling at the speed c, there is, relatively, smallness we can grasp and on which we can build. In the infinite world that results from light traveling at an infinite speed there is a vastness and fullness that is difficult to grasp.

The notions of time and space are something entirely different in both worlds, and it could be said that Space allows the full play of everything meant by Power, Knowledge, Harmony, and Presence to be seeded in it, and that Time allows that seeding to flower into fuller forms with its passage.

In other words, both space and time are not just abstract concepts but are essentially highly structured to allow physical-ness, vitality, mentality, and matter to become Presence, Power, Knowledge, and Harmony.

Space allows all the possibilities present in Presence, Power, Knowledge, and Harmony and seeded in vast diversity, to evolve into more fullness through the time stages of physical-ness, vitality, mentality, and increasingly harmonious matter.

Such a creation of space and time is synonymous with the creation of quanta. Quanta become the means for the possibilities inherent in the anterior worlds of Presence, Power, Knowledge, and Harmony to express themselves in a structured space and time. Quanta are therefore a passage into deeper worlds of Light, and a means for possibilities in these deeper realms to express themselves materially. A process of quantum computation involving anterior-layer codified possibilities, or "pre-genetic" information, and a range of shifting forces, arbitrates the expressions that manifest in space and time.   This will be described in detail in Section 3 that focuses on the mathematical foundations of the being-becoming cycle or partial-singularity dynamics.

But further, in this view it may also be proposed that space and time being so structured by the four properties of Light, need also to be quantized. That is, space and time must be experienced as quanta as well.

The Quantization of Space, Time, Matter, and Gravity

Chapter 1.1 proposed that quanta is related to the slowing down of light, and as such becomes the basis for material expression. In other words, for matter to express itself requires quanta.   But further it was also just proposed that space, consisting of seeds of the properties of light, would also require quanta to express itself. Philosophically this has to be in a world where light is traveling at a fraction of its possible speed. Further, it was also proposed that time is highly structured, expressing the growth of the seeds in space through definite phases. As such, growth through such phases can also be thought of as happening due to quantization that so allows phases to express themselves.

But if we take a deeper look at space, time, and matter, in light of the properties of Light, it can be deduced that space, consisting of a vast array of seeds derived from the properties of Light is itself an expression of Light's property of Knowledge.

Time, bringing forth the meaning contained in the seeds, regardless of circumstance, and even being opposed by circumstance, can be thought of as Light's property of Power.

Matter itself, being a container in which space and time can allow deeper properties of Light to become materially tangible, must be an expression of Light's property of Presence.

But it is also known from Einstein's General Theory of Relativity that gravity is associated with mass and space, in that, as the physicist John Wheeler has put it, it is none other than a mass's instruction telling space how to curve, and again is nothing else that space's instruction telling

39

mass how to move through it (Wheeler, 2000). As such, where mass and space exist, there gravity has to exist as well. Hence it may be inferred that gravity is none other than an expression of Light's property of Harmony, which fixes the collective relationship between object and object.

But if as proposed, matter, space, and time all need to be quantized in order to express themselves, then this must also be true of gravity.

But also, the emergence of space, time, matter, and gravity can be seen as a quantum computational outcome of light precipitating from the reality where it travels at an infinite speed to the reality where it travels at c.

Subsequent chapters will explore these claims in greater detail.

The previous chapters revealed four intrinsic properties that exist in Light at its native state. These are omnipresence, omnipotence, omniscience, and omninurturance, or simply Presence, Power, Knowledge, and Harmony, respectively. These four properties shed insight into the nature of Light and exist in a tight union. It is this union that reveals Love.

Light in its nature is therefore one with Love. Light in its fullness is nothing but love. As light begins to precipitate by projecting itself at slower speeds it will be seen that it maintains its essential nature of love regardless of the reality that is generated at a particular speed.

Therefore, for example, when light travels at c, an essential fourfoldness that operates together is revealed – that of past, present, future, and matter. Recall that the essential nature of this fourfoldness can be thought of as physicality, vitality, mentality, and connection, which is none other than Presence, Power, Knowledge, and Harmony seen from a different angle. In mathematical terms there is an essential symmetry that is maintained, and this has to be since the essential unity of light will always seek to maintain that unity. Hence, past, present, future, and matter are nothing other than a quadrumvirate construct of Light at c. But this revealing of a quadrumvirate that is one, is nothing other than an act of love.

As discussed in the previous chapter Light slowing down results in the Big Bang, which is the apparent starting point of the physical universe. In this starting point one of the first essential quadrumvirate constructs that is revealed is that of space-time-energy-gravity. This essential quadrumvirate is nothing other than an expression of love, and as will be discussed in much greater detail in subsequent sections of the book, in any light-based-singularity – that phenomenon where creation and operation of a new whole or being maintains integrity with all the previous precipitated layers of light and wholes or beings – the process of becoming is in fact seeking to express another construct of love to further enhance the light-based foundation of manifested existence.

As such the cycle or spiral of being and becoming contributes to Life through participating in the inherent dynamic of Love that is the deepest nature of Light.

The space-time-energy-gravity quadrumvirate can therefore be thought of as a being of cosmic proportions that exists in all of the physical universe. This quadrumvirate space-time-energy-gravity being can create additional beings from its substance where space-time-energy-gravity can behave in precise ways to support phenomenon such as black holes, stellar curvatures, and so on, as will be explored later in the book. As also suggested in the previous chapter, an essential operation of all materialization will be traced to space-time-energy-gravity quantization. This is an example of a being that participates in the becoming of another being, as will be elaborated.

So also at the electromagnetic field level, there is a sustainable fourfold wholeness that comes into place: archetypes materialized as different wave-lengths define the knowledge aspect, the range of frequencies that allows different energies to come into being define the power aspect, different mass-potentialities define the presence aspect, and this whole conglomerate travels at a speed that allows materialized creations to interact as seemingly isolated entities defining the harmony aspect. Without the participation of each of these aspects subsequent layers of matter could not emerge. But such complete interaction of different aspects working together to create a whole different possibility is nothing other than Love.

At the quantum level it can similarly be said that quarks or leptons or bosons or the Higgs-boson by themselves would not have been able to create the foundation for the next layer of complexity in Light's journey. If they acted independently there would have only been a flurry of endless flows. It is only the combined action of the four held together in a tight embrace that is able to create the reality of the atom. But such a tight embrace of distinct actors or personalities that causes cohesive and unified action is nothing other than Love.

Hence it is Love that pulls together the possibilities projected by Light to create a unity that must exist for any next layer of complexity to emerge.

Continuing with the exploration of the layer of quantum particles, it is the coordinated action of bosons that allows the richness, the possibilities, the dynamism of quantum particle interaction, and all the myriad results of this to come into being. Gluons allow the nucleus to form and therefore allow the sustained codification of what an element will mean in the scheme of things to come into being, hence precipitating the knowledge

42

aspect. W and Z bosons allow the release of energy to be recycled or used for other purposes of interaction, through decay of heavier quarks and leptons, hence precipitating the power aspect. Photons as the carrier particle of the electromagnetic spectrum allow this field to be ever-present in this reality, hence precipitating the presence aspect. Hypothetical gravitons would allow built-up objects to enter into a space-time relationship with each other: it bounds objects to behave in a unified way in a specified space and for a specified time, hence precipitating the harmony aspect. But all this holding together, and releasing, and recycling, and relating, this cohesiveness of the action of the force-carriers or bosons is only another instance of the action of Love. Any of these forces acting by themselves would not be able to create the foundation on which subsequent emergences are built.

Similarly, the complexity of molecules resulting in the diversity of possibility is only so because such a molecular foundation is built by Love: it is Love that causes a sustainable molecular foundation constructed from the s-Group or power aspect, p-Group or knowledge aspect, f-Group or harmony aspect, and d-Group or presence type atoms to come into being. A universe created by one type of atom only would quickly be crushed under the weight of an ever more unsustainable imbalance.

Cells themselves require a balance of lipids or the harmony aspect, polysaccharides or the power aspect, proteins or the presence aspect, and nucleic acids or the knowledge aspect to meaningfully sustain themselves. Such balance is the sign that Love has left its stamp.

So, where there is Light, Love too is. Where Light has projected itself into ever-more crystallized forms of matter and life, there too Love must be. This is so because Light in its wholeness is inseparable. The properties of Light when projected therefore, still have to act as though one because that is the innate law of their being. This continued action of oneness in crystal form is what Love is, and it acts so that the properties of Knowledge, Power, Harmony, and Presence can more fully manifest what Light really is.

The becoming of quadrumvirate stability reveals a being. Being can spawn sub-beings, and being builds on beings. Hence there is a double action of Love displayed as horizontal and vertical integrity in which more of the fullness of Light is progressively embraced. Life itself complexifies through this display of horizontal and vertical integrity solidifying as it were with the spiraling through being and becoming.

A Cosmology of Light is predicated on the notion that light can and does exist at different constant speeds. In such an approach Light's native state is assumed to be an existence where it travels infinitely fast. This, as preliminarily explored creates a fourfold reality of omnipresence, omnipotence, omniscience, and omninurturance. This reality has infinite potential coded into it as infinite information. In order to materialize this information Light projects itself at slower and slower constant speeds. Each of these projected constant speeds creates a particular kind of reality, and a cosmology comes into being as a result of the worlds so created. Hence, as suggested, light traveling at the constant speed of c creates the reality of our physical universe. Light projected so that it is stagnant, effectively with 0 speed, creates a veritable hell. As will be further explored light at different speeds similarly creates different worlds.

In such a cosmology each projected layer of light and the world so created exists in layers of light faster than it. Each projected layer of light existing in layers of light that precede it forms a light-based-singularity where in effect the integrity of the dynamics of preceding layers of light are maintained and can influence operations in worlds so created. Each light-based-singularity so created is in effect a Being. This is because there is an integrity and wholeness of something in potentiality in Light at its native state having projected itself into living form. Becoming is the process by which a Being becomes a Being. The existence of living beings creates Life. Life therefore continues to complexify with the birth of beings. The range of beings are infinite, and as will be explored subsequently include quadrumvirate creations of the nature of space-time-energy-gravity, past-present-future-matter, quarks-leptons-bosons-Higgsboson, amongst many others, that all derive their reality from Light's implicit omnipresent-omnipotent-omniscient-omninurturing nature.

Subsequent sections will construct a comprehensive mathematical model of being, becoming, and life, based on light-based singularities. Note that my previous book, The Second Singularity, explored the mathematics of light-based singularities in depth, and this book will leverage and enhance the mathematical models created there.

Light-based singularities can themselves be of different natures and will be the foundation for different types of beings, and different processes of becoming. Hence, the Seed-Singularity is a singular light-based structure created by light traveling at multiple constant speeds greater than c. This would include the layer of light in its native state traveling infinitely fast, and subsequent layers where light travels at some speed k, much greater than c, and at some speed n, somewhat greater than c. The seed-singularity is constructed to be constituted by light traveling at 3 different constant speeds and this allows a representative and mathematically manageable model to be created that yet illustrates the fullness in such a cosmology.

Hence, K-Beings are created when light travels at speed k, and N-Beings are created when light travels at speed n, and these beings have particular properties that impact all subsequent layers of light. K-beings and n-beings exhibit particular kinds of dynamics best understood by phenomenon such as entanglement and superposition that have surfaced in the contemporary study of quantum-level phenomenon. These will be elaborated in more detail in following sections.

When light travels at speed c, then as suggested earlier, the materialization of vast amounts of antecedent information contained in layers of light traveling faster than c, results in the phenomenon of the Big Bang, and defines too, a cosmogony, a process by which the physical universe comes into Being. By definition any process of materialization is on-going and prescribed by a process of time, since time is emergent as part of the space-time-energy-gravity quadrumvirate being that becomes when the Big Bang occurs, and measures out how possibility existing in seeds in space mature to arrive at their full potentiality. Hence, there is not a finality in the process of becoming, and any light-based-singularity, such as space-time-energy-gravity while capturing something stable and concrete is itself termed a "Partial-Singularity", to emphasize the notion that many subsequent partial-singularities will come into being.

There is envisioned to be an iterative process by which a being, or existing partial-singularity, serves as a foundation for subsequent beings or partial-singularities to become. There is a process of arbitration, as it were, in which all possibilities related to a particular becoming, and resident in different layers of light, are computed into existence. This computation occurs at the quantum level at the interface between layers of light, and involves quanta, bridge devices that accumulate information at a faster-moving layer of light and materialize that at a slower-moving layer of light. In effect there is envisioned to be a persistent quantum-level computation that therefore computes the material world into existence.

My book, Emperor's Quantum Computer, had covered this in some depth.

The output of any quantum computation is information, and more precisely, genetic-type information, that then becomes "law" for beings affected by it. Such law prescribes the set of behaviors and possibilities existing for a being. While such laws appear to be permanent for beings, it is possible to change them by affecting the process of quantum-level computation. This process was further elaborated in my book, The Origin and Possibilities of Genetics, and is at the heart of the process of becoming. Hence, beings can become other beings, enhancing life itself, through affecting quantum-level computation in effect adjusting or even creating genetic-type laws in place for a particular type of being.

Such a process of enhancing life, by iterating between being and becoming, is proposed to be a viable process for transhumanism. This is perhaps in addition to current approaches to transhumanism built off fragmented-singularities. In a Cosmology of Light, processes that inherently do not leverage life or existing light-based singularities but are based on first-order or second-order derivatives of existing beings, are termed "Fragmented-Singularities". First-order derivatives of the human species, themselves a partial-singularity because all the dynamism and logic of light is inherent in them, would be computer-based programs, for example. Second-order derivatives would be constructions of such computer-based programs such as created when using 3D printing, for example. Such fragmented-singularities are termed "singularities", as suggested in The Second Singularity, because of the popular notion that AI can create a singularity transcending human capability. They are termed "fragmented" because they in effect negate the inherent advantages available to any light-based-singularity. All the connection with the design, the laws, the antecedent layers of light, that had continued to exercise themselves through the long play of Space and Time to allow beings to become, will have been severed.

The Second Singularity, by contrast, is one in which the dynamics that arise in the layer where light travels at c are fully and consciously integrated with the positive dynamics of all antecedent light layers. Hence the Second Singularity, due to a complete vertical integration with antecedent layers and horizontal integration with all four categories of light in any endeavor, is a triumph of love, and one in which future becoming is based on the deepest positive founts of creation. Life and any transhumanism that were to remain connected with light and love in such a manner. would be of a fundamentally more robust kind than any derivatives or life-derivative hybrids founded on fragmented-singularities.

A composite light-based-singularity superstructure in which the seed-singularity, partial-singularities, and the Second Singularity occur is termed the Light-Based-Singularity Superstructure. The light-based-singularity superstructure, hence, can be thought of as the matrix, termed Light-Matrix, that gives rise to the different types of beings that will arise in a Cosmology of Light. The next Section will begin to explore the Light-Based-Singularity Superstructure or Light-Matrix and the various types of beings that arise in it in more detail.

SECTION 2: MODELING THE MATHEMATICAL STRUCTURE OF BEING & THE BECOMING OF THE P-TYPE BEING

This section explores the mathematical structure of being and the becoming of the P-type being in a cosmology of light.

There are five representative levels of light that are illustrated in this section, that are themselves determined by the play of four ubiquitous bases or properties of light. These levels spawn five types of beings corresponding to each layer of light. While the number of levels or types of beings may potentially be infinite, the five levels seem to represent important and perhaps even fundamental quantum-based shifts or "quantizations", as light projects itself at slower and slower speeds to progressively reveal materially more of the vast information implicit in it. The quantized appearance of a level relates to a fundamental genre of emergence or being replete with different kinds of dynamics. Note that "information" relates to the four fundamental properties intuited to occur in Light.

Chapter 2.1 explores the Light-Matrix or the Light-Based-Singularity Superstructure, comprising of the five representative levels of light, that is also the source of the types of beings that are envisioned to exist in our universe. The light-based-singularity superstructure as suggested in the previous section, is such because the seed-singularity, partial-singularities, and the Second Singularity occur in it. The light-based-singularity superstructure gives rise to the first apparent layer or type of being of the "ubiquitous-point-instant" and to subsequent layers culminating in the layer that results in the vast diversity of life. Each layer, and the subtle-libraries of information they automatically spawn, thereby materializing a set of laws that the corresponding being will generally be subject to, is then explored in more detail in subsequent chapters through the ubiquitous-point-instant or UPI-type being (Chapter 2.2), the architectural forces or the K-type being (Chapter 2.3), and uniqueness of organization or the N-type being (Chapter 2.4).

The fourth layer in the Light-Matrix is created by light traveling at speed c and is the foundation for the U-type being, where 'U' suggests 'untransformed'. This layer, hence, is where material reality emerges. But the nature of the emergences will themselves be influenced by conditions that are prevalent in space and time. It is the combined dynamics of the seed-singularity, comprised of the three antecedent layers of light, with the play of the emergences in space and time arbitrated through a persistent quantum-level computation that outputs genetic-type information, that constitutes the process of becoming, also detailing the dynamics of any partial-singularity and emergence of the P-type being, where 'P' stands for 'partial-singularity'. The P-type being is hence an instance of the U-type being that has iterated through a process of becoming to reveal a stability of some aspect of Light that originated in pure potentiality in Light at its native state.

Chapter 2.5 suggests a mathematical basis for a process of becoming involving mutations for the P-type being. To be clear, the very basis of mutation derives from the seed-singularity itself, as will be elaborated. This mutational sequence thereby also suggests the conditions by which a partial-singularity or the P-type being could eventually evolve into the Second Singularity or the S-type being, where 'S' stands for 'Second Singularity'. Chapter 2.6 attempts a broader systematization of the process of becoming of the P-type being.

The mathematical foundations explored in Section 2 will set up the basis for the creation and dynamics of chains of four-base ecosystems that encode logic and provide insight into the biography of becoming of partial-singularities or the P-type being, leading also to the culmination in the S-type being, as explored in Section 3.

Chapter 2.1: The Light-Based-Singularity Superstructure as the Matrix of Being

This chapter will explore the Light-Based-Singularity Superstructure aka the Light-Matrix, a mathematical expression of the codification in Light formed at the time of the Big Bang (Malik et al., 2018). Further a case will be made equating the Light-Matrix as the primary or downward-strand in a ubiquitous subtle-DNA (Malik, 2019a). In its wholeness the Light-Matrix contains within it the seed-singularity, and the mechanism by which all partial-singularities, and the Second Singularity arise. Therefore, it is also the matrix in which UPI-type beings, K-type beings, N-type beings, and U-type beings originate. As suggestion in the introduction to this section, through a process of becoming U-type beings can also give rise to relatively transformed P-type beings, and more fully transformed S-type beings. Generally speaking, 'transformation' refers to the state when a being embodies more of the full potentiality of light within it, as will become clear through the course of this book.

This Light-Matrix can be thought of as a light-based-singularity superstructure that provides insight into dynamics of the universe. The Light-Matrix codifies dynamics that may be experienced at the quantum borders of realities created through light traveling at different speeds, and further in the worlds that the quantum veil or window provides access to. Weaving together possible realities created through light traveling at different speeds, the Light-Matrix also provides a model of superposition and entanglement. The Light-Matrix also provides insight into the range of different types of beings that will arise in a cosmology of light.

Light slowing down or projecting itself at a particular speed is envisioned to create a particular type of universe that will exist in all universes where light is traveling at any faster speed. Hence, all universes will exist in that created where light is in its native state traveling at an infinite speed. This implies that all properties true of the native state will also be present in any sub-universe. It is the presence of such properties and dynamics true of supra-universes that that gives rise to the phenomenon of superposition and potential entanglement in any sub-universe.

But further, such superposition also creates a "strand" as it were, that intimately connects together these universes and the realities represented by them. Entanglement creates the dynamic of influence. So, it is the combined action of superposition and entanglement that also creates the notion of ubiquity of the structure of "subtle-DNA", an apt metaphor to understand a key dynamic that occurs in any light-based-singularity.

It is interesting to note that Stanford anthropologist, Jeremy Narby, in his book 'The Cosmic Serpent: DNA and the Origins of Knowledge' (Narby 1998) points first to a shift in vision experienced by shamans around the world after the ingestion of ayuhuasca, and second to the commonality of the ensuing vision marked primarily by two entwining serpents. His interpretation is that these shamans enter into the molecular realm and actually perceive DNA and accompanying structures at the cellular levels. This book also requires a shift in vision, one bought about by perceiving Light in a precipitating series of layers caused by different speeds of light. The result of such a perception is the view of a ubiquitous subtle-DNA and a vast subtle structure continually arbitrating the output of genetic-type information to create the realities we exist in materially. This book aims to bring the wholeness or singularity containing such an enterprise into view, and the gradations and possibilities of being and becoming that will naturally arise in such a singularity.

Nature of Light at U and the U-Type Being

As laid out in Section 1, it is perhaps fair to say that the speed of light has significant implications on the nature of reality. The finiteness, c, at 186,000 miles per second in a vacuum creates an upper bound to the speed with which any object may travel also implying, and as discussed in Section 1, that objective reality will be experienced as a past, a present, a future from the point of view of that object (Einstein, 1995). These characteristics – a past, a present, a future – are implicit in the nature of light and become part of objective reality because of the speed of light. So, the way we experience time seems to be determined by c.

Further, c also creates a lower bound when inverted ($1/c$) being proportional, or arguably even determining Planck's constant, h, that pegs the minimum amount of energy or quanta required for expression at the sub-atomic level. Planck's constant, h, pegging the amount of energy required for expression, therefore may allow matter to form as suggested by the physicist Lorentz a century ago (Lorentz, 1925). Note that Einstein postulated quanta as a fundamental property of light itself, rather than as something that arose in the interaction of light with matter as suggested by Max Planck (Isaacson, 2008). Hence, c contributes to or even establishes a reality of nature with a past, present, and future, to also be experienced as a phenomenon of connection between seemingly independent islands of matter. This characteristic of 'material connection' or 'connection' is therefore also proposed to be implicit in the nature of light and becomes part of objective reality because of the speed of light.

But as explored in Chapter 1.2, a 'past' can also be viewed as established reality as defined by what the eye or other lenses of perception can see. Hence, 'It is the perceivable result of all the work and effort that has taken place so far. It is the foundation upon which the present and future will be built. It represents a status quo, stability, and even rigidity, and given that it is the result of the long play of time, it will not easily be persuaded to become another thing. It can be thought of as that which the eye can see when it looks around it. There is "physicality" to what the eye can see and so the essence of the past is a kind of physical-ness. So, ingrained in light, is this ability to project or create physical-ness.'

Also as explored previously, the 'present' is 'the tremendous play of forces of all kinds to express themselves here and now. There is "vitality" that is present in this play and often it is the most energetic or forceful of the forces that will win out, as opposed to the most insightful or thoughtful. All the tremendous possibility of the future is seeking for expression now and so this essential vitality can also be thought of as a projection or possibility implicit in light.'

The 'future' is 'the inevitability of what will manifest. The great thoughts, the great ideas, the purpose, the possibilities will sooner or later express themselves in what we call the future. And the essence of this is thoughtfulness or a curiosity or a purpose that we can summarize as an essential "mentality". So embedded in light is this ability to project mentality.'

While in its native state these properties of light are merged together, the experience of the past, present, future, and matter that arise when light travels at c can be thought of as a means by which the essential oneness is disaggregated in a way that it becomes more plainly perceivable as qualities of physicality, vitality, mentality, and connection by fundamentally fragmented constructs and emergences.

Further, the past-present-future-matter or physicality-vitality-mentality-connection quadrumvirates reveals essential dynamics that characterize the U-type being. Since, as will be discovered, there are an infinite number

53

of U-type beings, this book will refer to any stable quadrumvirate realities, as beings. Hence, the practically dynamic past-present-future-matter quadrumvirate being reveals an essential physicality-vitality-mentality-connection quadrumvirate being, that are both U-type beings.

Note that the whole notion of fragmentation becoming whole again is the basis of light-based-singularity dynamics in which being and becoming iterates to create more and more transformed or light-filled beings, as will be explored in detail in this book. The constant quantum-level computation and generation of genetic-type information that compels emergences to behave in consistency with their class of genetic-type information is the essence of the being-becoming dynamic of any light-based-singularity.

These implicit characteristics of the U-type being or nature of light as experienced at the layer of reality so set up by a finite speed of light may hence be summarized by Equation 2.1.1, where c_U refers to the speed of light of 186,000 miles per second, that has created the perceived nature of reality, U, as expressed in Equation 2.1.1:

$$c_U: [Physical, Vital, Mental, Connection]$$

Eq. 2.1.1: Implicit Characteristics of the U-Type Being or Nature of Light at U

The notation 'U' suggests 'untransformed' as will become clearer with the discussion of the light-space-time emergence equation in Chapter 3.1.

Nature of Light at ∞ and the UPI-Type Being

Exploring further, it is thought however that at quantum levels the nature of reality is characterized by wave-particle duality. Light itself and matter may be experienced as both particles and waves. But for matter to be experienced as waves implies that 'h' must have become a fraction of itself, $h_{fraction}$, to allow the concentration or possibility of quanta to have dispersed into waveform. This further implies that c must have become greater than itself, c_N, such that the inequality specified by Equation 2.1.2 holds:

$$c_N > c_U$$

Eq. 2.1.2: Layer N, Layer U Inequality of Speed of Light

Note that what is implied here is that just as there is a nature of reality specified by U that is the result of the speed of light being 186,000 miles per second, so too there is another nature of reality specified by N that is the result of a speed of light greater than 186,000 miles per second.

This is akin to recent developments in physics with the notion of property spaces being separate from but influencing physical space as explored by Nobel Physicist Frank Wilczek in his book 'A Beautiful Question" (Wilczek, 2016). But further in "Slow Light" Perkowitz's recent treatment of today's breakthroughs in the science of light (Perkowitz, 2011) he states: "Although relativity implies that it's impossible to accelerate an object to the speed of light, the theory may not disallow particles already moving at speed c or greater."

So light traveling at c_N may be possible. Current instrumentation, experience, and normal modes of thinking though, having developed as a bi-product of the characteristics so created in the layer of reality U may be inadequate to access N without appropriate modification. The notion of wave-particle duality already challenges the notion of normal thinking perhaps because wave-like phenomena could be a function of faster than c motion, and particle-like phenomena a function of equal to c motion. That these may be happening simultaneously is reinforced by principles such as complementarity in which experimental observation may allow measurement of one or another but not of both as pointed out by Whitaker (Whitaker, 2006). But further the very notion of the Pilot-Wave Interpretation of Quantum Mechanics, as modeled by the physicists Bohm and DeBroglie (Holland, 1995), is that constructive interference of waves causes particles to move toward these areas. Hence in this interpretation both particle and wave always exist simultaneously.

But then taking this trend of a possible increase in the speed of light to its limit, this will result in a speed of light of infinite miles per second. The question is, what is the nature of reality when light is traveling at infinite miles per second? As explored in Section 1, in any continuum, light originating at any point will instantaneously have arrived at every other point. Hence light will have a full and immediate *presence* in that continuum. Further, that light will *know* everything that is happening in that continuum completely and instantaneously – that is know what is emerging, what is changing, what is diminishing, what may be connected to what, and so on - or have a quality of *knowledge*. It will connect every object in that continuum completely and therefore have a quality of connection or *harmony*. Finally, nothing will be able to resist it or set up a separate reality that excludes it and hence it will have a quality of *power*.

The resulting stability describes a ubiquitously present being, referred to as UPI-type being, for ubiquitous-point-instant type being that is felt in all space and time regardless of how it is configured. The stability is described as a presence-power-knowledge-harmony quadrumvirate being that is a UPI-type being.

These implicit characteristics of the UPI-type being or nature of light as experienced at the layer of reality so set up by an infinite speed of light may hence be summarized by Equation 2.1.3, where c_∞ refers to the speed of light of ∞ miles per second, that has created the perceived nature of reality, ∞:

$$c_\infty : [Presence, Power, Knowledge, Harmony]$$

Eq. 2.1.3: Implicit Characteristics of UPI-Type Being or Nature of Light at ∞ Speed

Transformation from ∞ to U and the Creation of K-Type and N-Type Beings

But by (2.1.3) it can also be noticed that 'physical' is related to Presence, 'vital' is related to Power, 'mental' is related to Knowledge, and 'connection' is related to Harmony.

If this is the case the question then, is how do these apparent qualities at ∞ precipitate or become the physical-vital-mental-connection based diversity experienced at U? This may be achieved through the intervention or action of a couple of mathematical transformations acting on the implicit characteristics of nature of light at ∞ speed as summarized by (2.1.3).

First, the essential characteristics of Presence, Power, Knowledge, Harmony that it is posited exists at every point-instant by virtue of the ubiquity of light at ∞ will need to be expressed as sets with up to infinite elements. Second, elements in these sets will need to combine together in potentially infinite ways to create a myriad of seeds or signatures that then become the source of the immense diversity experienced at U. Note that

56

these mathematical transformations suggest that light may gather itself in such a way so as to first release, as it were, from its essential nature such sets with infinite variation around the essential characteristics, and second, an infinite combination of such vast variation. This suggests that all that is seen and experienced at U may be nothing other than 'information' or 'content' of light and as such that there are fundamental mathematical symmetries at play where everything at U is essentially the same thing that exists at ∞. This then is also the primordial basis of genetics. Further, the heart of creation, in this view, is perhaps the result of persistent quantum-computations that interrelates various realities so set up by light traveling at different speeds. The essential "nodes" of such quantum-computation take place at distinct levels or realities created at an interface of light traveling at different speeds, and these nodes distinguish the precipitating structure of a downward-strand of ubiquitous subtle-DNA.

Assuming that the first transformation occurs at a layer of reality K where the speed of light is c_K, such that $c_U < c_K < c_\infty$, this may be expressed by Equation 2.1.4:

$$c_K : [S_{Pr}, S_{Po}, S_K, S_H]$$

Eq. 2.1.4: The K-Type Being or First Transformation to Sets at Layer K

S_{Pr} signifies 'Set of Presence', S_{Po} signifies 'Set of Power', S_K signifies 'Set of Knowledge', S_H signifies 'Set of Harmony/Nurturing' (note: Harmony and Nurturing will be used interchangeably through this book).

Assuming that the second transformation occurs at a layer of reality N where the speed of light is c_N, such that $c_U < c_N < c_K < c_\infty$, this may be expressed by Equation 2.1.5:

$$c_N : f(S_{Pr} \times S_{Po} \times S_K \times S_H)$$

Eq. 2.1.5: The N-Type Being or Second Transformation to Seeds at Layer N

The unique seeds are therefore a function, f, of some unique combination of the elements in the four sets S_{Pr}, S_{Po}, S_K, S_H.

The relationship between the layers of light may be modeled by the following matrix in Equation 2.1.6:

$$Light_{Matrix} = \begin{vmatrix} c_\infty : [Pr, Po, K, H] \\ (\downarrow R_{C_K} = f(R_{C_\infty})) \\ c_K : [S_{Pr}, S_{Po}, S_K, S_H] \\ (\downarrow R_{C_N} = f(R_{C_K})) \\ c_N : f(S_{Pr} \times S_{Po} \times S_K \times S_H) \\ (\downarrow R_{C_U} = f(R_{C_N})) \\ c_U : [P, V, M, C] \end{vmatrix}$$

Eq. 2.1.6: Light-Matrix & Downward-Strand of Subtle-DNA

The matrix should be read from the top row down to the bottom row as indicated by the $\downarrow$ between rows, and suggests a series of transformations leading from the ubiquitous nature of light implicit in a point – presence, power, knowledge, harmony - to the seeming diversity of matter observed at the layer of reality U which is fundamentally the same presence, power, knowledge, and harmony projected into another form of itself.

The first transformation summarized by (2.1.4) into light-based sets is the result of a quantization function that culls out set-based wholes from the infinite potentiality in Light, and is summarized by Equation 2.1.7:

$$R_{C_K} = f(R_{C_\infty})$$

Eq. 2.1.7: Quantization Function Resulting in Light-Based Sets

This is suggesting that the reality at the layer specified by the speed of light c_K, R_{C_K}, is a function of the reality at the layer specified by the speed of light c_∞. The function itself is a quantization-function that allows some essential characteristics in the point-nature of light to express itself more "materially", relatively speaking, in sets described by (2.1.4), by light traveling at a relatively slower speed. This quantization-function can be thought of as a node in which a higher-level quantum-computation occurs by which infinite possibility begins to disaggregate itself.

The second transformation summarized by (2.1.5) into light-based seeds, is summarized by Equation 2.1.8:

$$R_{C_N} = f(R_{C_K})$$

This is suggesting that the reality at the layer specified by the speed of light c_N, R_{C_N}, is a function of the reality at the layer specified by the speed of light c_K. This function combines elements of the light-based sets into unique seeds and is envisioned as also being a quantization-function that allows what is expressed by the light-based sets to gather into unique light-based seeds.

Note that (2.1.3), (2.1.4), (2.1.5), (2.1.7), and (2.1.8) specify the dynamics of the seed-singularity. There is a light-based wholeness in the seed-singularity that results in a first meaningful sphere of cohesion with the creation of myriad light-based seeds. Each of these light-based seeds will give rise to a particular being through a subsequent process of becoming played out in space and time.

The quantization-function that interrelates layers N and U puts in place the essential dynamics of partial-singularities and is summarized by Equation 2.1.9:

$$R_{C_U} = f(R_{C_N})$$

Eq. 2.1.9: Quantization Function Resulting in Material Diversity

This is suggesting that the reality at the layer specified by the speed of light c_U, R_{C_U}, is a function of the reality at the layer specified by the speed of light c_N. This transformation builds on the unique seeds suggested by (2.1.5) to create the diversity of U as specified by (2.1.1). This too is envisioned as being a quantization-function that further allows all the combination of possibility to express itself in more material form. Note that the entire upward-strand to be discussed soon, also plays a key part in the dynamics of any partial-singularity.

In this framework the notion of wave-particle duality hence may become complementary block-field-wave-particle "quadrality" where block refers to phenomenon resident to ∞, field to phenomenon resident to K, wave to phenomenon resident to N, and particle to phenomenon resident to U. The block is all the reality always present behind the surface and is captured by (2.1.3) or the top line in the Light-Matrix as expressed in (2.1.6). The field is captured by (2.1.4) or the second major (non-parenthetical) line from the top in (2.1.6) and can be thought of as layers of possibility existing in each of the sets. The wave is captured by the creation of seeds represented by (2.1.5) or by the third major (non-parenthetical) line in (2.1.6). The particle that is apparently disconnected from the whole is captured by (2.1.1) or the bottom line in (2.1.6).

So, what we arrive at is a fundamental Light-Matrix or the downward-strand of a ubiquitous subtle-DNA that suggests key dynamics for different layers of Light and key types of being. This Light-Matrix also summarizes the essential dynamics of a Light-Based-Singularity Superstructure in which the seed-singularity, all partial-singularities, and the Second Singularity emerge. Note that the Second Singularity will necessitate the fulfilling of all conditions in a 'Space-Matrix" or upward-strand to be elaborated in Section 3.

Layer Zero and the 0-Type Being

For the sake of completeness, the thought-experiment to do with light existing at a zero-speed must now be brought up. In Chapter 1.1 a possible reality that would result if light were to exist at this speed was described as opposite to the reality were light to travel at an infinite speed.

Hence, ubiquitous Presence of Light would become the ubiquitous Absence of Light, or Darkness. The aspect of Power would become utter Weakness. The aspect of Knowledge would become complete Ignorance. The aspect of Harmony would become total Chaos. The resulting darkness-weakness-ignorance-chaos quadrumvirate being, hence, would be an instance of the 0-type being. And all this because Light would be unable to travel from where it is, in complete opposition to its known nature of traveling at a speed of c, and perhaps its true nature when traveling at a speed of ∞. Light, hence, would be hidden in itself, so as to speak, in some sort of a negative infinity. An Equation, 2.1.10, Nature of Light at Speed Zero and the 0-Type Being, would hence be:

c_0: $[Darkness, Weakness, Ignorance, Chaos]$

Eq. 2.1.10: Nature of Light at Speed Zero and the 0-Type Being

Further, (2.1.6), the Light-Matrix, would be modified to become Equation 2.1.11, Light-Matrix with Zero-Limit, where c_0 implies light at zero-speed, and D, W, I, and C imply Darkness, Weakness, Ignorance, and Chaos respectively:

$$Light_{Matrix}(0_{limit}) = \begin{vmatrix} c_\infty: [Pr, Po, K, H] \\ (\downarrow R_{C_K} = f(R_{C_\infty})) \\ c_K: [S_{Pr}, S_{Po}, S_K, S_H] \\ (\downarrow R_{C_N} = f(R_{C_K})) \\ c_N: f(S_{Pr} \times S_{Po} \times S_K \times S_H) \\ (\downarrow R_{C_U} = f(R_{C_N})) \\ c_U: [P, V, M, C] \\ \Uparrow \\ c_{0:[D,W,I,C]} \end{vmatrix}$$

Eq. 2.1.11: Light-Matrix with Zero-Limit

The implication of the bottom, zero-limit line, is that just as there is proposed to be an influence from the upper layers of light on U, as will be explored in subsequent chapters, so too there is a subtle influence from this lower layer of "light" that perhaps is responsible for the obstinacy of the untransformed nature of practical reality at U. Such obstinacy will be further explored in Chapters 2.5, 2.6, and 3.7 and reinforces the need for the notion of 'transformation' in the first place. Further, such obstinacy also has a bearing on the phenomena of genetic mutation by which iterative P-type beings with greater spheres of influence, and the S-type beings can come into being, as will also be discussed in greater detail subsequently. Note that since any P-type being, in reality a partial-singularity, is a biography, subsequent partial-singularities or P-type beings will always contain previous P-type beings within themselves, and hence will become P-type beings with greater spheres of influence.

Superposition, Entanglement, Quantum Computation, Genetic Mutation, Singularities, Beings, and Becoming

As may be apparent in the Light-Matrix (2.1.6) and its complete form (2.1.11), circumstance at U is the outcome of a number of influences from layers ∞, K, N, U itself, and Zero (0). But further since each of these layers is itself a reality caused by a different speed of light, the interface between these layers occurs at the quantum-levels. Hence the quantum-levels are replete with superposition emanating from **layers** ∞, K, N, U itself, and Zero (0). There is simultaneity and multiple possibilities that will determine what manifests at U. But as will be explored in more detail in subsequent chapters there is a logic or process to what manifests. This logic and process provides insight into the reality of genetic mutation or becoming, which itself may be seen as a process leading to the reality of P-type beings or partial-singularities with greater spherical influence, and to a possible Second Singularity or birth of S-type beings.

The notion that a quantum-object – an entity at the quantum-level – can be in an infinite number of superposed states that randomly collapses into a more precise state as is presumed in several leading interpretations of quantum mechanics, is called into question in the interpretation offered in this book. This was the subject of a previous book on quantum computation (Malik, 2018c). There may in fact be infinite number of possibilities vying for manifestation, but there is likely a more precise process by which this manifestation takes place. In other words, genetic mutation or the process of becoming is not random, but follows process involving multiple layers of light, or more precisely interaction between different types of beings.

Further, the process of entanglement by which quantum-objects are made to relate to each other, may be otiose since in the interpretation offered in this book entanglement exists ab initio at the layer ∞. This entanglement will be referred to as ∞-entanglement and will be further explored in Chapter 2.2. Further, there are additional forms of entanglement that occur even before a quantum-object becomes perceivable at layer U. There is entanglement at layer K by virtue of field-type action of light-based architectural sets. This will be referred to as K-entanglement and will be further explored in Chapter 2.3. There is an entanglement at layer N by virtue of the light-based wave-type action of seeds. This will be referred to as N-entanglement and will be further explored in Chapter 2.4.

Note that any process of entanglement is relevant from the point of view of genetic mutation or process of becoming, as it can be thought of as a source of becoming. Further, quantum computation, integral to the precipitation of quantization-function as described in this chapter, is therefore a reality in the manifestation of the minutest of circumstances at U and is also integral to the process of becoming. Such becoming, relying on a process of genetic mutation in which repositories of instruction change, allows the wholeness implicit in beings generated by the seed-singularity to materialize as the many P-type beings with greater and greater spheres of influence, culminating in the S-type being we will encounter later in this book.

Chapter 2.2: The UPI-Type Being and its Dynamics

The 'ubiquitous-point-instant' captures the inherent nature that appears to exist in the system and is represented by the top-line in Equation (2.1.6) and (2.1.11), different forms of the Light-Matrix, and is referred to as the UPI-type being. The prevalent dynamics characteristic of the UPI-type being are a function of that reality where light travels at ∞ speed and is envisioned as being infinitely entangled by virtue of light being omnipresent-omnipotent-omniscient-omninurturing in that realm. An omnipresent-omnipotent-omniscient-omninurturing quadrumvirate being can therefore be thought of as an instance of the UPI-type being, the essence of which is a presence-power-knowledge-harmony quadrumvirate being as suggested by (2.1.3) reproduced here for convenience:

$$c_\infty : [Presence, Power, Knowledge, Harmony]$$

The 'point' aspect of the UPI-type being suggests the space-dimension and as discussed in Chapter 1.3, that space is seeded with the possibilities inherent in the properties of Presence, Power, Knowledge, and Harmony or the presence-power-knowledge-harmony quadrumvirate being. The 'instant' aspect suggests the time-dimension and gives insight into the process of emergence that the possibilities in space progressively surface as. Note that the conception of UPI or ubiquitous-point-instant is different from the point-instant coined by Samuel Alexander in 'Space, Time and Deity' (Alexander, 1920). His notion was that matter is made up of point-instant motions of Space-Time and emergence happens successively from Space-Time which is considered to be the basic stuff of reality. This is in contrast to the view being modeled in this book where Light is the substance of existence and has infinite information or potentiality within it.

Therefore, the 'point' or space-aspect provides insight into structural aspects of genetic-type information. The 'instant' or time-aspect provides insight into the mutational aspects of genetics related processes. "Genetics" when viewed in this manner is a lens into the type and process of information-richness associated with a range of emergences and further, the process by which partial-singularities or P-type beings develop greater spheres of influence culminating in the possibility of the Second Singularity or S-type beings. If there is a history of singularities, or a biography that relates the seed-singularity with partial-singularities

64

with the Second Singularity, then such genetics can be thought of as the language of becoming in which it is written. Note though that such separation into point and instant aspects is just a way of trying to unpack some of the dynamics in that realm where everything, even space and time, are all one thing.

In its ubiquitous-point-instant, Presence-Power-Knowledge-Harmony wholeness, Presence allows emergences to continue to develop as per the possibilities implicit in the past-present-future or physical-vital-mental pathway. Beginning to translate this into an equation, the notation $System_{Pr}$ is given to system-presence. Note that the derivation of this and many subsequent equations was part of my doctoral work at University of Pretoria (Malik, 2017a). This system-presence is true across any considered Time-Space continuum starting from a time-space boundary '0' to a time-space boundary 'N'. This notion is characterized by the notation $TS_{0 \rightarrow N}$. Within that boundary from 0 to 'N', the 'presence' is such that it will always seize an opportunity to cause a shift from the physical-leading to the vital-leading, and from the vital-leading to the mental-leading. Research shows (Malik, 2009) that greater degrees of freedom are afforded by such a shift.

The notion that the 'presence' seizes on 'opportunity' is characterized by the notation:

$$Presence \\ \downarrow \\ Opportunity$$

The shift from physical-leading (P_L) to vital-leading (V_L) and vital-leading (V_L) to mental-leading (M_L) is characterized by:

$$P_L \rightarrow V_L$$
$$V_L \rightarrow M_L$$

Hence in this approach it is suggested that:

$$System_{Pr} \equiv TS_{0 \rightarrow N} \begin{bmatrix} Presence \\ \downarrow \\ Opportunity \end{bmatrix} \begin{bmatrix} P_L & \rightarrow & V_L \\ V_L & \rightarrow & M_L \end{bmatrix}$$

But there is something else about this Presence as well. All other developments take place in it. That is, it provides a container of sorts in which the plays of system-power, system-knowledge, and system-harmony/system-nurturing can take place. This notion is summarized by the notation:

$$Container \begin{bmatrix} System_P \\ System_K \\ System_N \end{bmatrix}$$

Hence, combining these various components, an equation for 'system-presence', Equation 2.2.1, arises:

$$System_{Pr} \equiv TS_{0 \to N} \begin{bmatrix} Presence \\ \downarrow \\ Opportunity \end{bmatrix} \begin{bmatrix} P_L & \to & V_L \\ V_L & \to & M_L \end{bmatrix} \& Container \begin{bmatrix} System_P \\ System_K \\ System_N \end{bmatrix} $$

Eq 2.2.1: System Presence

In its ubiquitous-point-instant, Presence-Power-Knowledge-Harmony wholeness, Power allows emergences to continue to happen in spite of tremendous oppositions of all kinds; this too, regardless of field or area. Constructing an equation for system-power, the notation $System_P$ is used to represent system-power. Any endeavor will always be met with resistances of various kinds. The resistances that arise along the physical dimension are referred to as P_R. The resistances that arise along the vital dimension are referred to as V_R. The resistances that arise along the mental dimension are referred to as M_R. In the fruition of any endeavor one or all of these types of resistances may arise. Further, resistance of one kind often feeds on resistance of another kind, and to generalize the resistances encountered in an endeavor, these may be characterized as the product of the three types of resistance:

$$P_R * V_R * M_R$$

These resistances arise across any considered Time-Space boundary from 0 to 'N', and therefore it may be said that the power of the system is such that:

$$power > \sum_{TS=0}^{N} P_R * V_R * M_R$$

An equation for 'system-power', Equation 2.2.2, hence, is the following:

$$System_P \equiv power > \sum_{TS=0}^{N} P_R * V_R * M_R$$

Eq 2.2.2: System Power

In its ubiquitous-point-instant, Presence-Power-Knowledge-Harmony wholeness, Knowledge orchestrates emergences to continue to happen by leveraging the right instruments and circumstances. Translating this into an equation, the notation, $System_K$, is used for system-knowledge. This $System_K$ is such that it leverages the right instrumentation and circumstance to bring about the progress that is possible. This concept of 'instrumentation' is denoted by the subscript 'I'. The concept of 'circumstance' is denoted by the subscript 'C'. Both instrumentation and circumstance can be of a physical, vital, or mental type and this possibility is denoted by:

$$\begin{bmatrix} P_{I,C} \\ V_{I,C} \\ M_{I,C} \end{bmatrix}$$

Further, the notion that the 'knowledge' is such that it 'leverages' the right instrumentation and circumstance is depicted by:

Knowledge
↓
Leverage

This act of leveraging results in a fundamental shift so that the physical-leading yields to the vital-leading, and the vital-leading yields to the mental-leading. Hence:

$$\begin{matrix} Knowledge \\ \downarrow \\ Leverage \end{matrix} \begin{bmatrix} P_{I,C} \\ V_{I,C} \\ M_{I,C} \end{bmatrix} \rightarrow \begin{bmatrix} P_L & \rightarrow & V_L \\ V_L & \rightarrow & M_L \end{bmatrix}$$

Since this behavior may exist across any Time-Space continuum an equation for system-knowledge, Equation 2.2.3, is suggested:

$$System_K \equiv TS_{0 \rightarrow N} \begin{bmatrix} Knowledge \\ \downarrow \\ Leverage \end{matrix} \begin{bmatrix} P_{I,C} \\ V_{I,C} \\ M_{I,C} \end{bmatrix} \rightarrow \begin{bmatrix} P_L & \rightarrow & V_L \\ V_L & \rightarrow & M_L \end{bmatrix} \end{bmatrix}$$

Eq 2.2.3: System Knowledge

In its ubiquitous-point-instant, Presence-Power-Knowledge-Harmony wholeness, Harmony or Nurturing allows emergences to continue to happen with more and more degrees of freedom coming to the surface. The characteristic of this implicit-nurturing may be referred to as 'system-nurturing'. Like the other characteristics it is suggested to exist across a Time-Space continuum. This is depicted by:

$$TS_{0 \to N}$$

There is an action of nurturing such that any state is always advanced to a higher level. This is depicted by:

$$\coprod_{Nurturing} \begin{pmatrix} P_- & M_+ \\ V_- & V_+ \\ M_- & P_+ \end{pmatrix}$$

Hence, there is a 'union', depicted by 'U' that 'nurtures' the negatives towards their positives.

Further, there is an increasing action of nurturing such that the possibility of integration is always increased to form a larger and larger basis. This increasing basis is depicted as being modulated by the polar coordinates 'r' and 'θ', where r is the radius which increases from an initial value of '0', and 'θ' is an angle from '0' to '360'.

Hence, the equation of system-nurturing, Equation 2.2.4, is depicted as:

$$System_N \equiv TS_{0 \to N} \left(\coprod_{Nurturing} \begin{pmatrix} P_- & M_+ \\ V_- & V_+ \\ M_- & P_+ \end{pmatrix} mod\ (r, \theta) \right)$$

Eq 2.2.4: System Nurturing

It is suggested that these four characteristics exist across any system, and to denote this it is generalized that every point-instant in any system is embedded with this four-fold intelligence. In other words, there is in effect a library of information represented by Equations 2.2.1 – 2.2.4 whose operation is governed by ∞-entanglement. Further, there is a category of constructive genetic mutation, ∞ -entanglement mutation rooted in

Equations 2.2.1 – 2.2.4 that may be activated to bring about change at the material level.

In effect the ubiquitous-point-instant pre-genetic ∞-entanglement library provides insight into the dynamics of the UPI-type being native to the seed-singularity, and further to the kind of constructive ∞-entanglement genetic mutation that will subsequently be available at the material level.

The following chapters in this section explore additional categories of entangled pre-genetic libraries implicit in the seed-singularity, summarizing dynamics of K-type and N-type beings, that also spawn additional categories of constructive genetic mutation at the material level. It is such constructive genetic mutation that alters the information by which various emergences exist, thereby allowing a biography toward the S-type being to in effect be recorded.

Note further, that libraries created at subsequent levels of the downward-strand, and the very process of genetic mutation involving the material level, are influenced by this subtlest level UPI-type being and its ∞-entanglement library.

Chapter 2.3: The K-Type Being and its Dynamics

The characteristics of the UPI-type being, summarized in (2.1.3), suggests a possibility that is hard to fathom. One can only glimpse this extraordinary nature. As previously referred to this extraordinary nature is responsible for a broader set of architectural forces that exists behind the visible face of things and comes into view in a reality where light travels at the speed c_K (above c, and closer to ∞, as defined in Chapter 2.1) also spawning or creating the K-type being. Such architectural forces can be envisioned as emanating from a pre-genetic library. Due to the process of K-entanglement this pre-genetic library can also be thought of as constructively influencing genetic mutation and therefore all singularity-biographies. Recall that any singularity-biography is essentially a process of becoming by which the light-matrix materializes possibility contained in the seed-singularity through the iteration of being and becoming to create P-type beings. An iterative being-becoming process will eventually build on a foundation of P-type beings to create more transformed S-type beings.

Hence, system-presence, system-power, system-knowledge, and system-nurturing that define the nature of every point in our system, become more tangible as a broader set of architectural forces that emanate from each of them.

Considering system-presence, here is a characteristic that appears to be everywhere (Malik, 2015) at the service of all the constructs that develop within it. There is a diligence and perseverance by which any opportunity for progress is seized. Further, if one considers the extraordinary detail that appears in any construct, whether an atom, a body, a planet, or a galaxy, one is struck by the high degree of perfection that surfaces in this presence. So, if one contemplates the nature of this system-presence there is a set of forces that surface. Depicting such a set as $S_{System_{Pr}}$, one can arrive at elements such as Service, Perfection, Diligence, Perseverance, amongst others, that are part of this set. Hence, the set can be described by Equation 2.3.1:

$$S_{System_{Pr}} \ni [Service, Perfection, Diligence, Perseverance, ...]$$

Eq 2.3.1: Set of System Presence

But from the point of view of the speed of light, K is a reality set up by light traveling at c_K, where $c_U < c_N < c_K < c_\infty$. Conversely assuming an inverse proportionality with 'h', as c_K is closer to ∞, h will be closer to zero, and 'matter' or form will be highly dispersed. This dispersal is presumed to be field-like so that in effect any element of the set described in (2.3.1) or in to be discussed sets (2.3.2-4) will have a field-like reality and express K-entanglement such that all emergences emanating or comprised of that element will automatically partake in an entanglement with every other unique emergence having that element as part of its foundation. K-entanglement is therefore different from ∞-entanglement and every emergence will at least have both these types of entanglements subtly coordinating or influencing its action.

Similarly, considering the characteristic of system-power, one can hypothesize that there is a family of forces that emanates from it. The kinds of forces may be thought of as Power, Courage, Adventure, Justice, amongst others. The set for system-power can hence be depicted by Equation 2.3.2:

$$S_{System_P} \ni [Power, Courage, Adventure, Justice, ...]$$

Eq 2.3.2: Set of System Power

Similarly, considering the system-knowledge as the root of various powers that emanate from it, one may characterize the set for system-knowledge by Equation 2.3.3:

$$S_{System_K} \ni [Wisdom, Law Making, Spread of Knowledge ...]$$

Eq 2.3.3: Set of System Knowledge

The set for system-nurturing is depicted by Equation 2.3.4:

$$S_{System_N} \ni [Love, Compassion, Harmony, Relationship ...]$$

Eq 2.3.4: Set of System Nurturing

Equations 2.3.1 – 4 describe potentially infinite-element sets that comprise the Architectural Force K-entanglement Pre-Genetic Library existing in a K-type being. Further, there is a category of constructive genetic mutation, K-entanglement mutation, that may be activated to bring about change at the material level. As already suggested, this category of genetic mutation will also influence the narrative that defines any singularity-biography.

Chapter 2.4: The N-Type Being and its Dynamics

The seed-singularity mathematical model that leads from unity to unique diversity is perhaps justified by observation of phenomena. Such observation generates hypotheses that every organization, whether an atom, cell, person, team, corporation, market, or country is unique and that this uniqueness can be specified in terms of elements of the derived light-based sets for power, knowledge, presence, and nurturing.

At the sub-atomic level, hence, Nobel Laureate Wolfgang Pauli's 'Pauli Exclusion Principle' states that no two similar fermions, which include fundamental particles with half-integer spin such as protons, neutrons, and electrons, can occupy the same quantum states simultaneously (Pauli, 1964). Spin has to do with the angle that the particle has to rotate through before being symmetrical with its original state. Half-integer spin particles need to rotate through 720 degrees before being symmetrical with their original state. The implication of the Pauli Exclusion Principle is that fundamental structure and consequently stability comes into being at the atomic level, which as is evident in the Periodic Table also allows the separation of function related to form. This stability related to the underlying structure of atoms implies the basis of uniqueness and diversity. In the absence of the Exclusion Principle matter would just be a dense soup (Hawking, 1988) with particles occupying overlapping space.

At the observable level uniqueness is evident from the immense diversity of distinct species on earth (Mora, 2011) estimated to be over 2 million, and further the uniqueness of every member of each species. This member-level uniqueness is suggested by the difference in non-coding regions of the DNA that may vary in their sequence by about 1 to 4 percent, which in turn result in unique protein binding sequences of each human (Snyder, 2010), as an example, which in turn results in unique observable qualities.

At the astronomical level Einstein's Special Theory of Relativity (Einstein, 1995) suggests that every coordinate system potentially has its own space-time rendering as opposed to there being one absolute space and time. This implies the notion of uniqueness as an implicit property of space.

The four properties explored in Chapter 2.1 and elaborated in Chapter 2.2 define the source of that uniqueness. From this source emanate 4 sets of forces that suggest the boundaries of that uniqueness as explored in Chapter 2.3.

Assuming then that the fount of uniqueness is system-presence, a general equation for organizations that belong to the family of system-presence can be derived. Such uniqueness can be depicted as Sig_x where the subscript 'x' refers to the source family, and 'Sig' or signature to 'uniqueness'. Hence the uniqueness of an organization in the family of system-presence would be notated by Sig_{Pr}.

In line with the development of properties of a point and the precipitating architectural forces as discussed in Chapter 2.2 and 2.3 respectively, an approach to constructing such uniqueness is to assume a primary factor X that drives the uniqueness that belongs to the set $S_{System_{Pr}}$. Further, assume that the uniqueness is qualified by a number of secondary factors Y that may belong to any of the 4 sets - $S_{System_{Pr}}, S_{System_P}, S_{System_K}, S_{System_N}$. The primary factor X would have a greater weightage than any of the secondary factors Y. The weightage of X hence could be depicted by the number 'a', and the weightage of Y a number 'b_{0-n}', such that a > b. Further, the secondary element can repeat from '0 – n' times and is hence depicted as $\overline{Yb_{0-n}}$.

The equation, Equation 2.4.1, hence for a unique organization derived from the family of system-presence is:

$$Sig_{Pr} = Xa + \overline{Yb_{0-n}} \quad where \begin{bmatrix} X \in [S_{System_{Pr}}] \\ Y \in [S_{System_{Pr}}, S_{System_P}, S_{System_K}, S_{System_N}] \\ a, b \ are \ integers; a > b \end{bmatrix}$$

Eq 2.4.1: System Presence Based Unique Organization

But from the point of view of the speed of light, N is a reality set up by light traveling at c_N, where $c_U < c_N < c_K < c_\infty$. Conversely assuming an inverse proportionality with 'h', as c_N is closer to but greater than c_U, h_U will be less than h, and 'matter' or form will be unable to accumulate as it does at U, instead tending to disperse like a wave. The wave-like nature implies an entanglement, N-entanglement, such that any emergence will always be unique. Uniqueness could not be unless there was a dynamic such as N-entanglement in place. Further, such dynamics of uniqueness and entanglement at layer N are representative or the result of the N-type being.

Similarly, an equation, Equation 2.4.2, for a unique organization derived from the family of system-power is:

$$Sig_P = Xa + \overline{Yb_{0-n}} \quad where \left[\begin{array}{c} X \in [S_{System_P}] \\ Y \in [S_{System_{Pr}}, S_{System_P}, S_{System_K}, S_{System_N}] \\ a, b \ are \ integers; a > b \end{array} \right]$$

Eq 2.4.2: System Power Based Unique Organization

An equation, Equation 2.4.3, for a unique organization derived from the family of system-knowledge is:

$$Sig_K = Xa + \overline{Yb_{0-n}} \quad where \left[\begin{array}{c} X \in [S_{System_K}] \\ Y \in [S_{System_{Pr}}, S_{System_P}, S_{System_K}, S_{System_N}] \\ a, b \ are \ integers; a > b \end{array} \right]$$

Eq 2.4.3: System Knowledge Based Unique Organization

An equation, Equation 2.4.4, for a unique organization derived from the family of system-nurturing is:

$$Sig_N = Xa + \overline{Yb_{0-n}} \quad where \left[\begin{array}{c} X \in [S_{System_N}] \\ Y \in [S_{System_{Pr}}, S_{System_P}, S_{System_K}, S_{System_N}] \\ a, b \ are \ integers; a > b \end{array} \right]$$

Eq 2.4.4: System Nurturing Based Unique Organization

The four preceding equations can be generalized by Equation 2.4.5 that also summarizes uniqueness dynamics for the N-type being:

$$Sig = Xa + \overline{Yb_{0-n}} \quad where \left[\begin{array}{c} X \in [S_{System_{Pr}}, S_{System_P}, S_{System_K}, S_{System_N}] \\ Y \in [S_{System_{Pr}}, S_{System_P}, S_{System_K}, S_{System_N}] \\ a, b \ are \ integers; a > b \end{array} \right]$$

Equations 2.4.1 – 5 describe potentially infinite-element sets that comprise the Organizational-Uniqueness Pre-Genetic N-entanglement Library.

The N-entanglement level, like the ∞-entanglement and K-entanglement levels, is also hypothesized as being that part of the backbone of subtle-DNA, where distinct elements of the deeper fourfold-based sets combines, to create unique seeds. These seeds are the primary organizing force at the center of any and every organization and shed light into a key dynamic of the seed-singularity. Hence the subtle-library behind any emergence of matter and life derives primarily from the action on the N-entanglement level and dynamics of the N-type being, and further, there is a category of constructive genetic mutation, N-entanglement mutation, that may be activated to bring about change at the material level.

The next two chapters turn away from the structural or space-aspect of genetics implicit in the seed-singularity, and focus on the emergent, mutational, or time-aspect of genetics implicit in any light-based partial-singularity.

Hence Chapter 2.5 will focus on possible mutation-based narrative that describes dynamics related to the creation of P-type beings. Essentially the deep structural basis of genetics hinted at by the ∞-entanglement, K-entanglement, and N-entanglement levels, also becomes the basis for mutational-emergence. In other words, the possibilities inherent in these levels can change emergence through genetic-type mutation to prescribe or narrate a biography of P-type beings aka light-based partial-singularities.

Note that massively significant mutations such as took place on the SRY gene and led to the evolution of human from the chimpanzee species (Ridley, 1999) are architected by a layer of light traveling faster than c and likely involved one of the three entanglement schemes just described - ∞-entanglement, K-entanglement, or N-entanglement.

Chapter 2.6 will focus on the further mathematical systematization of the mutational-sequence discussed in Chapter 2.5, thus more generally describing the pathway of emergence for any P-type being.

Chapter 2.5: A Process of Becoming Involving Mutations for the P-Type Being

While the uniqueness of organizations made possible through the N-type being and as represented by the Signature is a seed, like any seed there is a process for its emergence (Kaufmann, 1995; Portugali, 2012; Yates, 2012), and the uniqueness will often be hidden or very much behind the scene until certain conditions are fulfilled (Malik, 2009). This chapter suggests a process of becoming involving mutations by which the ambiguity of possibility progressively becomes more precise resulting in the emergence of a P-type being.

The process of becoming is envisioned as involving the alteration of information fields associated with the emergence of uniqueness. Necessarily information fields influencing material entities will involve the mechanisms of genetics and will entail mutations. This mutational mechanism be discussed in greater detail throughout the book. Such mutations may either tap into the subtle-libraries created through the structures already set up by the ∞-entanglement, K-entanglement, and N-entanglement levels, or involve direct interaction with the UPI-type, K-type, or N-type beings.

Note that the libraries so set up by the ∞-entanglement, K-entanglement, and N-entanglement levels would be quite different from the Library of Babel, referred to in Beinhocker's Origin of Wealth (Beinhocker, 2006). The Library of Babel is imagined containing all the possible 500-page books in the English language and would be vastly larger than the universe. A vast majority of these books would be gibberish with characters strung together in random fashion. By contrast the subtle-libraries at the N, K, or ∞ layers are a play on qualities or functions related to the four-foldness implicit in Light. These libraries potentially contain an infinite number of positive and useful functions that will be subtly available to inform any play involving genetic-type information.

The implicit nature of Time and Space suggests a universal developmental model that provides a cue as to the process for such emergence. In this model the four sets of architectural forces and the combination of their elements form a pool in space, as discussed in Section 1, from which possibility arises. Possibility itself is unique across Space and is governed by the Equation for Uniqueness (Equations 2.4.1 through 2.4.5) described in the previous chapter.

In other words, Space contains seeds, as suggested in Section 1, and the implication of the discussion in the previous chapters in Section 2, is that seeds can be thought of as the result of the superposition of ever-present ∞-entanglement, K-entanglement, and N-entanglement libraries. Thus, it could be said that Space is filled with multiple levels of superposed entanglements in static form.

Time on the other hand appears to be the working out of the possibilities implicit in these superposed entanglements. The inevitable trajectory implicit in these superposed entanglements will typically follow a mutational-sequence to be described in this chapter. What is variable is the amount of time for the mutational-sequence to express itself, as this will depend on the strength of the different forces or influences of the UPI-type, K-type, and N-type beings operating from each of the different layers so set up by light.

In equation form Space could be described by Equation 2.5.1:

$Space = STATIC(superposition\ (\infty - entanglement, K - entanglement, N - entanglement))$

Eq 2.5.1: Space

In equation form Time could be described by Equation 2.5.2:

$Time = DYNAMIC(superposition\ (\infty - entanglement, K - entanglement, N - entanglement))$

Eq 2.5.2: Time

Hence it is observed that initially the mutational-sequence takes a 'physical' form, moving on to a 'vital' form, and then onto a 'mental' form, as will be illustrated in greater detail subsequently. Relating these forms to the equation-segment in (2.1.11), the shaded lines in Illustration 2.5.1 suggest which set of dynamics tend to be more active. Hence in the process of becoming at this level it is hypothesized that it is due to the influence of the 0-type being that movement tends to be slower from one form to another:

$$c_\infty: [Pr, Po, K, H]$$
$$(\downarrow R_{C_K} = f(R_{C_\infty}))$$
$$c_K: [S_{Pr}, S_{Po}, S_K, S_H]$$
$$(\downarrow R_{C_N} = f(R_{C_K}))$$
$$c_N: f(S_{Pr} \times S_{Po} \times S_K \times S_H)$$
$$(\downarrow R_{C_U} = f(R_{C_N}))$$
$$c_U: [P, V, M, C]$$
$$\Uparrow$$
$$c_{0:[D,W,I,C]}$$

Illustration 2.5.1: 0-Type Being Influencing Initial P-V-M Stages of Mutational Sequence

Once the characteristics implicit in each of the Physical, Vital, and Mental phases are assimilated, then the mutational-sequence takes on a more integral form. At this stage it is the influence of the N-type being that is active as illustrated in Illustration 2.5.2:

$$c_\infty: [Pr, Po, K, H]$$
$$(\downarrow R_{C_K} = f(R_{C_\infty}))$$
$$c_K: [S_{Pr}, S_{Po}, S_K, S_H]$$
$$(\downarrow R_{C_N} = f(R_{C_K}))$$
$$c_N: f(S_{Pr} \times S_{Po} \times S_K \times S_H)$$
$$(\downarrow R_{C_U} = f(R_{C_N}))$$
$$c_U: [P, V, M, C]$$
$$\Uparrow$$
$$c_{0:[D,W,I,C]}$$

Illustration 2.5.2: N-type Being Influencing Integral Stage of Mutational Sequence

The integral form due to the influence of the N-type being is a threshold phase, and allows the uniqueness suggested by the Signature to emerge in fuller force or in its 'force' form. At this stage it is the influence of the K-type being that becomes more active as suggested by Illustration 2.5.3:

$$c_\infty: [Pr, Po, K, H]$$
$$(\downarrow R_{C_K} = f(R_{C_\infty}))$$
$$c_K: [S_{Pr}, S_{Po}, S_K, S_H]$$
$$(\downarrow R_{C_N} = f(R_{C_K}))$$
$$c_N: f(S_{Pr} \times S_{Po} \times S_K \times S_H)$$
$$(\downarrow R_{C_U} = f(R_{C_N}))$$
$$c_U: [P, V, M, C]$$
$$\Uparrow$$
$$c_{0:[D,W,I,C]}$$

Illustration 2.5.3: K-Type Being Influencing Force Stage of Mutational Sequence

The final phase is the 'contextual form' that allows the signature to act with impunity within a considered context. The dynamics active in the contextual form are suggested by the shaded lines in Illustration 2.5.4, which also has a more overt influence of the UPI-type being:

$$c_\infty: [Pr, Po, K, H]$$
$$(\downarrow R_{C_K} = f(R_{C_\infty}))$$
$$c_K: [S_{Pr}, S_{Po}, S_K, S_H]$$
$$(\downarrow R_{C_N} = f(R_{C_K}))$$
$$c_N: f(S_{Pr} \times S_{Po} \times S_K \times S_H)$$
$$(\downarrow R_{C_U} = f(R_{C_N}))$$
$$c_U: [P, V, M, C]$$
$$\Uparrow$$
$$c_{0:[D,W,I,C]}$$

Illustration 2.5.4: UPI-Type Being Influencing Contextual Stage of Mutational Sequence

Mathematically, if an organization exists at the physical phase, it may be suggested that its signature or uniqueness is modulated by the constant 'π'. π is the seed of a circle or sphere and can be thought of as defining behavior that is tightly bound. Within such a tightly bound volume it will likely not even be apparent what the uniqueness of an organization necessarily is. Assuming the uniqueness to be defined by the derived equation *Sig*, the physical-level (P) behavior can be described by the following equation-segment where 'mod' signifies modulated-by:

$P: Sig * mod\,(\pi)$

79

If an organization exists at the vital level, it may be suggested that its uniqueness is modulated by the Euler-constant 'e'. e is at the root of exponential behavior. The vital, by definition, is about assertive and aggressive growth the symbol of which is 'e'. Hence vital-level (V) modulation (represented by 'mod') can be described by the following equation-segment:

V: $Sig * mod\ (e)$

If an organization exists at the mental level, it may be suggested that its uniqueness is modulated by the Gaussian Distribution 'G'. G summarizes rational behavior with a key direction followed by most, and directions more on the edge followed by outliers. Mental-level dynamics are arguably quite similar, and it can be suggested are best modeled by such a distribution. Mental-level (M) modulation (mod) can hence be described by the following equation-segment:

M: $Sig * mod\ (G)$

The physical, the vital, and the mental levels as descriptive of the nature of light at c as discussed in Section 1, are also at the root of orientations that emerge later in which patterns of perceiving, being, behaving are set in certain ways. Each pattern has its purpose and its limitation, and it can be argued that being able to learn from each orientation and yet being able to move beyond that, is the next logical step in any developmental model. The integral level, hence, is about being able to leverage each of the patterns that naturally arise at the three preceding levels at will, and about further, being able to integrate these and arrive at new ways of perceiving and being.

Mathematically such behavior may be represented as being an integrative function ($\int x$) where 'x' is the ability to move between the patterns emanating from G, e, π, at will, represented by $\overline{G, e, \pi}$. Integral-level (I) modulation (mod) of uniqueness (*Sig*) can hence be represented by the following equation-segment:

I: $Sig * mod\ (\int \overline{G, e, \pi}\ )$

The condition of overcoming any fixed and limiting patterns is the prerequisite for the emergence of 'Force' or for entering into the force-level. At this level the uniqueness behind the particular development being considered can emerge in its purity and become a truly creative dynamic. This aspect of creativity that is in a sense not bound by circumstance may be represented by the constant 'c', the speed of light in a vacuum, which is an upper limit of the layer that systems practically operate in. This is also likely the level at which N-entanglement is overtly active. Force-level (F) modulation (mod) of uniqueness (*Sig*) can hence be represented by the following equation-segment:

$$F: Sig * mod (c)$$

Once the signature of an organization arises and continues to exercise itself in its purity, it achieves contextual-mastery (C) and is able to enforce itself as though the context it is acting in, that can vary in scale and complexity, were all of the same substance as itself. This is likely the level at which K-entanglement is active. This equality may be represented by the integrative function '$\int = 1$'. The equation-segment that notates this contextual-level (C) modulation (mod) applied to organizational uniqueness (*Sig*) is hence:

$$C: Sig * mod \left(\int = 1 \right)$$

Piecing all the equation-segments together the equation for the emergence of uniqueness (*Sig$_E$*), where 'X' can be any of the discussed modulations at the respective development-model levels (P, V, M, I, F, C), is hence summarized by Equation 2.5.3:

$$Sig_E = X \begin{vmatrix} C: Sig * mod \left(\int = 1 \right) \\ F: Sig \ mod (c) \\ I: Sig \ mod \left(\int \overline{G, e, \pi} \right) \\ M: Sig * mod (G) \\ V: Sig * mod (e) \\ P: Sig * mod (\pi) \end{vmatrix}$$

Eq 2.5.3: Mutational-Sequence Active in Becoming of P-Type Being

What is implied by (2.5.3) is that there is a bias in the phenomenon of mutation so that an implicit sequence is followed such that more and more of the uniqueness of an organization may emerge. **Material change is** therefore tightly tied in with implicit function. But this has to be if all is just a play of Light. Further, for the active influence of the N-type, K-type, UPI-type beings to become overt implies a much greater vertical integration with deeper levels of light. As will be discovered later, such a greater span of vertical integration into deeper levels of light is in itself supported by a corresponding and in a sense impersonal display or dynamic love. The love required to penetrate deeply, to want to penetrate deeply, so that the deepest dynamics that architect things can become active at the U-level is nothing other than a triumph of love.

Such a mutational-sequence also implies an upward-strand to the ubiquitous subtle-DNA. The general mathematics related to the upward-strand, and therefore to a key aspect of the dynamics of the becoming of P-type beings will be further elaborated in the subsequent chapter.

Chapter 2.6: The Systematization of the Process of Becoming of the P-Type Being

So far, the inherent creativity in Light as summarized by the influence of the UPI-type being, with its dynamics of four overarching properties in the nature of a point-instant and its attendant '∞-entanglement', has been considered. Further, how this deep fount of creativity is present everywhere, and how light-based sets with their attendant 'K-entanglement' dynamics representative of the influence of the K-type being that further materialize the range of creative forces, has also been considered. These architectural forces elaborate the possibility inherent in any system. Leveraging these sets of forces by virtue of 'N-entanglement', representative of the influence of the N-type being, an equation for the uniqueness of an organization, regardless of scale, was also arrived at.

In some sense the precipitation of creativity from the barely perceptible nature of the ubiquitous point-instant, to how this reveals a play of forces, to how organizations take their seed and grow from these forces, giving insight too into the dynamics of entanglement and superposition in space and time, has been traced. It is proposed that such creativity is what also spawns subtle-libraries that elaborate the structure of subtle-DNA, also providing insight into being and becoming as key light-based-singularity dynamics.

In Chapter 2.5 the process of becoming of the P-type being involving a mutational-sequence arising in time was explored, thereby also beginning to articulate the language of partial-singularity biographies. While such a sequence is intricately tied with the precipitating layers of light, and in fact can be thought of as a reversal of the downward precipitation, thereby reinforcing the notion of subtle-DNA since DNA at the material level is composed of two tightly related strands in opposite directions, this chapter will further explore a systematization of such a process of becoming of the P-type being.

Such an upward-strand is organic in nature suggesting likely paths of the process of becoming of the P-type being involving mutational-sequences, itself determined by what sets of influences are active. Yet the possibilities are wholly determined by the downward-strand and the overarching organizational principles put in place as light projects itself at varying constant speeds.

Hence, starting with the physical, which recall is suggested as being a projection of Light's property of Presence, an equation, Equation 2.6.1, is summarized as:

$$Physical = \begin{bmatrix} M_3 \rightarrow System_{Pr} \\ (\uparrow F \rightarrow I) \\ M_2 \rightarrow S_{System_{Pr}} \\ (\uparrow Sig \rightarrow F) \\ M_1 \rightarrow Sig_P \\ (\uparrow > P_P) \\ U \rightarrow Physical_U \end{bmatrix} TC \rightarrow Physical_T$$

$$Where \begin{bmatrix} Physical_U \ni [inertia, lethargy, status\ quo, ...] \\ Physical_T \ni [adaptability, durability, strength, ...] \end{bmatrix}$$

Eq 2.6.1: *Process of Physical-Type Systems Becoming of the P-Type Being*

Essentially this equation is laying out the conditions of moving from the untransformed or negative physical state represented by $Physical_U$ to the transformed or positive physical state represented by $Physical_T$.

The matrix should be read from the bottom to the top:

$$\begin{bmatrix} M_3 \rightarrow System_{Pr} \\ (\uparrow F \rightarrow I) \\ M_2 \rightarrow S_{System_{Pr}} \\ (\uparrow Sig \rightarrow F) \\ M_1 \rightarrow Sig_P \\ (\uparrow > P_P) \\ U \rightarrow Physical_U \end{bmatrix}$$

Hence, at the bottom is the starting point '$U \rightarrow Physical_U$' which identifies the default or untransformed (U) level of the physical. The next row up, $(\uparrow > P_P)$, states that when the patterns of the untransformed physical (P_P) have been overcome (>), movement to the next level ($\uparrow$) is facilitated. Breaking through to the next level, $M_1 \rightarrow Sig_P$, allows its dynamics to become active with the influence of the N-type being. Hence, the signature or uniqueness of the physical (Sig_P) becomes active at meta-level 1 (M_1). As this signature becomes more like a Force ($Sig \rightarrow F$), the conditions for breakthrough ($\uparrow$) to the next level are achieved with the active influence of the K-type being. This next level is referred to as meta-level 2 (M_2), and indicates that the architectural forces represented by the set of system-presence ($S_{System_{Pr}}$) have become more consciously active. When this Force becomes Integral ($F \to I$) then the conditions for breakthrough ($\uparrow$) to the next level are achieved with the active influence of the UPI-type being. The next level is notated as M_3 for meta-level 3, and the dynamics here indicate that the equation for system-presence becomes active. Becoming active basically means that the respective meta-level being, and its dynamics begin to act at the once 'untransformed' level (U) further modifying it. Modification or transformation began when M_1 became active. Transformation is accelerated when M_2 becomes active, and even further accelerated when M_3 becomes active.

The rate of the transformation can be better envisioned when considering action of the Transformation Circle, or TC. The TC can be thought of as 4 concentric circles, with M_3 at the center. M_3 is surrounded by M_2, which is surrounded by M_1. The outer circle is U. If TC is considered to be a clock, then at time 't = 0', the 'physical' can be thought of as being entirely in U. The clock starts ticking only when some initial patterns P_P are overcome ($>P_P$). From this point on as time proceeds the conditions for breakthrough become riper, and a sinusoidal wave begins to integrate more of the concentric circles together. The sinusoid wave (sin) is itself modulated by an euler function, e^x, where 'x' is determined by the strength to overcome patterns ($\uparrow$) which will likely vary over time but will likely tend to be positive once the clock has started ticking because of the naturally building creative expression with progressive movement. Being that the limit is the outer boundary of the concentric circles, there is further modulation by π until the 4 concentric circles have been integrated. TC, hence, may be represented by Equation 2.6.2:

$$TC \equiv (> P_P) \to \mathrm{mod}\,(\sin,\, e^x,\, \pi)$$

Eq 2.6.2: Transformation Circle

Hence, the initial nature of the physical that may be characterized by the set comprising of elements such as, lethargy, acceptance of the status quo, amongst other such elements, is represented by:

$$(Physical_U \ni [inertia, lethargy, status\ quo, ...])$$

This transforms into a physical more characterized by elements such as adaptability, durability, strength, and so on. That is:

$$(Physical_T \ni [adaptability, durability, strength, ...])$$

This transformation or process of becoming represents the inherent creativity-dynamic driving any mutation-sequence within physical-type systems. Such transformation as discussed in the previous chapters will implicitly involve the mechanisms of superposition and entanglement as emergence takes place.

Similarly, the equation for the 'Vital', Equation 2.6.3, which recall is suggested as being a projection of Light's property of Power, also shows the built-in transformation that represents the innovation-dynamic within the vital:

$$Vital = \begin{bmatrix} M_3 \to System_P \\ (\uparrow F \to I) \\ M_2 \to S_{System_P} \\ (\uparrow Sig \to F) \\ M_1 \to Sig_V \\ (\uparrow > P_V) \\ U \to Vital_U \end{bmatrix} TC \to Vital_T,$$

$$Where \begin{bmatrix} Vital_U \ni [aggression, self\ centeredness, exploitation, ...] \\ Vital_T \ni [energy, support, adventure, enthusiasm, ...] \end{bmatrix}$$

Eq 2.6.3: Process of Vital-Type Systems Becoming of the P-Type Being

The equation for the 'Mental', Equation 2.6.4, which recall is suggested as being a projection of Light's property of Knowledge, is similarly summarized as:

$$Mental = \begin{bmatrix} M_3 \to System_S \\ (\uparrow F \to I) \\ M_2 \to S_{System_S} \\ (\uparrow Sig \to F) \\ M_1 \to Sig_M \\ (\uparrow > P_M) \\ U \to Mental_U \end{bmatrix} TC \to Mental_T$$

$$Where \begin{bmatrix} Mental_U \ni [fixation, fundamentalism, fragmentation, ...] \\ Mental_T \ni [understanding, imagination, inspiration, ...] \end{bmatrix}$$

Eq 2.6.4: Process of Mental-Type Systems Becoming of the P-Type Being

The equation for the 'Integral', Equation 2.6.5, suggested as being a projection of Light's property of Harmony, is similarly summarized as:

$$Integral = \begin{bmatrix} M_3 \rightarrow System_N \\ (\uparrow F \rightarrow I) \\ M_2 \rightarrow S_{System_N} \\ (\uparrow Sig \rightarrow F) \\ M_1 \rightarrow Sig_I \\ (\uparrow > P_I) \\ U \rightarrow Integral_U \end{bmatrix} TC \rightarrow Integral_T$$

$$Where \begin{bmatrix} Integral_U \ni [possession, usurpation, hidden\ agendas, \dots] \\ Integral_T \ni [appreciation, shift\ POV, MPV, synthesis, \dots] \end{bmatrix}$$

Eq 2.6.5: Process of Integral-Type Systems Becoming of the P-Type Being

The preceding equations can be generalized by Equation 2.6.6:

$$P - Type\ Becoming_{orientation-x} = \begin{bmatrix} M_3 \rightarrow System_X \\ (\uparrow F \rightarrow I) \\ M_2 \rightarrow S_{System_X} \\ (\uparrow Sig \rightarrow F) \\ M_1 \rightarrow Sig_X \\ (\uparrow > P_X) \\ U \rightarrow x_U \end{bmatrix} TC \rightarrow x_T, where \begin{bmatrix} x_U \ni [\dots] \\ x_T \ni [\dots] \end{bmatrix}$$

Eq 2.6.6: Process of Becoming of the P-Type Being

In this generalized equation, $P - Type\ Becoming_{orientation-x}$, refers to the inherent innovation within a specific orientation. Orientation refers to the physical, the vital, the mental, or the integral.

Further, the notion of a core-matrix can be summarized by the following equation, Equation 2.6.7, and summarizes the essential conditions and levels common to the different orientations in the journey through time. Hence it also summarizes the upward-strand of subtle-DNA:

87

$$Core\_matrix = \begin{bmatrix} M_3 \rightarrow System_X \\ (\uparrow F \rightarrow I) \\ M_2 \rightarrow S_{System_X} \\ (\uparrow Sig \rightarrow F) \\ M_1 \rightarrow Sig_X \\ (\uparrow > P_x) \\ U \rightarrow x_U \end{bmatrix}$$

Eq 2.6.7: Core Matrix or Upward-Strand of Subtle-DNA

Equations 2.6.1 through 2.6.7, hence, further systematize the dynamics of mutation that are integral to the becoming of the P-type being. While these equations are tied to the time-dimension, they also define logical conditions that will allow more of the possibility resident in any seed to emerge. These conditions, as will be explored in more detail later in the book, also influence the process of quantization whereby further possibility can emerge in the material reality typified by light traveling at c.

SECTION 3: FURTHER ELABORATION OF THE MATHEMATICS OF THE BECOMING OF THE P-TYPE BEING

The previous section outlined an essential mathematical structure of being and a mathematical structure of becoming of the P-type being. Each being thus existing in the Matrix of Being or the light-based-singularity superstructure is essentially a light-based-singularity. The UPI-type, K-type, and N-type beings exist in the seed-singularity portion of the downward-strand of subtle-DNA. Each of these beings is related to layers of light traveling faster than c. As light slows down to c the material layer U comes into existence. Dynamics at U are the start of the becoming of the vast range of the P-type being.

The downward-strand is caused by light slowing down in quantized-decelerations, as it were, progressively concretizing more of the information in light. The becoming of beings in this downward-strand are therefore envisioned as precise "quantized" accumulations architected to bring about unfolding possibilities in our material universe. Other sets of decelerated-quantizations may serve other purposes and create other kinds of universes that would in effect exist simultaneously with ours. As such the UPI-type, K-type, and N-type beings are envisioned to play a significant role in any effort at transhumanism.

The upward-strand is envisioned as prescribing a time-variable sequence wholly determined by the nature of the beings in the downward-strand. The time-variability is due to the result of the interplay of the active influences of the UPI-type, K-type, and N-type beings in the downward-strand. But further, as already alluded to, each of these beings spawn subtle-libraries of pre-genetic information as part of their dynamics. These subtle-libraries are effectively infinitely large, encoding many different possibilities. Such genetic-type information is the language in which biographies of light-based singularities and all light-based beings, including the P-type being, are written.

This section will illustrate some of the interplay involving the subtle-libraries and a first projection of their possibilities in a space-time-energy-gravity quadrumvirate construct. The space-time-energy-gravity construct, itself one of the first instances of the P-type being, is positioned as being fundamental in altering any genetic information at the material level. Further, numerous other constructs that have their logic determined in 'four-base logic-encoding ecosystems' (FBLEE), envisioned as existing in the quantum layer antecedent to the material layer, will also be explored in subsequent sections. The process of becoming of the P-type being, hence, involves encoding at the quantum level as a prerequisite to further materialization of becoming. It is such ecosystems that can change due to interplay with the material layer. In the interplay or call from 'below' in which emergences find that their forms are inadequate in expressing the fullness of Light, as it were, there can be thought to be a response from 'above', from more of the fullness that Light is, that may precipitate one of the infinite functions already created by the subtle-libraries that are part of the UPI-type, K-type, and N-type beings. Such a call and response system can be thought of as operating like a lock-and-key mechanism.

As such, this section will reinterpret some basic equations and existing quantum mechanics fundamentals based on the Cosmology of Light view adopted in this treatise. Further, it will focus on the derivation of the Light-Space-Time Emergence or P-Type Being Generator equation leveraged in subsequent quantization analyses. The Light-Space-Time Emergence equation models the basis for quantization suggesting emergent reality for all phenomena from Light. Schrodinger's wave equation and Heisenberg's uncertainty principle are also interpreted from the point of view of the Light-based Interpretation of quantum mechanics central to this treatise, to reinforce the notion that even when considered from these points of view the existence of multiple layers of light is feasible. The chapter on quantization of space, time, matter and gravity, models how these phenomena are related to Light and subsequently also models how these fundamentals work together to potentially impact any genetic-type information.

This section therefore elaborates dynamics central to the becoming of the P-type being.

Chapters in this section will focus on:

- The Light-Space-Time Emergence aka P-Type Being Generator Equation
- Speed of Light and Quanta
- Interpreting Schrodinger's Equation
- Interpreting Heisenberg's Uncertainty Principle

- A Deeper Look at Quantization of Space, Time, Matter, and Gravity
- Effect of Levels of Light on the Becoming of the P-Type Being
- Application of Qualified Determinism on the Becoming of the P-Type Being

Chapter 3.1: The Light-Space-Time Emergence or P-Type Being Generator Equation

Equation 2.6.6, Process of Becoming of the P-Type Being, can be restated as an evolving form true for all time, as in Equation 3.1.1:

$$P-Type\ Becoming_{orientation-x}$$

$$= \left(\begin{bmatrix} M_3 \to System_x \\ (\uparrow F \to I) \\ M_2 \to S_{System_x} \\ (\uparrow Sig \to F) \\ M_1 \to Sig_x \\ (\uparrow > P_x) \\ U \to x_U \end{bmatrix} TC \to x_T, where \begin{bmatrix} x_U \ni [...] \\ x_T \ni [...] \end{bmatrix} \right)_{\langle x_U | x_T \rangle}$$

Eq 3.1.1: Evolving Form of Becoming of the P-Type Being

The added notation of $\langle x_U | x_T \rangle$ implies that the output of the previous iteration of the equation of innovation, x_T, where the subscript T implies relatively-transformed, now becomes the input, x_U, for the next iteration of the equation, where U implies relatively-untransformed. Hence through time there is greater and greater transformation that pushes experienced reality to greater and greater levels of functional-richness.

But further, given that quanta are proposed to be a doorway to deeper worlds of Light, that in fact allow aspects of those worlds or layers to become active at the surface layer U, the question is when have those aspects become active in manifest time. The following timeline based on generally accepted models of universal history (Particle Data Group, 2015) suggests when. Note too that the subsequent sections of exploring the generation of pre-genetic and genetic information as part of the developing biography of light-based singularities at the levels of the electromagnetic spectrum, matter, and life, will explore in far greater detail some of the statements made in the following timeline:

- At time, $t \le 0$ seconds, only M_3 is active, and then remains active for all $t < \infty$. Recall that M_3 represents the UPI-type being in its four-fold reality present in every ubiquitous-point-instant.

- At time, $0 \geq t > \infty$, M_2 the set of architectural forces continually gets added to, thereby increasing the size of the sets of forces. This implies the action of the K-type being.

- At time, $t \geq 0$, space, time, energy, gravity, the first clear expression of the four-fold order, emerges. This emergence marks the generation of the Big-Bang P-type being, as will be explored in subsequent chapters. This first expression is significant because it sets in motion the interplay between the antecedent quantum-layer and the layer where matter will materialize. Note that the antecedent quantum-layer is envisioned to house the continually-enhanced four-base logic-encoding ecosystems (FBLEE) critical to the biography of becoming of the P-type being as scripted in the language of genetic-type information.

- At time, $0 > t \geq 10^{-36}$ seconds, the equation for P-type becoming, $P - Type\ Becoming_{orientation-x}$, is such that M_1 also becomes active. The activation of M_1 implies the materialization of unique expressions or signatures and the active influence of the N-type being. The activation is due to the materialization of the ubiquitous electromagnetic-spectrum (EM Spectrum). The EM Spectrum as a vehicle of the four-fold order implies the material increase of the action of unique signatures from this point in time on. This also marks the generation of the EM Spectrum P-type being.

- At time, $t \sim 10^{-10}$ seconds, fundamental particles emerge as an essential material basis of the four architectural forces that frame all further development. As in the case of the EM Spectrum this implies the activity of M_1, with the increasing influence of the N-type being, and then also of U. This also marks the generation of the quantum particle P-type being.

- At time, $t \sim 3 \times 10^5$ years light atoms emerge, and at time $t \sim 10^9$ years heavier atoms in the stars emerge. These also imply the continued activity of M_1 and U and the generation of atom-based P-type beings. Since the action of previous P-type beings is the foundation for the atom-based P-type being this also implies the increase in the sphere of influence of this P-type being – a concept that will be elaborated later.

- At time, $t \sim 13.8 \times 10^9$ years, a further clear expression of the same fourfold order as the bases of an even more complex organization, that of cellular life and all that is founded on it comes into being. This time-point will be represented by the notation $t \sim E_{Cell}$, where 'E' stands for emergence. This too implies the activity of M_1 and the increasing influence of the N-type being. Note that with the generation of the living-cell P-type being the sets of architectural forces specified by M_2 continue to increase their cardinality as the complex interaction between the layers continues.

- At time $t > 13.8 \times 10^9$ years, human-beings, and more complex social organizations emerge. Here TC acts with an implicit direction of operation from U to M_3. This time-point will be represented by $t \sim E_{Human}$ and marks the generation of the human-based P-type being. Note that with the reversal of operation from the meta-levels (M) to the untransformed layer (U), the P-type being is further qualified as the PUM-type being implying the need for much greater and conscious effort on the part of the P-type being, as distinguished from the PMU-type being in which meta-layers organize activities, as it were, automatically.

Note that the emergence of the space-time-energy-gravity quadrumvirate PMU-type being, and subsequently the generation of the electromagnetic spectrum PMU-type being, quantum particles PMU-type being, and atoms-based PMU-type being, implies that the pre-genetic information in their attendant FBLEE must also be present in some form in genes as appear later with the advent of cellular life. This must be the case since the logic of cosmic fundamentals contained in PMU-type beings, has to be deeply ingrained in all things, inanimate and animate. This feature of subsequent emergences containing biographies for all previous emergences is fundamental to light-based emergences and to the dynamics of any light-based-singularity. PMU-type biographies have to be a part of PUM-type biographies. A vast and actionable information-base therefore animates any light-based emergence or PUM-type being.

Based on the aforementioned timeline and description Equation 3.1.2 for Emergence true of any space-time scale may be generalized as the following:

$$Emergence_{space-time} = \left| \begin{array}{c} \begin{bmatrix} M_3 \; \to \; System_X \\ (\uparrow F \; \to \; I) \\ M_2 \; \to \; S_{System_X} \\ (\uparrow Sig \; \to \; F) \\ M_1 \; \to \; Sig_X \\ (\uparrow > P_{x)} \\ U \; \to \; x_U \end{bmatrix}_{Space} \\ \left| U \; \to \; \begin{bmatrix} M_3 : \; -\infty \; \leq t \; \leq \; \infty \\ \downarrow \\ M_2 : 0 \; \geq t \; > \; \infty \\ \downarrow \\ M_1 : \; 0 > t \; > \; \infty \\ \downarrow \\ \begin{array}{c} t \; \leq \; E_{Cell}; \text{TC:} \; M_3 \; \to \text{U} \\ t \sim E_{Human}; \text{TC:} \; U \; \to M_3 \end{array} \\ TC \; \to x_T \, , where \; \begin{bmatrix} x_U \ni [\ldots] \\ x_T \ni [\ldots] \end{bmatrix} \end{bmatrix}_{Time} \right| \end{array} \right|_{\langle x_U | x_T \rangle}$$

Eq 3.1.2: Space-Time Emergence

An implication of this equation, brought out more explicitly through the elaboration of the 'Time' component, is that the layers U, M_1, M_2, and M_3 exist simultaneously. Further the advancement of time also causes the generation of P-type beings with increasing spheres of influence. Adding the Light-Matrix derived in Chapter 2.1 enhances Equation 3.1.2 to the Light-Space-Time Emergence form as represented by:

$Emergence_{light-space-time} =$

$$
\begin{bmatrix}
c_\infty: [Pr, Po, K, H] \\
(\downarrow R_{C_K} = f(R_{C_\infty})) \\
c_K: [S_{Pr}, S_{Po}, S_K, S_H] \\
(\downarrow R_{C_N} = f(R_{C_K})) \\
c_N: f(S_{Pr} \times S_{Po} \times S_K \times S_H) \\
(\downarrow R_{C_U} = f(R_{C_N})) \\
c_U: [P, V, M, C] \\
\Uparrow \\
c_0: [D,W,I,C]
\end{bmatrix}_{Light}
\begin{bmatrix}
M_3 \to System_X \\
(\uparrow F \to I) \\
M_2 \to S_{System_X} \\
(\uparrow Sig \to F) \\
M_1 \to Sig_X \\
(\uparrow > P_X) \\
U \to x_U
\end{bmatrix}_{Space}
$$

$$
\begin{bmatrix}
U \to
\begin{array}{c}
M_3 : -\infty \le t \le \infty \\
\downarrow \\
M_2 : 0 \ge t > \infty \\
\downarrow \\
M_1 : 0 > t > \infty \\
\downarrow \\
t \le E_{Cell}; TC: M_3 \to U \\
t \sim E_{Human}; TC: U \to M_3
\end{array}
\end{bmatrix}_{Time}
\qquad TC \to x_T \qquad \langle x_U | x_T \rangle
$$

Eq 3.1.3: Light-Space-Time Emergence or P-Type Being Generator

In (3.1.3) there is a 1:1 mapping between the Light and Space matrices in that M_3 reflects the ever-present C_∞, M_2 reflects C_K, M_1 reflects C_N, and U reflects C_U. Hence there are fundamental mathematical symmetries at play that suggests that the same underlying reality is only being perceived from a different point of view. The Time matrix simply gives estimates of the time at which each of the layers became active. But also (3.1.3) can, in this analysis, be thought of as the equation that generates P-type beings. Hence, (3.1.3) is also a P-type being generator equation.

Chapter 3.2: Speed of Light and Quanta

As discussed conceptually in Section 1 and also mathematically in Chapter 2.1, since c is finite and therefore there is past, present, and future implied by it, this also implies that at U a point has to become quanta. This is implicit in the notion of finiteness. Since light takes a finite amount of time to get from A to B, a "unit" of light can be thought of as related to the finite time to traverse that. Quanta at the subatomic level can be thought of as related to this finite time and distance for a unit of light to be expressed.

Planck's discovery that energy at the subatomic level requires a minimum threshold 'quantum' to express itself therefore makes sense. It is to be noted though that Planck's treatment of quanta was more as a mathematical convenience that allowed the derivation of an equation that explained the curve of radiation wavelengths at varying temperatures of a heated black body (Isaacson, 2008). Einstein though postulated quanta as a fundamental property of light itself, rather than as something that arose in the interaction of light with matter as Planck thought. Einstein's theory produced a law of the photoelectric effect where the energy of emitted electrons would depend on the frequency of light. Einstein received the Nobel Prize for this discovery (Isaacson, 2008).

Summarizing, if c is the upper limit of the layer U, then it makes sense that the lower limit h (Planck's constant) should be inversely proportional to c. Hence:

$$h \propto \frac{1}{c}$$

This relationship is substantiated by combining two well-known equations: the first is the electromagnetic equation connecting speed of light with wavelength and frequency, and the second is Einstein's photoelectric equation connecting energy with frequency of light:

(1) $C = v\lambda$
(2) $E = hv$

Yields:

$$h = \frac{E\lambda}{C}$$

About h, H.A. Lorentz the Dutch scientist has commented in The Science of Nature (Lorentz, 1925): "We have now advanced so far that this constant not only furnishes the basis for explaining the intensity of radiation and the wavelength for which it represents a maximum, but also for interpreting the quantitative relations existing in several other cases among the many physical quantities it determines. I shall mention a few only, namely the specific heat of solids, the photo-chemical effects of light, the orbits of electrons in the atom, the wavelengths of the lines of the spectrum, the frequency of the Roentgen rays which are produced by the impact of electrons of given velocity, the velocity with which gas molecules can rotate, and also the distances between the particles which make up a crystal. It is no exaggeration to say that in our picture of nature nowadays it is the quantum conditions that hold matter together and prevent it from completely losing its energy by radiation."

So just as c sets up the past-present-future experience and reality of U, h suggests that this experience will take place in shells of matter. In the absence of the limit h, as pointed out by Lorentz, only radiation, and no matter would exist. This 'past-present-future-matter' dynamic or quadrumvirate being reinforces the notion of the four-foldness implicit in the nature of Light as already discussed in Sections 1 and 2.

The suggested variance in the speed of light by meta-layer may also throw some further light on the quantum realm. First, summarizing:

1. At U the speed of light in a vacuum, c_U, is finite at 186,000 miles/sec. This finiteness creates the reality and experience of past-present-future, and further a sense of fragmentation and separation. Further, assuming that c_U is a fundamental upper-limit at U, the inverse of it, $\frac{1}{c_U}$, must define some fundamental lower limit at U. This is indeed the case as Planck's constant, h, can be perceived as being proportional to this. 'h' allows for matter to be sustained, as it fundamentally limits the dispersion of energy as suggested by Lorentz. Such upper and lower limits at U can be thought of as dynamics representative of the U-type being, as already suggested.

2. At M_3, the speed of light, c_{M_3}, is suggested as being ∞ miles/sec. This allows a reality of 'oneness' and the possibility of a suggested fourfold-intelligence existing in every ubiquitous-point-instant as the UPI-type being, as already discussed.

3. As also already suggested the quantum world, here designated by Q, because it is at boundary of U, accesses and interrelates with the meta-levels. As such, the speed of light, c_Q, will appear as a hybrid as in the following figure. Note though that it is really the speed of light at the native or resident layer that becomes active, and that this is simply being represented as c_Q for convenience:

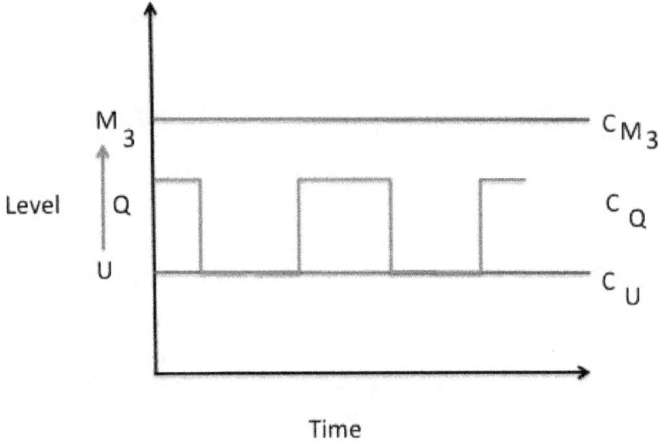

Figure 3.2.1 Speed of Light at Quantum Level

Note that research on the speed of light also indicates that it may go faster than c_U. While the speeds suggested currently through experimental research, and summarized below, may be only incrementally higher than c_U the notion that c_U can be exceeded appears to be put in place:

1. The Heisenberg Uncertainty Principle (to be discussed subsequently in Chapter 3.4) already suggests that photons can travel at any speed, even exceeding c_U, for short periods.
2. Notion of different space-time realities, also known as meta-levels in this treatise, suggests that light can travel differently in a layer different from the four-dimensional space-time that apparently defines our observable world (Hawking, 1988).
3. In his book QED Feynman (Feynman, 1985) says "...there is also an amplitude for light to go faster (or slower) than the conventional speed of light. You found out in the last lecture that light doesn't go only in straight lines; now, you find out that it

doesn't go only at the speed of light! It may surprise you that there is an amplitude for a photon to go at speeds faster or slower than the conventional speed, c." In research conducted at Humboldt University (Chown, 1990), Scharnhorst has made calculations using the theory of quantum electrodynamics to reveal the possible existence of "faster-than-light" photons. This is known as the Scharnhorst effect.

4. As reviewed in Chapter 2.1, Perkowitz makes the point that the theory of relativity does not disallow particles already moving at c or greater.

The point is that the reality at Q is going to be different than the reality at U. This should be apparent from considering the relation of c_X, where 'x' is '∞', 'K', 'N', U', and 0 respectively, to the consequent reality as elaborated in Chapter 2.1. In Q the fundamental lower limit, h, which allows matter to sustain itself, is itself going to fluctuate. Hence, as X tends to M_3, c_X will tend to infinity, and h will tend to 0. As it becomes a fraction of itself the quantization effect will be lowered, and matter will get dispersed more and more easily to in effect take on a wave-like, or field-like, or even block-like appearance as also suggested in Chapter 2.1.

Chapter 3.3: Reinterpreting Schrodinger's Equation in a Cosmology of Light

Schrodinger's equation, which seeks to model how a quantum state of a quantum system changes with time, or in other words seeks to model matter as a wave rather than as a particle (Stewart, 2012), is depicted in Equation 3.3.1:

$$i \frac{h}{2\pi} \frac{\partial}{\partial x} \psi = \hat{H} \psi$$

Eq 3.3.1: Schrodinger's Equation

ψ depicts a wave form and can be thought of as a probable cloud of possible states. $\hat{H}$ is the Hamiltonian operator, which is a focusing function, and in its essence what the equation may be suggesting is that the way a wave form changes over time is equivalent to some expressible state of the possibilities inherent in the cloud of possible states.

But the cloud of possible states is another way of saying that behind the layer U, form is represented in another way than at U. If the existence of the meta-levels, and in this case, of Sig_x at M_1, is considered possible, then it is far more reasonable to admit that form is configured by function and the very dynamics of what may appear to have been random, as some interpretations of quantum phenomena suggest (Wimmel, 1992), may now appear to be far more logical. There is now more *context* to interpreting observation at the quantum level.

Schrodinger's equation, a cornerstone in quantum theory, alludes to the mystery that accompanies the lifting of the material veil. Different interpretations of the quantum world have arisen to suggest the meaning of potentiality implicit in this equation. As summarized by Kleinman in his book "The Four Faces of the Universe" (Kleinman, 2006), at one end are theories to do with 'hidden variables' that bring into focus our ignorance of the deeper aspects of the system being studied, to David Bohm's notion of the 'implicate order' in which any element contains enfolded within it the totality of the universe (Bohm, 1983), to a 'many-worlds' interpretation asserting that every observation of a quantum system splits the universe into parallel and disconnected worlds.

Interestingly Schrodinger himself had misgivings about the applicability of this equation that seemed to apply at the quantum level, to the macro-world (Stewart, 2012). To bring his misgivings to light he invented a thought experiment concerning a cat. This cat would be in a superposed state in a quantum black-box. A radioactive particle, a decaying-particle detector, and a flask of poison were the other inhabitants of the black-box. At some point the particle will decay, be detected, and as in the thought-experiment at that point, triggered by the decaying particle, the poison in the flask would be released. The cat would then die. But in the meanwhile, the cat would be in superposed states of being both dead and alive. Only when the box was opened would the wave function collapse and a single definite state emerge. This thought experiment is also considered the origin of the notion of 'superposition'. The mathematical model presented in this book though has already interpreted 'superposition' differently as discussed previously.

Schrodinger was hoping to highlight the absurdity of the application of having a cat in both a dead and alive state at the macro-level. Instead physicists found this thought experiment to be sensible, even at the macro-level, and began to generalize the quantum theory based on this. Hence, for example the idea of superposition at the physical level began to be thought of as real. It is interesting to note that in his lectures on Schrodinger's equation Feynman (Gottlieb, 2013) has stated: "Where did he get that [equation] from? Nowhere. It is not possible to derive it from anything you know. It came out of the mind of Schrödinger".

The Copenhagen Interpretation of Quantum Mechanics in fact does away with the real questions about what reality is altogether, by simply focusing on the observed universe. This orientation perhaps stems from one of its founders, Neils Bohr who claimed that it was not the purpose of physics to answer questions about the nature of reality (Kleinman, 2006).

Considering Schrodinger's equation in the light of the discussion on Q in Chapter 3.2, 'i' is a complex number and suggests the interplay of two dimensions, one being real, and one being 'imaginary'. But the 'imaginary' dimension could be thought of as none other than the meta-levels implicit in the mathematical model in this treatise, and suggested to be real at Q. Further, $\frac{h}{2\pi}$ is in line with the suggestion also made in Chapter 3.2 that h will have to become a fraction of itself as c increases. Hence, the change in the wave function, $\frac{\partial}{\partial x}\psi$, is intimately related to i and $\frac{h}{2\pi}$, and perhaps more fully makes sense when considered in the context of i x $\frac{h}{2\pi}$ x $\frac{\partial}{\partial x}\psi$, which has to be the case when dealing with the integration of dynamics of multiple levels of light.

Further, the change in the wave function, $\frac{\partial}{\partial x}\psi$, is related to $\hat{H}\psi$, and suggests that there is some system "energy", represented by the Hamiltonian, $\hat{H}$, that when applied to the existing wave, ψ, will indicate how the wave will be expressed going forward.

But as discussed, at Q the dynamics of Sig_x representative of the N-type being become real, and in fact is a fundamental organizing principle for all organization at U, and starting at the space-dimension of h.

Chapter 3.4: Reinterpreting Heisenberg's Uncertainty Principle in a Cosmology of Light

In his book, The Little Book of String Theory, Princeton University's Gubser (Gubser, 2010) describes the effect on approaching absolute zero temperature on molecules. He takes the example of water molecules and relates that one cannot make the water molecules colder than absolute zero, -273.15 Celsius, because there is no more thermal energy to suck out at that temperature. However, quantum uncertainty, the phenomenon which relates the momentum and location of electrons in atoms necessitates that the water molecules will still vibrate. Gubser suggests this by considering Heisenberg's uncertainty relation, reproduced in Equation 3.4.1:

$$\Delta p \; X \; \Delta x \; \geq \; \frac{h}{4\pi}$$

Eq 3.4.1: Heisenberg's Uncertainty Relation

In Equation 3.4.1, Δp is the uncertainty in a particle's momentum, Δx is the uncertainty in the particle's location, and h is the Planck's constant. In frozen water crystals it is precisely known where the water molecules are, and therefore Δx is fairly small. This means that Δp has to be considerably larger, and therefore that the water molecules are still vibrating even though they are at absolute zero. This innate vibration, known as 'quantum zero-point' energy, expresses the phenomenon of quantum fluctuations.

The Planck's constant order of magnitude (10^{-34}) though, suggests the boundary between U and M_1 and the quantum fluctuations, the uncertainty relation, and the quantum zero-point energy could be an expression of the essential Signature function, Sig_x, that is posited as a key formative force behind organization at U. Sig_x, as suggested earlier is a key dynamic of the N-type being, and encapsulates uniqueness precipitated as light projects itself in quantized-decelerations from its native state of traveling infinitely fast, to its scientifically recognized speed of c, at layer U.

In this interpretation the thermal energy describes the essential energy at U, while the uncertainty relation may suggest the phenomenon of function-precipitation from other layers of Light, "physically" linking M_1 and U. In this case it may be suggested that integration of meta-levels with the surface level, I_U^M, is indicated by the uncertainty relation, as in Equation 3.4.2:

$$I_U^M \rightarrow \Delta p \; X \; \Delta x \geq \frac{h}{4\pi}$$

Eq 3.4.2: Integration of Levels (Leveraging Heisenberg's Uncertainty Relation)

This also casts a different interpretation on the cosmological phenomenon of quantum vacuum, reinforcing in line with this treatise, that the "emptiness" of space is in fact a veil to infinitely more information seeking precipitation at the material level.

But further, it may also be suggested that the uncertainty principle itself is only valid at U, and that too, because of the finiteness of c. This finiteness as already suggested implies h, which implies that if the position of a particle is going to be observed by shining light on it, the light has to have at least a quantum of energy. But to determine the position of a particle accurately, light of a shorter wavelength would have to be used (Hawking, 1988) which would have to have a minimum amount of energy, which in turn would interfere with the velocity and hence momentum of the particle. The uncertainty in measuring the momentum could therefore be thought of as a consequence of the finiteness of the speed of light, c.

If c_U were to approach c_Q though, which as suggested in Chapter 3.2 could be anywhere between c and ∞ miles per second, the quantum would be smaller and the uncertainty in measuring position or momentum would be reduced. At c_{M_3} there would be no uncertainty since light would accurately tell both position and momentum definitively.

Hence, the uncertainty principle may be further qualified, as in Equations 3.4.3, 3.4.4, and 3.4.5:

$$@C_U: \Delta p \; X \; \Delta x \geq \frac{h}{4\pi}$$

Eq 3.4.3: Uncertainty Principle at U

$@C_Q: \Delta p \times \Delta x \to 0$

Eq 3.4.4: Uncertainty Principle at Q

$@C_{M_3}: \Delta p \times \Delta x = 0$

Eq 3.4.5: Uncertainty Principle at M3

The trend toward a minimization of uncertainty as the speed of light increases to infinity, also makes sense from the point of view of eliminating the genre of dynamics originating from the 0-type being, that due to its nature of ignorance, myopia, and stagnation, amongst other such dynamics, will always have an essentially 'uncertain' influence on any dynamics at U. This is in contrast to dynamics originating from the UPI-type, K-type, and N-type beings that will on the contrary always have a more certain and precise influence on dynamics at U.

The notion of position and momentum becoming finite at U also may imply that space, time, and quanta are emergent rather than absolute properties, as also suggested in Section 1. This is also the conclusion of Arkani-Hamed of the Institute of Advanced Studies in the following thought experiment (Wolchover, 2013):

'Locality says that particles interact at points in space-time. But suppose you want to inspect space-time very closely. Probing smaller and smaller distance scales requires ever higher energies, but at a certain scale, called the Planck length, the picture gets blurry: So much energy must be concentrated into such a small region that the energy collapses the region into a black hole, making it impossible to inspect. "There's no way of measuring space and time separations once they are smaller than the Planck length," said Arkani-Hamed. "So we imagine space-time is a continuous thing, but because it's impossible to talk sharply about that thing, then that suggests it must not be fundamental — it must be emergent."

Unitarity says the quantum mechanical probabilities of all possible outcomes of a particle interaction must sum to one. To prove it, one would have to observe the same interaction over and over and count the frequencies of the different outcomes. Doing this to perfect accuracy would require an infinite number of observations using an infinitely large measuring apparatus, but the latter would again cause gravitational collapse into a black hole. In finite regions of the universe unitarity can therefore only be approximately known.'

Chapter 3.5: The Space-Time-Energy-Gravity Quadrumvirate Micro- and Macro-Beings

The Light-Space-Time Emergence equation (3.1.3) derived in Chapter 3.1, and reproduced here for convenience, offers some insight into the process of quantization. Reproducing the equation:

$$Emergence_{light-space-time} =$$

$$
\begin{bmatrix}
\begin{bmatrix}
C_\infty : [Pr, Po, K, H] \\
(\downarrow R_{C_K} = f(R_{C_\infty})) \\
c_K : [S_{Pr}, S_{Po}, S_K, S_H] \\
(\downarrow R_{C_N} = f(R_{C_K})) \\
c_N : f(S_{Pr} \times S_{Po} \times S_K \times S_H) \\
(\downarrow R_{C_U} = f(R_{C_N})) \\
c_U : [P, V, M, C] \\
\Uparrow \\
C_{0:[D,W,I,C]}
\end{bmatrix}_{Light}
\begin{bmatrix}
M_3 \rightarrow System_X \\
(\uparrow F \rightarrow I) \\
M_2 \rightarrow S_{System_X} \\
(\uparrow Sig \rightarrow F) \\
M_1 \rightarrow Sig_X \\
(\uparrow > P_x) \\
U \rightarrow x_U
\end{bmatrix}_{Space} \\
\begin{bmatrix}
U \rightarrow
\begin{matrix}
M_3 : -\infty \le t \le \infty \\
\downarrow \\
M_2 : 0 \ge t > \infty \\
\downarrow \\
M_1 : 0 > t > \infty \\
\downarrow \\
t \le E_{Cell}; \, TC: M_3 \rightarrow U \\
t \sim E_{Human}; \, TC: U \rightarrow M_3
\end{matrix}
\end{bmatrix}_{Time}
\quad TC \rightarrow x_T
\end{bmatrix} \langle x_U | x_T \rangle
$$

The Light-Matrix, the top left-hand matrix, suggests that there are particular kinds of quantization that could occur. Along the vertical realm these can be thought of as inter-relating one layer of the matrix with the previous layers, or c_0 with c_U. Hence the kind of quantization that is relevant to the physical layer, U, would relate c_U with c_x where $x \in (N, K, \infty, 0)$, one of the possible speeds of light as per this mathematical treatise. This quantization can be represented by h_U, where h stands for Planck's constant. Note that in this model there are several fundamental quantization possible that may inter-relate a specific light-layer with other layers of light. For the sake of simplicity, it will be assumed that h_U is the quanta experienced in the inter-relation between U and the collective-set of light-layers behind or above U.

In looking at the Light-Matrix it is also clear that each subsequent layer below the top layer, describing the reality set up by an infinite speed, is emergent. Hence Space, Time, Energy, and Gravity, that arise when Light slows down from c_N to c_U, can be thought of as emergent phenomena. The emergence itself can be thought of as a function of the Light-Matrix. At layer U, and as discussed in Chapter 1.2, Space is related to Light's property of Knowledge, Time is related to Light's property of Power, Matter or Energy is related to Light's property of Presence, and Gravity is related to Light's Property of Harmony or Nurturing. Each of these will emerge in a particular way and it is possible that there are multiple h_U's.

Hence, the Planck's constant for matter or energy, which we are already familiar with from past scientific discoveries, can be depicted as h_{UPr}, since it is related to Presence (Pr). But similarly, quantization for space, time, and gravity could potentially be governed by other similar constants, referred to as h_{UK}, h_{UP}, and h_{UH}, and related to Knowledge (K), Power (P), and Harmony (H), respectively. Since the relationship between these constants, in absolute terms, is uncertain, this can be represented by the equality-inequality as depicted by Equation 3.5.1:

$$h_{UPr} \lesseqgtr h_{UK} \lesseqgtr h_{UP} \lesseqgtr h_{UH}$$

Eq 3.5.1: Equality-Inequality Relationship Between Different 'Planck' Constants

Further, the "quantization-window", that is, the window that quanta provides to layers of light behind U, as it were, potentially allows the precipitation of, or inter-relation with, or creation of a cohesive and compelling meta-function or signature as modeled by the c_N (or M_1) layer. This quantization-window is positioned as being key in allowing a phenomenon of "quantum-certainty" to occur. Quantum certainty can be thought of as allowing space-time-energy-gravity quantization to occur and is described in detail in the book Quantum Certainty (Malik, 2017e), part of the 6-book Cosmology of Light series (Malik, 2017-2018).

Subsequent Sections in this book will describe several examples of the activation of the multi-dimensional quantization that ensures change at the material-level initiated at FBLEE at the quantum-levels. Such change at the FBLEE and material levels is also the basis of genetic mutation. If the mutation is influenced by the layers set up by c_K, c_N, or c_∞, then the mutation is constructive because the material outcomes of Light are being more integrated with Light itself. If the mutation is influenced by c_0, then the mutation is destructive because the material outcomes of Light are being further separated or fragmented from Light itself.

Such quantization requires patterns to be overcome, and so long as these are, this will result in the phenomenon of quantum-certainty. The 'Time' matrix, in (3.1.3) reproduced above, indicates the general default direction under which such quantum-certainty can occur at Layer U. Generally, at the pre-human level it may proceed more automatically by the dynamics of the system itself, as represented by the segment: $t \leq E_{Cell}$; TC: $M_3 \rightarrow U$. Beyond this level it will generally occur through the action of a cohesive will, as represented by the segment: $t \sim E_{Human}$; TC: $U \rightarrow M_3$. The opening of this quantization-window is modeled by the 'Space' matrix and will happen when habitual patterns are overcome so that 'will' or 'want' becomes cohesive.

Such inter-dependence between human want or will and quantum dynamics is a possible basis of Human-Quantum computational models, as discussed in greater detail in 'The Emperor's Quantum Computer' (Malik, 2018c) and is envisioned to play a critical role in genetic mutation and evolution.

The specific quantization that are occurring along the space, time, energy/matter, and gravity dimensions are modeled by the following equations, that are all based on the Sig_x equations derived in Chapter 2.4. As a reminder (2.4.5) is reproduced here for convenience:

$$Sig = Xa + \overline{Yb_{0-n}}$$

$$where: \begin{bmatrix} X \in [S_{System_{Pr}}, S_{System_P}, S_{System_K}, S_{System_N}] \\ Y \in [S_{System_{Pr}}, S_{System_P}, S_{System_K}, S_{System_N}] \\ a, b \ are \ integers; a > b \end{bmatrix}$$

The Quantization and Structure of Space

The quantization of Space, which holds the seeds of knowledge of all that will emerge, hence, is modeled in the following way as specified by Equation 3.5.2:

$$Space_{quantization} = h_{UK}\left(Xa + \overline{Yb_{0-n}}\right)$$

$$\text{where:} \begin{bmatrix} X \in [S_{System_K}] \\ Y \in [S_{System_{Pr}}, S_{System_P}, S_{System_K}, S_{System_N}] \\ a, b \text{ are integers; } a > b \end{bmatrix}$$

Eq 3.5.2: Space Quantization

In this model 'space' as an emergent property of Light, is structured by infinite seeds of knowledge. But because the emergence is taking place in a layer of reality generally itself structured by a finite speed of light, c, it has to be quantized. The quantization assures that the knowledge is not dissipated, but can accumulate to create seeds, and therefore, the structure of space itself. But such quantization as suggested earlier, also ensures that something of the possibility available at light traveling at a higher speed is materialized at the layer created when light slows down. In this treatise the only reason that Light slows down is to allow such materialization to occur. Such a structure of space appears consistent with recent models on the structure of space as explored in Rovelli's 'Reality Is Not What It Seems' (Rovelli, 2017). It is assumed that there is some kind of 'Planck's constant' in effect, that is modeled as being specific to the way knowledge or space may be quantized – hence, h_{UK}.

The structure of Space itself would be the summation of infinite seeds, as in Equation 3.5.3, Structure of Space:

$$Space_{Structure} = \sum_{i,j=1}^{\to \infty} h_{UK}\left(X_i a + \overline{Y_j b_{0-n}}\right)$$

$$\text{where:} \begin{bmatrix} X_i \in [S_{System_K}] \\ Y_j \in [S_{System_{Pr}}, S_{System_P}, S_{System_K}, S_{System_N}] \\ a, b, i, j \text{ are integers; } a > b \end{bmatrix}$$

Eq 3.5.3: Structure of Space

The Quantization and Structure of Time

The quantization of Time, which holds the inevitability of the seeds or knowledge emerging in a phased maturity, hence, is modeled in the following way as specified by Equation 3.5.4, Time Quantization:

$$Time_{quantization} = h_{UP}\left(Xa + \overline{Yb_{0-n}}\right)$$

$$where: \begin{bmatrix} X \in [S_{System_P}] \\ Y \in [S_{System_{Pr}}, S_{System_P}, S_{System_K}, S_{System_N}] \\ a, b \ are \ integers; a > b \end{bmatrix}$$

Eq 3.5.4: Time Quantization

In this model 'time' as an emergent property of Light, is structured by an inevitable process of maturity, which due to its inevitability, is related to power. But because the emergence is taking place in a layer of reality generally itself structured by a finite speed of light, c, it has to be quantized. The quantization assures that the power is not dissipated, but can accumulate to express phased maturity, and therefore, the structure of time itself. It is assumed that there is some kind of 'Planck's constant' in effect, that is modeled as being specific to the way power or time may be quantized – hence, h_{UP}.

The infinity of Time itself would be the summation of the maturation of infinite seeds, as in Equation 3.5.5, Infinity of Time:

$$Time_{Infinity} = \sum_{i,j=1}^{\to \infty} h_{UP}(X_i a + \overline{Y_j b_{0-n}})$$

$$where: \begin{bmatrix} X_i \in [S_{System_P}] \\ Y_j \in [S_{System_{Pr}}, S_{System_P}, S_{System_K}, S_{System_N}] \\ a, b, i, j \ are \ integers; a > b \end{bmatrix}$$

Eq 3.5.5: Infinity of Time

The Quantization and Aggregation of Energy

Energy, which through a process of containment can result in Matter, hence, is modeled in the following way as specified by Equation 3.5.6:

$$Energy_{quantization} = h_{UPr}(Xa + \overline{Y b_{0-n}})$$

$$where: \begin{bmatrix} X \in [S_{System_{Pr}}] \\ Y \in [S_{System_{Pr}}, S_{System_P}, S_{System_K}, S_{System_N}] \\ a, b \ are \ integers; a > b \end{bmatrix}$$

Eq 3.5.6: Energy Quantization

In this model 'energy' as an emergent property of Light, results in the reality of matter. But because the emergence is taking place in a layer of reality generally itself structured by a finite speed of light, c, it has to be quantized. The quantization assures that the energy is not dissipated but can accumulate to create matter. Planck's constant is referred to as - h_{UPr}.

The Aggregation of Energy itself would be the cumulative dynamism in the play of infinite seeds, as in Equation 3.5.7, Aggregation of Energy:

$$Energy_{Aggregation} = \sum_{i,j=1}^{\rightarrow \infty} h_{UPr}\left(X_i a + \overline{Y_j b_{0-n}}\right)$$

$$where: \begin{bmatrix} X_i \in [S_{System_{Pr}}] \\ Y_j \in [S_{System_{Pr}}, S_{System_P}, S_{System_K}, S_{System_N}] \\ a, b, i, j \text{ are integers}; a > b \end{bmatrix}$$

Eq 3.5.7: Aggregation of Energy

Quantization and Universal Gravity

Gravity, which holds seemingly distinct objects in the layer of reality created by c together in a harmony, hence, is modeled in the following way as specified by Equation 3.5.8:

$$Gravity_{quantization} = h_{UH}\left(Xa + \overline{Yb_{0-n}}\right)$$

$$where: \begin{bmatrix} X \in [S_{System_N}] \\ Y \in [S_{System_{Pr}}, S_{System_P}, S_{System_K}, S_{System_N}] \\ a, b \text{ are integers}; a > b \end{bmatrix}$$

Eq 3.5.8: Gravity Quantization

In this model 'gravity' as an emergent property of Light, results in a harmonious collectivity of seemingly independent objects. But because the emergence is taking place in a layer of reality generally itself structured by a finite speed of light, c, it has to be quantized. The quantization assures that the harmony is not dissipated but can accumulate to express more and more complex collectivities on large-scale. It is assumed that there is some kind of 'Planck's constant' in effect,

113

that is modeled as being specific to the way harmony or gravity may be quantized – hence, h_{UH}.

Universal Gravity itself would be the cumulative harmony of distinct infinite seeds, as in Equation 3.5.9, Universal Gravity:

$$Gravity_{Universal} = \sum_{i,j=1}^{\to\infty} h_{UH}\left(X_i a + \overline{Y_j b_{0-n}}\right)$$

$$where: \begin{bmatrix} X_i \in [S_{System_N}] \\ Y_j \in [S_{System_{Pr}}, S_{System_P}, S_{System_K}, S_{System_N}] \\ a, b, i, j \ are \ integers; a > b \end{bmatrix}$$

Eq 3.5.9: Universal Gravity

The preceding analysis also suggests that Space is a repository of pre-genetic information or of pre-genetic libraries. Time, Gravity, and Energy work with Space to bring about material possibilities. Change to any material possibility requires therefore the action of space, time, energy, and gravity. At the margin there is a fourfold quantization that brings about such change.

Such composite space-time-energy-gravity quantization is hence a key light-based-singularity dynamic. This quadrumvirate can be thought of as a space-time-energy-gravity micro-being essential to the becoming of any P-type being. Such becoming would in fact not be, unless space-time-energy-gravity "gave" itself to such a process to alter the very genetic-type script necessary to enhancing the biography of the P-type being. Such an act of giving where one type of being offers its dynamics to another type of being, in this case the space-time-energy-gravity quadrumvirate micro-being giving itself to the P-type being, is nothing other than a dynamic that can be called Love.

It is proposed that any process of transhumanism has to involve larger and larger acts of love. Hence, effective transhumanism in a Cosmology of Light could be perceived as the result of beings such as the UPI-type, or K-type, or N-type, integrating with the PUM-type being, with the space-time-energy-gravity micro-being as scribe and witness, to in effect maintain the integrity of light-based-singularities.

Further, all the integration of light-based-singularities itself takes place in the macro-container of Space-Time-Energy-Gravity at their larger aggregated levels. This latter reality will be referred to as the Space-Time-Energy-Gravity Macro-Being.

Chapter 3.6: The Effect of Levels of Light on the Becoming of the P-Type Being

The Light-Space-Time Emergence or P-Type Being Generator equation (3.1.3) being iterative can be used to model emergence from simpler to more complex four-fold manifestations. In other words (3.1.3) suggests a computational approach to the development of the universe. Pre-genetic and genetic information may be thought of as the output of such computation. This computational approach involves multiple layers of reality as suggested by (3.1.3) and is driven more by a process of qualified determinism, to be explored in more detail in the next chapter, than probability and statistics. This notion of qualified determinism is in contrast to the prevalent probabilistic view that has been erected as the cornerstone of quantum theory, **and to MIT's Seth Lloyd's Copenhagen-like** quantum superposition, probability-based computational approach to the development of the universe **as described in his book** 'Programming the Universe' (Lloyd, 2007).

Examples of the P-type generation-based computation will be reviewed in Section 4, to provide an overview of the process that results in a series of more and more complex P-type beings. The output of such computations is genetic-type information that also creates the laws or script that determines the biographies by which P-type beings exist. Sections 5, 6, 7, and 8 will further elaborate the process of becoming, by reviewing in detail the Light-Space-Time based computation that drives the change in genetic-type script and the P-type being.

However, in keeping with the current quantum-analysis paradigm of using wavefunction to explore outcomes at a particular space and time, this chapter will develop two general approaches summarized by equations, to understanding the effect that different layers of light may have on the process of becoming of the P-type being. The first approach will view (3.1.3) from a wavefunction lens, while the second approach will follow the natural line of development already being followed in this book to reinforce an essentially iterative-type computation.

(3.1.3) is reproduced below for convenience and a more concise form of it, the Simplified Version of Light-Space-Time Emergence, Equation 3.6.1, shall be utilized through the rest of this chapter.

$$Emergence_{light-space-time} =$$

$$\left| \begin{bmatrix} \begin{bmatrix} c_\infty: [Pr, Po, K, H] \\ (\downarrow R_{C_K} = f(R_{C_\infty})) \\ c_K: [S_{Pr}, S_{Po}, S_K, S_H] \\ (\downarrow R_{C_N} = f(R_{C_K})) \\ c_N: f(S_{Pr} \times S_{Po} \times S_K \times S_H) \\ (\downarrow R_{C_U} = f(R_{C_N})) \\ c_U: [P, V, M, C] \\ \Uparrow \\ c_{0:[D,W,I,C]} \end{bmatrix}_{Light} \quad \begin{bmatrix} M_3 \rightarrow System_X \\ (\uparrow F \rightarrow I) \\ M_2 \rightarrow S_{System_X} \\ (\uparrow Sig \rightarrow F) \\ M_1 \rightarrow Sig_X \\ (\uparrow > P_x) \\ U \rightarrow x_U \end{bmatrix}_{Space} \\ \begin{bmatrix} M_3 : -\infty \le t \le \infty \\ \downarrow \\ M_2 : 0 \ge t > \infty \\ \downarrow \\ M_1 : 0 > t > \infty \\ \downarrow \\ U \rightarrow \begin{matrix} t \le E_{Cell}; TC: M_3 \rightarrow U \\ t \sim E_{Human}; TC: U \rightarrow M_3 \end{matrix} \end{bmatrix}_{Time} \quad TC \rightarrow x_T \end{bmatrix} \right|_{\langle x_U | x_T \rangle}$$

Simplifying each of the main matrices in (3.1.3), hence, yields Equation 3.6.1, the Simplified Version of the Light-Space-Time Emergence equation:

$$Emergence_{light-space-time} = |[L][S][T]TC \rightarrow x_T|_{\langle x_U | x_T \rangle}$$

Eq. 3.6.1: Simplified Version of Light-Space-Time Emergence

As detailed in Chapter 2.6 on the systematization of the becoming of the P-type being, while relatively transformed organizations have opened to the active influence of meta-levels, relatively untransformed organizations remain under the influence of untransformed physical, vital, mental, and integral dynamics at U. This modeling can also be applied to micro-level dynamics such as mutation. Keep in mind that genetic-mutation as a level of granularity is important because this is the script or language in which partial-singularity or the P-type being biographies are written. Hence, as discussed in Chapter 2.6, the notion of relatively transformed organizations will translate to that of constructive-mutation. The notion of relatively untransformed organizations will translate to that of destructive-mutation.

117

Effect of Levels of Light on Mutation Using Wavefunction Form of Light-Space-Time Emergence

In Chapter 3.3 we were introduced to Schrodinger's wavefunction equation (3.3.1), reproduced below for convenience, which seeks to model matter as a wave rather than as a particle:

$$i\frac{h}{2\pi}\frac{\partial}{\partial x}\psi = \hat{H}\psi$$

This wavefunction equation deals with dynamics behind the surface, and particularly of as many as infinite potential superposed states that then collapse into a material possibility in time and space at U. Such superposition is another way of saying that there is wholeness behind the surface. In (3.6.1) such wholeness is represented by L, S, T - the Light, Space, and Time matrices - respectively.

As explained by Neil Turok in his book The Universe Within (Turok, 2012), Euler's formula reproduced below, can be used to model many naturally occurring phenomena because of its sinusoidal oscillation between narrow bounds as x increases. Reproducing Euler's formula:

$$e^{ix} = \cos x + i\sin x$$

A key characteristic of this equation is that the sum of the squares of the ordinary and complex parts, on the right side of the equation, is one. In many contemporary interpretations of quantum theory this ensures that the probabilities for all possible outcomes add up to one. Dealing in probabilities become important when considering as many as infinite superposed states.

Further, in modeling material reality a modified notation of the Schrodinger wavefunction as interpreted by Feynman is leveraged. This version features the integral sign, $\int$, meaning that all terms to the right of it have to be summed up for all space and time till the moment when the wavefunction is required to be known.

Hence, combining (3.6.1) with Feynman's interpretation of the Schrodinger wavefunction, with the Euler formula yields the following equations, 3.6.2-7 for genetic mutation:

$$\psi_{destructive-mutation} = \left|\left| \int e^{i \int |[L][S][T]|} \right|\right|_{U}$$

Eq. 3.6.2: Wavefunction for Destructive-Mutation

Note that in (3.6.2) the formulation specifying iteration, $\langle x_U | x_T \rangle$, becomes redundant since iteration is implied by the integral across space and time, and hence is removed. Further, (3.6.2) suggests that so long as the basis of an organization is untransformed, specified by the U following the vertical-brackets, the outcome is going to be one of destructive mutation, specified by $\psi_{destructive-mutation}$.

Further, as specified by Equation 3.6.3, the Probability-View of Destructive-Mutation, the driver of such destructive-mutation is going to be either the untransformed physical (P_U), the untransformed vital (V_U), the untransformed mental (M_U), or the untransformed integral (I_U):

$$|\psi_{destructive-mutation}|^2$$
$$= P_U^2 + V_U^2 + M_U^2 + I_U^2 = 1$$

Eq. 3.6.3: Probability-View for Destructive-Mutation

As specified by (3.6.3) the probability that any of these bases will be leveraged in destructive mutation adds up to one. Note that the physical, the vital, the mental, and integral were discussed in Chapter 2.6 focusing on the systematization of the process of becoming of the P-type being.

Equation 3.6.4, Wavefunction for Random-Mutation, suggests the mixed bases for organizations, as specified by dynamics of both the untransformed (U) and the meta-levels (M_x), which therefore results in random mutation.

$$\psi_{ransom-mutation} = \left|\left| \int e^{i \int |[L][S][T]|} \right|\right|_{U \& M_x}$$

Eq. 3.6.4: Wavefunction for Random-Mutation

As specified by Equation 3.6.5, Probability-View for Random-Mutation, the probability that any of the untransformed (U) and transformed (T) bases will be leveraged in random mutation adds up to one:

119

$$|\psi_{random-mutation}|^2 = P_U^2 + P_T^2 + V_U^2 + V_T^2 + M_U^2 + M_T^2 + I_U^2 + I_T^2 = 1$$

Eq. 3.6.5: Probability View for Random-Mutation

Equation 3.6.6, Wavefunction for Constructive-Mutation, suggests some transformed bases for an organization, as specified by dynamics of the meta-levels (M_x), which therefore can result in a constructive mutation.

$$\psi_{constructive-mutation} = \left| \int e^{i \int |[L][S][T]|} \right|_{M_x}$$

Eq. 3.6.6: Wavefunction for Constructive-Mutation

As specified by Equation 3.6.7, Probability-View for Constructive-Mutation, the probability that the transformed (T) bases will be leveraged in constructive mutation adds up to one:

$$|\psi_{constructive-mutation}|^2 = P_T^2 + V_T^2 + M_T^2 + I_T^2 = 1$$

Eq. 3.6.7: Probability View for Constructive-Mutation

Now the question is how would genetic information be impacted based on the levels of light active, and what might an equation that captured that look like?

Equation 3.6.8, Potential Effect of Levels of Light on Genetic-Type Information (Wavefunction Form), is such an equation:

$$\psi_{Potential\ Effect\ of\ Levels\ of\ Light\ on\ Pre\ Genetic\ or\ Genetic\ Information} =$$

$$\begin{bmatrix} \left(\left| \int e^{i \int |[L][S][T]|} \right|_{Y=(U,U\&M_x,M_x)} \right) \\ \times \\ \left((Y > U: Z_Q) \vee (Y \leq U: Z_F) \vee (Y = U: Z_R) \right) \\ \ni \\ \left(\begin{array}{c} Z \in \mathbb{U}\ (\text{Space, Time, Energy, Gravity}) \\ Q: Quantization; F: Fragmentation; R: Random \end{array} \right) \end{bmatrix} \rightarrow$$

$$h \ni h \in \left(\begin{array}{c} \text{Constructive zone,} \\ \text{Constructive zone} \wedge \text{Constructive mutation,} \\ \text{Destructive mutation,} \\ \text{Random mutation} \end{array} \right)$$

Eq. 3.6.8: Potential Effect of Levels of Light on Genetic-Type Information (Wavefunction Form)

Taking each line in the equation separately: Line 1 from the top combines (3.6.2), (3.6.4), and (3.6.6) to essentially state that the wave-function effect on genetic information while governed by the Light-Space-Time matrices in (3.1.3) will vary based on the bases that are active, hence $Y = (U, U\&M_x, M_x)$, where Y equates to the family of bases that may be active: untransformed (U), a mix (U & M_x), or M_x. Since the primary part of Line 1 falls under the $\int$ it means that the operation is iterative since dynamics are summed up for space and time to that point where the wavefunction is sought.

Line 1 is then subjected ($\times$) to a determination of the dominant levels of light that may be active, designated by Line 2, '$\big((Y > U: Z_Q) \vee (Y \leq U: Z_F) \vee (Y = U: Z_R) \big)$'. Unpacking this, 'Y > U' implies meta-levels are active and as a result it is possible that Z_Q is going to take place (the subscript 'Q' implies quantum-level action). In other words the "quantization-window" referred to in the previous chapter has been opened. This also implies activation and potential change of FBLEE. The call from below, as it were, may invoke some function that already exists in the subtle-libraries 'above', so that some already existing function may influence FBLEE through ∞-entanglement, K-entanglement, or N-entanglement. This may be thought of as a key-and-lock mechanism, where a deep enough visceral urge from below acting as the key, opens an entangled lock to alter FBLEE as per the visceral urge. The alteration of FBLEE is what is referred to as "quantum-certainty" in the previous chapter because such certain alteration is happening at the quantum-level. 'Y $\leq$ U' implies that only the untransformed levels are active, and therefore also the sub-level where the speed of light is 0 is active and as a result Z_F is going to take place (the subscript 'F' implies 'fragmentation'). 'Y = U' implies that all levels are active and as a result Z_R is going to take place (the subscript 'R' implies 'random').

Line 3 elaborates the significance of Z_Q, Z_F, and Z_R. Hence Z is the union of potential quantum-operations of space, time, energy, and gravity, designated by '$Z \in \mathbb{U}\,(Space, Time, Energy, Gravity)$'. But the nature of the operations is designated by 'Q: $Quantization$; F: $Fragmentation$; R: $Random$'. Z_Q, then, implies that the full quantization originating from updated four-base logic-encoding ecosystems (FBLEE) can take place. Z_F implies that the essential set will be fragmented and that only libraries at the level of local cellular-level DNA or precipitated material-fabric logic can potentially be altered. Z_R also precludes full quantization, and that some partial local-library constructive or destructive mutation may take place.

The '$\rightarrow h \ni h \in$' segment resolves the outcome of the operations implied by Lines 1 – 3, suggesting that the outcome will be 'h' such that ($\ni$) 'h' is an element ($\in$) of the set specified by the members '[Q]: *Constructive zone*', '[Q]: *Constructive zone AND Constructive mutation*', '[F]; *Destructive mutation*', and '{R}: *Random mutation*'. '[Q]: *Constructive zone*' implies that the in-built buffer has been crossed and that access to the deeper four-base logic-encoding ecosystem (FBLEE) has been granted. '[Q}' in this segment implies that there is the possibility that full-quantization as specified by Line 3 of the previous matrix can take place. Access to this zone is a prerequisite for constructive mutation to occur, as designated by the element '[Q}: *Constructive zone* ∩ *Constructive mutation*', which implies that full-quantization is going to take place and will result in material change. The '[F]' specifies the relationship between 'Fragmentation' in Line 3 of the previous matrix and destructive mutation. The '[R]' specifies the relationship between 'Random' in Line 3 of the previous matrix and random mutation.

Effect of Levels of Light on Mutation Using Iterative Version of Light-Space-Time Emergence

Equations 3.6.9 through 3.6.11 summarize three types of mutation using (3.6.1). These are the destructive-mutation with only untransformed bases active, random-mutation with mixed bases, and constructive-mutation with bases relatively transformed due to the active influence of meta-levels, respectively. Hence:

$$Destructive - Mutation = \left\| |[L][S][T]TC \rightarrow x_T|_{\langle x_U | x_T \rangle} \right\|_U$$

Eq. 3.6.9: Destructive-Mutation

$$Random - Mutation = \left\| |[L][S][T]TC \rightarrow x_T|_{\langle x_U | x_T \rangle} \right\|_{U \& M_x}$$

Eq. 3.6.10: Random-Mutation

$$Consgtructive - Mutation = \left\| |[L][S][T]TC \rightarrow x_T|_{\langle x_U|x_T \rangle} \right\|_{M_x}$$

Eq. 3.6.11: Constructive-Mutation

Now the question is to what extent will the antecedent, and therefore the more permanent four-base logic-encoding ecosystem (FBLEE) be altered? For in the model based on Light it is this category of library that is important in heredity or the longer-term scheme of things. Destructive or random mutation may alter local copies of libraries, such as exist in DNA at the cellular-level, but it is only constructive mutation that can alter the deeper level subtle libraries that are preserved in a different kind of way in longevity.

This has to be the case since there is an in-built buffer due to the quantization that occurs as light slows down. The conditions for entering into that region where light is traveling faster must require a special, non-trivial set of conditions to be active. The conditions for traversing this in-built buffer will be explored in greater detail in subsequent chapters. Traversing the buffer implies that meta-levels are active, and that they have to become active in a special way for FBLEE to be affected.

Equation 3.6.12, Potential Effect of Levels of Light on Genetic-Type Information, summarizes how this may happen:

$$Potential\ Effect\ of\ Levels\ of\ Light\ on\ Genetic - Type\ Information =$$

$$\begin{bmatrix} STATIC\ \langle |[L][S][T]TC \rightarrow x_T|_{\langle x_U|x_T \rangle} \rangle \\ \times \\ \left((Y > U: Z_Q) \vee (Y \leq U: Z_F) \vee (Y = U: Z_R) \right) \\ \ni \\ \left(\begin{array}{c} Z \in \mathbb{U}\ (Space, Time, Energy, Gravity) \\ (Q: Quantization; F: Fragmentation; R: Random) \end{array} \right) \end{bmatrix} \rightarrow h \ni$$

$$h \in \left(\begin{array}{c} [Q]: Constructive\ zone, \\ [Q]: Constructive\ zone \wedge Constructive\ mutation, \\ [F]: Destructive\ mutation, \\ [R]: Random\ mutation \end{array} \right)$$

Eq. 3.6.12: Potential Effect of Levels of Light on Genetic-Type Information

Line 1 from the top in the matrix is simply a static form of (3.6.1) the Simplified Light-Space-Time Emergence equation. The static form is designated by 'STATIC' and implies that fundamental operations true of (3.6.1) are being highlighted in (3.6.12). In other words (3.6.1) already has all the operations highlighted in (3.6.12) in it, but by 'freezing' it by making it static, the essential dynamics leading to possible mutations at the genetic level can more clearly be highlighted.

Line 1 is then subjected ($\times$) to a determination of the dominant levels of light that may be active, designated by Line 2, $'\big((Y > U: Z_Q) \vee (Y \leq U: Z_F) \vee (Y = U: Z_R) \big)'$. Unpacking this, '$Y > U$' implies meta-levels are active and as a result it is possible that Z_Q is going to take place (the subscript 'Q' implies quantum-level action). In other words the "quantization-window" referred to in the previous chapter has been opened. This also implies activation and potential change of FBLEE. The call from below, as it were, may invoke some function that already exists in the subtle-libraries 'above', so that some already existing function may influence FBLEE through ∞-entanglement, K-entanglement, or N-entanglement. This may be thought of as a key-and-lock mechanism, where a deep enough visceral urge from below acting as the key, opens an entangled lock to alter FBLEE as per the visceral urge. The alteration of FBLEE is what is referred to as "quantum-certainty" in the previous chapter because such certain alteration is happening at the quantum-level. '$Y \leq U$' implies that only the untransformed levels are active, and therefore also the sub-level where the speed of light is 0 is active and as a result Z_F is going to take place (the subscript 'F' implies 'fragmentation'). '$Y = U$' implies that all levels are active and as a result Z_R is going to take place (the subscript 'R' implies 'random').

Line 3 elaborates the significance of Z_Q, Z_F, and Z_R. Hence Z is the union of potential quantum-operations of space, time, energy, and gravity, designated by '$Z \in \mathbb{U}$ ($Space, Time, Energy, Gravity$)'. But the nature of the operations is designated by 'Q: $Quantization$; F: $Fragmentation$; R: $Random$'. Z_Q, then, implies that the full quantization originating from updated four-base logic-encoding ecosystems (FBLEE) can take place. Z_F implies that the essential set will be fragmented and that only libraries at the level of local cellular-level DNA or precipitated material-fabric logic can potentially be altered. Z_R also precludes full quantization, and that some partial local-library constructive or destructive mutation may take place.

The '→ h ∋ h ∈' segment resolves the outcome of the operations implied by Lines 1 – 3, suggesting that the outcome will be 'h' such that (∋) 'h' is an element (∈) of the set specified by the members '[Q]: *Constructive zone*', '[Q}: *Constructive zone AND Constructive mutation*', '[F}; *Destructive mutation*', and '{R}: *Random mutation*'. '[Q]: *Constructive zone*' implies that the in-built buffer has been crossed and that access to the deeper four-base logic-encoding ecosystem (FBLEE) has been granted. '[Q}' in this segment implies that there is the possibility that full-quantization as specified by Line 3 of the previous matrix can take place. Access to this zone is a prerequisite for constructive mutation to occur, as designated by the element '[Q}: *Constructive zone* ∩ *Constructive mutation*', which implies that full-quantization is going to take place and will result in material change. The '[F]' specifies the relationship between 'Fragmentation' in Line 3 of the previous matrix and destructive mutation. The '[R]' specifies the relationship between 'Random' in Line 3 of the previous matrix and random mutation.

Chapter 3.7: Application of Qualified Determinism in the Becoming of the P-Type Being

The previous chapter explored the possible impact of different layers or levels of light on genetic mutation. Recall that genetic mutation results in altering or creating new genetic-type information, which is positioned as being the biographical language in the becoming of the P-type being. The exploration was at a high level and broadly considered a mapping between active layers of light and the type of mutation that may result. Further, the exploration was framed in terms of the light-space-time matrices, central to the mathematics in this treatise, and additionally in terms of a wavefunction form, central to the probabilistic foundation in quantum theory.

Consistent with the mathematics in this treatise though, this chapter further elaborates a non-probabilistic approach by application of a process of Qualified Determinism to genetic mutation. In this process structural forces influencing mutation are disaggregated into vertical and horizontal components. The vertical component is resolved by determining the most active layer of light. The horizontal component is resolved by determining the most influential base of change amongst the physical, vital, mental, and integral possibilities.

This chapter, hence, explores a mathematical notion of qualified determinism - Dynamic Interaction (DI) that has a 'vertical' and a 'horizontal' component. The vertical component is designated as DI_V and the horizontal component as DI_H. Several equations to capture the inherent dynamism manifest as the likely process of becoming of the P-type being have already been derived in Chapter 2.6. These included equations for the process of becoming propelled by the physical, the vital, the mental, and the integral bases. The derived equations propose a model to give insight into how the process of becoming for the P-type being occurs by changing the fundamental states that a microorganism, as an instance of organization in general, is subject to. Several scientists, such as (Prigogine, 1977) and others, are proposing that a system can bifurcate in unpredictable ways to create an emergent property that cannot be predicted. DI is going to propose that in fact there is a 'qualified determinism' as opposed to randomness that occurs (Malik et al, 2017).

DI is going to propose an alternative to the paradigm of statistics and probability thought to govern quantum and other nano- and micro-level objects.

This qualified determinism is the result of the relative strengths of the levels within the core-matrix as summarized by Equation 2.6.7 and reproduced here for convenience:

$$\begin{bmatrix} M_3 \rightarrow System_X \\ (\uparrow F \rightarrow I) \\ M_2 \rightarrow S_{System_X} \\ (\uparrow Sig \rightarrow F) \\ M_1 \rightarrow Sig_x \\ (\uparrow > P_{x)} \\ U \rightarrow x_U \end{bmatrix}$$

The application of the vertical component of the new function being proposed, DI_V, to this core matrix will yield the nature or 'strength' of the state (x) or orientation under consideration. If the untransformed or U layer is strongest, implying that the habitual patterns that keep an organization locked into its untransformed way of operation are still very active, then the nature of the output of DI_V, notated by x-state, will be x_U. If the habitual patterns have been overcome then the strength of the x-state increases since it is the dynamics of M_1 or Sig_x representative of the N-type being that are now active. In this case the x-state will be Sig_x. If the unique 'signature' has become a 'force', then the conditions for activation of M_2 have been put in place and the x-state will be even higher, S_{System_x}, representative of the K-type being. The architectural forces active in M_2 are by definition more powerful than Sig_x that is a derivation of a set of such architectural forces. If the 'force' so acting becomes impersonal so that an organizational ego-state is overcome, then the x-state will have the most strength and is characterized by $System_x$ active at M_3 and representative of the UPI-type being Hence, DI_V applied to a core-matrix will yield the 'strength' in terms of the x-dynamic that is active. Recall also, that each of these upward-strand levels is in fact determined by the parallel light-level that exists in the downward-strand. This is illustrated by the following equation, Equation 3.7.1, which can be considered to be a deductive proof in the context of this model:

$$DI_V \begin{bmatrix} M_3 \rightarrow System_X \\ (\uparrow F \rightarrow I) \\ M_2 \rightarrow S_{System_X} \\ (\uparrow Sig \rightarrow F) \\ M_1 \rightarrow Sig_X \\ (\uparrow > P_{X)} \\ U \rightarrow x_U \end{bmatrix} =>$$

x-state $\in \left(x_U, Sig_X, S_{System_X}, System_X \right)$

Where: $Strength(System_X) > Strength(S_{System_X}) > Strength(Sig_X)$
$$> Strength(x_U)$$

Eq 3.7.1: Illustrating Action of Dynamic Interaction – vertical component

What is to be noted here is that while the action of DI_V yields a relative strength and therefore a 'single' value for the core- or x-matrix under consideration yet each x-matrix in itself could have an infinite number of possibilities. This should be clear in looking at how x_U, Sig_X, S_{System_X}, and $System_X$, were initially defined.

Hence, taking the example where x = physical:

$Physical_U \ni [inertia, lethargy, status\ quo, ...]$

As can be seen $Physical_U$, defined in Chapter 2.6, is practically speaking already an infinite set with qualities similar to the ones specified.

Similarly, Sig_{Pr}, defined in Chapter 2.4, also has an infinite variation:

$$Sig_{Pr} = Xa + \overline{Yb_{0-n}} \quad where \begin{bmatrix} X \in [S_{System_{Pr}}] \\ Y \in [S_{System_{Pr}}, S_{System_P}, S_{System_K}, S_{System_N}] \\ a, b\ are\ integers; a > b \end{bmatrix}$$

$S_{System_{Pr}}$, defined in Chapter 2.3, is also an infinite set with forces of the nature specified in the following equation:

$S_{System_{Pr}} \ni [Service, Perfection, Diligence, Perseverance, ...]$

And recall that in Chapter 2.2, $System_{Pr}$ has been defined as:

$$System_{Pr} \equiv TS_{0 \to N} \begin{bmatrix} \begin{matrix} Presence \\ \downarrow \\ Opportunity \end{matrix} \begin{bmatrix} P_L & \to & V_L \\ V_L & \to & M_L \end{bmatrix} & Container \begin{bmatrix} System_P \\ System_K \\ System_N \end{bmatrix} \end{bmatrix}$$

So in essence DI_V is really giving us a summary assessment of the 'level' of the x-matrix under consideration with all its infinite potentiality. An example will follow shortly.

The other component of DI, as suggested earlier in this chapter, is the horizontal component, DI_H. Just as DI_V yields a summary assessment of the level that an x-matrix is operating at, similarly DI_H yields a summary assessment of the direction that a system or organization under consideration is going to continue its development in, considering the physical, the vital, the mental, and the integral bases to be the choices.

Assuming that any organization or system is inherently unique, as this mathematical model proposes, and as the phenomena of entanglement proves, and assuming that the infinite sets of x_U and S_{System_X} applied across the physical, vital, mental, and integral bases or orientations respectively will account for any state that a nano- or micro-organization can experience, then at a certain point in time any such organization under consideration is going to have a direction-bias in one of the possible physical, vital, mental, or integral directions. Hence, DI_H will yield the summary direction that is going to lead an organization into its future given the current states active in it.

This summary direction is going to be yielded by considering the relative strengths of the separate core x-matrices – the physical, the vital, the mental, and the integral. The assumption is that there will be one core-matrix that will be stronger than the others.

Hence, as an example, first applying DI_V across all four x-matrices may, for example, yield the following results, with the strongest level within each x-matrix highlighted and bolded:

$$\begin{bmatrix} System_{Pr} \\ \mathbf{S_{System_{Pr}}} \\ Sig_P \\ Physical_U \end{bmatrix} \begin{bmatrix} System_P \\ S_{System_P} \\ Sig_V \\ \mathbf{Vital_U} \end{bmatrix} \begin{bmatrix} System_K \\ S_{System_K} \\ Sig_M \\ \mathbf{Mental_U} \end{bmatrix} \begin{bmatrix} System_N \\ S_{System_N} \\ \mathbf{Sig_I} \\ Integral_U \end{bmatrix}$$

Since by definition the strength of $System_x$ is greater than S_{System_x}, which is greater than Sig_x, which is greater than x_U, applying DI_H, as in Equation 3.7.2, across these x-matrices, as in the example following it will then yield the strongest direction, which in this example is the Physical:

$$DI_H \left(\begin{bmatrix} System_{Pr} \\ S_{System_{Pr}} \\ Sig_P \\ Physical_U \end{bmatrix} \begin{bmatrix} System_P \\ S_{System_P} \\ Sig_V \\ Vital_U \end{bmatrix} \begin{bmatrix} System_K \\ S_{System_K} \\ Sig_M \\ Mental_U \end{bmatrix} \begin{bmatrix} System_N \\ S_{System_N} \\ Sig_I \\ Integral_U \end{bmatrix} \right) = Orientation_{Strongest}$$

Eq 3.7.2: Illustrating Action of Dynamic Interaction - horizontal component

Example:

$$DI_H \left(\begin{bmatrix} System_{Pr} \\ \mathbf{S_{System_{Pr}}} \\ Sig_P \\ Physical_U \end{bmatrix} \begin{bmatrix} System_P \\ S_{System_P} \\ Sig_V \\ \mathbf{Vital_U} \end{bmatrix} \begin{bmatrix} System_K \\ S_{System_K} \\ Sig_M \\ \mathbf{Mental_U} \end{bmatrix} \begin{bmatrix} System_N \\ S_{System_N} \\ \mathbf{Sig_I} \\ Integral_U \end{bmatrix} \right) = Physical$$

Hence, DI function will yield the following direction of mutation, as in Equation 3.7.3, where 'x_matrix' is used interchangeably with 'orientation' or 'bases':

$$Mutation\_Dir = DI \left(\begin{bmatrix} M_3 \rightarrow System_{Pr} \\ (\uparrow F \rightarrow I) \\ M_2 \rightarrow S_{System_{Pr}} \\ (\uparrow Sig \rightarrow F) \\ M_1 \rightarrow Sig_P \\ (\uparrow > P_{P)} \\ U \rightarrow Physical_U \end{bmatrix} \begin{bmatrix} M_3 \rightarrow System_P \\ (\uparrow F \rightarrow I) \\ M_2 \rightarrow S_{System_P} \\ (\uparrow Sig \rightarrow F) \\ M_1 \rightarrow Sig_V \\ (\uparrow > P_{V)} \\ U \rightarrow Vital_U \end{bmatrix} \right.$$
$$\left. \begin{bmatrix} M_3 \rightarrow System_S \\ (\uparrow F \rightarrow I) \\ M_2 \rightarrow S_{System_S} \\ (\uparrow Sig \rightarrow F) \\ M_1 \rightarrow Sig_M \\ (\uparrow > P_{M)} \\ U \rightarrow Mental_U \end{bmatrix} \begin{bmatrix} M_3 \rightarrow System_N \\ (\uparrow F \rightarrow I) \\ M_2 \rightarrow S_{System_N} \\ (\uparrow Sig \rightarrow F) \\ M_1 \rightarrow Sig_I \\ (\uparrow > P_{I)} \\ U \rightarrow Integral_U \end{bmatrix} \right) \rightarrow$$

$x\_matrix_{strongest} @ level_{strongest}$

Eq 3.7.3: Direction of Mutation

Generalizing, as in Equation 3.7.4, where the direction of mutation is going to be synonymous with the direction of becoming of the P-type being, P-Type_Becoming_Dir is direction of becoming of the P-type being:

$$P - Type\_Becoming\_Dir = DI \left(\begin{bmatrix} M_3 \rightarrow System_X \\ (\uparrow F \rightarrow I) \\ M_2 \rightarrow S_{System_X} \\ (\uparrow Sig \rightarrow F) \\ M_1 \rightarrow Sig_x \\ (\uparrow > P_P) \\ U \rightarrow x_U \end{bmatrix}_{x=p,v,m,i} \right) \rightarrow$$

$x\_matrix_{strongest} @ level_{strongest}$

Eq 3.7.4: Generalized Equation for Direction of P-Type Becoming

Hence, this mathematical model is suggesting that any situation, including micro and nano-level situations, rather than having a random outcome, has a 'qualified deterministic' outcome. In the introduction to his book "Where is Science Going?" (Planck, 1933), James Murphy points out that the reason Planck spent so much of his time giving lectures on causation was because of the trend of physicists at the time, which has continued to the modern day, to overthrowing the principle of causation following the development of quantum theory, which he felt was misplaced. "Planck would claim", he wrote, "and so would Einstein, that it is not the principle of causation itself which has broken down in modern physics, but rather the traditional formulation of it." Murphy also quotes James Jeans (Jeans, 1932) to suggest the issue associated with causation and determinism: "Einstein showed in 1917 that the theory founded by Planck appeared, at first sight at least, to entail consequences far more revolutionary than mere discontinuity", and here he is referring to the finding that radiant energy is not emitted in a continuous flow, but in integral quantities, or quanta, which can be expressed in integral numbers. Continuing: "It appeared to dethrone the law of causation from the position it had therefore held as guiding the course of the natural world. The old science had confidently proclaimed that nature could follow only one road, the road which was mapped out from the beginning of time to its end by the continuous chain of cause and effect; state A was inevitably succeeded by state B. So far the new science has only been able to say that state A may be followed by state B or C or D or by innumerable other states. It can, it is true, say that B is more likely than C, C than D, and so on; it can even specify the relative probabilities of B, C, and D. But, just because it has to speak in terms of probabilities, it cannot speak with certainty which state will follow which; this is a matter which lies on the knees of the gods – whatever gods there may be."

While under the apparent dynamics at the quantum, micro, or nano level there may appear to be randomness and a dethroning of the principle of causation, the notion of a multiplicity of levels, each having its impact on the strength of an orientation or base and further on the consequent direction from a multiplicity of possible orientations, is being suggested here as determining the direction of any system, while still allowing infinite variation in the details that may define it. Hence, the positions of Planck and Einstein are vindicated when considering Equation 3.7.4.

Further, assuming any system where multiple elements are active, connected, interdependent, and emergent, it may be possible to understand, through application of calculus, as to which level is the source for change.

Hence, where N may be source of change, the rate of change of N will resolve into one of P_U, V_U, M_U, I_U, P_T, V_T, M_T, or I_T. This may be summarized by Equation 3.7.5, where y is either U or T:

$$\frac{dN}{dt} \rightarrow \begin{bmatrix} P_U & P_T \\ V_U & V_T \\ M_U & M_T \\ I_U & I_T \end{bmatrix} \rightarrow x_y, where\ y \in (U,T)$$

Eq 3.7.5: Establishing the Nature of the Change

If T, implying that the action of one of the meta-levels has caused the transformation or mutation, then application of one of the following integrals will determine which level is the likely source for change.

Hence, for M_1, representative of the N-type being, if the integral of $\frac{\partial(x_U \rightarrow x_T)}{\partial t}$ across a limited area 'a' in the vicinity of the change, is greater than some threshold value $Threshold_{Signature}$, then the signature dynamics are likely the source of change, where 'a' can be thought of as some kind of gene-footprint. This is summarized by Equation 3.7.6:

$$\int_0^a \frac{\partial(x_U \rightarrow x_T)}{\partial t} > Threshold_{Signature}$$

Eq 3.7.6: Signature Dynamics as the Source of Change

For M_2, representative of the K-type being, if the integral of $\frac{\partial(x_U \rightarrow x_T)}{\partial t}$ across a larger area 'b' extending beyond the vicinity of the change, is greater than some threshold value $Threshold_{Architectural\ Forces}$, then the architectural forces are likely the source of change, , where 'b' can be thought of as some kind of gene-footprint larger than 'a'. This is summarized by Equation 3.7.7:

$$\int_0^b \frac{\partial(x_U \rightarrow x_T)}{\partial t} > Threshold_{ArchitecturalForces}$$

Eq 3.7.7: Architectural Forces as the Source of Change

For M_3, representative of the UPI-type being, if the double integral of $\frac{\partial(x_U \rightarrow x_T)}{\partial t}$ across the system specified by 'A', and across some time 't', is greater than some threshold value $Threshold_{System\ Property}$, then the system properties are likely the source of change, , where 'A' can be thought of as some kind of gene-footprint larger than both 'a' and 'b'. This is summarized by Equation 3.7.8:

$$\int_0^t \int_0^A \frac{\partial(x_U \rightarrow x_T)}{\partial t} > Threshold_{SystemProperty}$$

Eq 3.7.8: System Properties as the Source of Change

SECTION 4: OVERVIEW OF GENERATION OF LIGHT-BASED-SINGULARITY BEINGS

In the mathematics presented in the previous sections, the persistent quantum-level computation arbitrates a process of becoming that is envisioned to culminate in a series of P-type beings with increasing spheres of influence. The process of becoming generates a stream of genetic-type information that articulates the biography and becomes "law" for the being it is generated for.

This section briefly summarizes the generation of different types of beings and some of the associated genetic-type code resulting from the persistent Light-Space-Time Matrix based computations ranging from an era pre-dating the Big Bang to modern day Global Civilization, and beyond.

The implications of this are that all light-based beings are imbued with code. It is just the way the code is housed that changes. This also implies that there is likely code within all living cells, as an example, that has to do with space, time, energy, gravity, the functioning of the electromagnetic spectrum, and all the stages of material elaboration that precede the emergence of a living cell. But this has to be, in a process of love, in which emergent P-type beings give themselves or are the foundation for the subsequent emergences of P-type beings with greater spheres of influence.

It is the sharing of such large swathes of code that precede an emergent form, with the emergent form, that typifies the notion of a light-based-singularity. In fact, it is a "singularity" because the most current emergent form has all previous code embedded or available to it thereby inherently abiding with all "laws" that have thus far emerged in previous singularities. This means that so long as an emergence is a light-based-singularity it inherently follows the law of love in that more and more of the expressed nature of light is integrated into a single, more-material whole. This also implies the notion that the seed-singularity itself will have gone through a process of becoming articulated by a progressive biography and resulting in partial-seed-singularities along the way, until the seed-singularity has itself emerged.

Chapter 4.1: Generation of Beings from the Pre-Big Bang to Big Bang Era

The Light-Matrix component of the Light-Space-Time Emergence equation (3.1.3), reproduced below for convenience, essentially provides an algorithm by which pre-Big Bang existence proceeds to a reality of the Big Bang due to precipitating light. The process generates beings through the pre-Big Bang stages, culminating in the generation of the space-time-energy-gravity micro- and macro-beings at the time of the Big Bang.

$Light - Space - Time\ Emergence_T =$

$$
\left[\left[\begin{array}{c} c_\infty : [Pr, Po, K, H] \\ (\downarrow R_{C_K} = f(R_{C_\infty})) \\ c_K : [S_{Pr}, S_{Po}, S_K, S_H] \\ (\downarrow R_{C_N} = f(R_{C_K})) \\ c_N : f(S_{Pr} \times S_{Po} \times S_K \times S_H) \\ (\downarrow R_{C_U} = f(R_{C_N})) \\ c_U : [P, V, M, C] \\ \Uparrow \\ c_{0 : [D, W, I, C]} \end{array} \right]_{Light} \left[\begin{array}{c} M_3 \to System_X \\ (\uparrow F \to I) \\ M_2 \to S_{System_X} \\ (\uparrow Sig \to F) \\ M_1 \to Sig_x \\ (\uparrow > P_x) \\ U \to x_U \end{array} \right]_{Space} \right.
$$

$$
\left. \left[U \to \begin{array}{c} M_3 : -\infty \le t \le \infty \\ \downarrow \\ M_2 : 0 \ge t > \infty \\ \downarrow \\ M_1 : 0 > t > \infty \\ \downarrow \\ t \le E_{Cell}; TC: M_3 \to U \\ t \sim E_{Human}; TC: U \to M_3 \end{array} \right]_{Time} \ TC \to x_T \right] \langle x_U | x_T \rangle
$$

Generation of Partial-Seed-Singularity Pre-Big Bang Pre-Precipitation Pre-Material-Fabric UPI-Type Being and Associated Pre-Genetic Code

Before the Big Bang, as per discussions in Section 1 and 2, it can be assumed that Light precipitated in stages to the point where it slowed down to c. Equation 4.1.1 highlights the generation of the pre-precipitation pre-material-fabric UPI-Type being and associated pre-genetic code. Hence (3.1.3) collapses to the active highlighted portion only.

$Light - Space - Time\ Emergence_{UPI-Type\ Being} =$

$$\left[\begin{bmatrix} \boxed{c_\infty : [Pr, Po, K, H]} \\ \left(\downarrow R_{C_K} = f(R_{C_\infty})\right) \\ c_K : [S_{Pr}, S_{Po}, S_K, S_H] \\ \left(\downarrow R_{C_N} = f(R_{C_K})\right) \\ c_N : f(S_{Pr} \times S_{Po} \times S_K \times S_H) \\ \left(\downarrow R_{C_U} = f(R_{C_N})\right) \\ c_U : [P, V, M, C] \\ \Uparrow \\ c_{0:[D,W,I,C]} \end{bmatrix}_{Light} \begin{bmatrix} M_3 \to System_X \\ (\uparrow F \to I) \\ M_2 \to S_{System_X} \\ (\uparrow Sig \to F) \\ M_1 \to Sig_x \\ (\uparrow > P_x) \\ U \to x_U \end{bmatrix}_{Space} \quad \Rightarrow \right.$$

$$\left. \begin{bmatrix} U \to \begin{matrix} M_3 : -\infty \leq t \leq \infty \\ \downarrow \\ M_2 : 0 \geq t > \infty \\ \downarrow \\ M_1 : 0 > t > \infty \\ \downarrow \\ t \leq E_{Cell}; TC: M_3 \to U \\ t \sim E_{Human}; TC: U \to M_3 \end{matrix} \end{bmatrix}_{Time} \quad TC \to x_T \quad \langle x_U | x_T \rangle \right.$$

$$\left(System_{Pr} \equiv TS_{0 \to N} \begin{bmatrix} Presence \\ \downarrow \\ Opportunity \end{bmatrix} \begin{bmatrix} P_L & \to & V_L \\ V_L & \to & M_L \end{bmatrix} \& Container_{\begin{bmatrix} System_P \\ System_K \\ System_N \end{bmatrix}} \right]$$

$$System_P \equiv power > \sum_{TS=0}^{N} P_R * V_R * M_R$$

$$System_K \equiv TS_{0 \to N} \begin{bmatrix} Knowledge \\ \downarrow \\ Leverage \end{bmatrix} \begin{bmatrix} P_{I,C} \\ V_{I,C} \\ M_{I,C} \end{bmatrix} \to \begin{bmatrix} P_L & \to & V_L \\ V_L & \to & M_L \end{bmatrix}$$

$$System_N \equiv TS_{0 \to N} \left(\coprod_{Nurturing} \begin{pmatrix} P_- & M_+ \\ V_- & V_+ \\ M_- & P_+ \end{pmatrix} mod\,(r, \theta) \right)$$

Eq 4.1.1: Generation of Partial-Seed-Singularity Pre-Big Bang Pre-Precipitation Pre-Material-Fabric UPI-Type Being and Associated Pre-Genetic Code

Note the equation-segments following the '⇒' were derived in Chapter 2.2, The UPI-Type Being and its Dynamics. The generated pre-genetic code belongs to the pre-material-fabric genre since it will remain in a subtle domain exercising its influence through ∞-entanglement where possible. Further, the generated pre-genetic code also narrates the biography for a light-based-singularity in that it is whole and complete, even though it is describing only a part of the seed-singularity. Hence, strictly speaking (4.1.1) is describing a partial-seed-singularity.

Generation of Partial-Seed-Singularity Pre-Big Bang Pre-Precipitation Pre-Material-Fabric 0-Type Being and Associated Pre-Genetic Code

As discussed in the Chapter 2.1, The Light-Based-Singularity Superstructure and the Matrix of Being, Light could project itself at speed zero. This would create a reality opposite to that in its native state. This 0-state reality could have been projected prior to the gradual precipitation from light traveling infinitely fast, that is the corner stone of this treatise on Light. The reality created when light exists at zero speed, as already discussed in Sections 2 and 3, can be seen to play a significant part in destructive mutation, and is seen to influence the reality modeled by the layer U, where light moves at c. Because obstinate patterns can be better explained with a 'speed zero' reality in the background, it will be assumed that this pre-exists the precipitation of Light down to c.

Equation 4.1.2 highlights the generation of the pre-precipitation pre-material-fabric 0-type being and pre-genetic code. Hence (3.1.3) collapses to the active highlighted portion only.

$Light - Space - Time\ Emergence_{0-Type\ Being} =$

$$
\left\| \begin{bmatrix} \begin{array}{l} c_\infty : [Pr, Po, K, H] \\ \left(\downarrow R_{C_K} = f(R_{C_\infty}) \right) \\ c_K : [S_{Pr}, S_{Po}, S_K, S_H] \\ \left(\downarrow R_{C_N} = f(R_{C_K}) \right) \\ c_N : f(S_{Pr} \times S_{Po} \times S_K \times S_H) \\ \left(\downarrow R_{C_U} = f(R_{C_N}) \right) \\ c_U : [P, V, M, C] \\ \Uparrow \\ C_{0:[D,W,I,C]} \end{array} \end{bmatrix}_{Light} \begin{bmatrix} M_3 \to System_X \\ (\uparrow F \to I) \\ M_2 \to S_{System_X} \\ (\uparrow Sig \to F) \\ M_1 \to Sig_x \\ (\uparrow > P_x) \\ U \to x_U \end{bmatrix}_{Space} \right\| \Rightarrow
$$

$$
\left\| U \to \begin{bmatrix} M_3 : -\infty \leq t \leq \infty \\ \downarrow \\ M_2 : 0 \geq t > \infty \\ \downarrow \\ M_1 : 0 > t > \infty \\ \downarrow \\ t \leq E_{Cell}; TC: M_3 \to U \\ t \sim E_{Human}; TC: U \to M_3 \end{bmatrix}_{Time} \quad TC \to x_T \right\| \langle x_U | x_T \rangle
$$

$\langle Partial - Seed - Singularity\ Pre - Big - Bang\ Pre - Precipitation,$
$\ Pre - Material - Fabric\ UPI - Type\ Being\ and\ Pre - Genetic\ Code \rangle +$

$(c_0: [Darkness, Weakness, Ignorance, Chaos])$

Eq 4.1.2: Generation of Partial-Seed-Singularity Pre-Big Bang Pre-Precipitation Pre-Material-Fabric 0-Type Being and Associated Pre-Genetic Code

Note that the last equation-segment following the '$\Rightarrow$' was derived in Chapter 2.1, The Light-Based-Singularity Superstructure and the Matrix of Being. This equation is stating that when light is projected at zero-speed the only active code-segments are due to ∞-entanglement as discussed in the previous sub-section, and zero-speed of light. The fact that code segments due to ∞-entanglement are still active implies that this too can be thought of as a partial-seed-singularity.

Generation of Partial-Seed-Singularity Pre-Big Bang Pre-Material-Fabric K-Type Being and Associated Pre-Genetic Code

The first stage of precipitation prior to the Big Bang is modeled by K-entanglement as discussed in Chapter 2.3, The K-Type Being and Its Dynamics. Equation 4.1.3 highlights the generation of the pre-Big Bang pre-material-fabric K-type being and pre-genetic code active at this point. Hence (3.1.3) collapses to the active highlighted portion only.

$Light - Space - Time \; Emergence_{k-Type \; Being} =$

$$
\begin{bmatrix}
\begin{bmatrix}
c_\infty: [Pr, Po, K, H] \\
\left(\downarrow R_{C_K} = f(R_{C_\infty}) \right) \\
c_K: [S_{Pr}, S_{Po}, S_K, S_H] \\
\left(\downarrow R_{C_N} = f(R_{C_K}) \right) \\
c_N: f(S_{Pr} \times S_{Po} \times S_K \times S_H) \\
\left(\downarrow R_{C_U} = f(R_{C_N}) \right) \\
c_U: [P, V, M, C] \\
\Uparrow \\
c_{0:[D,W,I,C]}
\end{bmatrix}_{Light}
\begin{bmatrix}
M_3 \rightarrow System_x \\
(\uparrow F \rightarrow I) \\
M_2 \rightarrow S_{System_x} \\
(\uparrow Sig \rightarrow F) \\
M_1 \rightarrow Sig_x \\
(\uparrow > P_x) \\
U \rightarrow x_U
\end{bmatrix}_{Space}
\\
\begin{bmatrix}
U \rightarrow
\begin{bmatrix}
M_3 : -\infty \le t \le \infty \\
\downarrow \\
M_2 : 0 \ge t > \infty \\
\downarrow \\
M_1 : 0 > t > \infty \\
\downarrow \\
t \le E_{Cell}; TC: M_3 \rightarrow U \\
t \sim E_{Human}; TC: U \rightarrow M_3
\end{bmatrix}_{Time}
\end{bmatrix}
\begin{matrix}
TC \rightarrow x_T \\
\\
\langle x_U | x_T \rangle
\end{matrix}
\end{bmatrix} \Rightarrow
$$

$$\left(\begin{array}{c} Partial - Seed - Singularity\ Pre - Big - Bang\ Pre - Precipitation \\ Pre - Material - Fabric\ UPI - Type\ Being\ and\ Pre - Genetic\ Code \end{array} \right) +$$

$$\left(\begin{array}{c} Partial - Seed - Singularity\ Pre - Big - Bang\ Pre - Precipitation \\ Pre - Material - Fabric\ 0 - Type\ Being\ and\ Pre - Genetic\ Code \end{array} \right) +$$

$$\left(\begin{array}{l} \sum S_{System_{Pr}} \ni [Service, Perfection, Diligence, Perseverance, \ldots] \\ \quad \sum S_{System_{P}} \ni [Power, Courage, Adventure, Justice, \ldots] \\ \sum S_{System_{K}} \ni [Wisdom, Law\ Making, Spread\ of\ Knowledge \ldots] \\ \quad \sum S_{System_{N}} \ni [Love, Compassion, Harmony, Relationship \ldots] \end{array} \right)$$

Eq 4.1.3: Generation of Partial-Seed-Singularity Pre-Big Bang Pre-Material-Fabric K-Type Being and Associated Pre-Genetic Code

Note that the last equation-segment following the '⇒' was derived in Chapter 2.3, **The K-Type Being and Its Dynamics.** The '$\sum x$' signifies all the possibilities for the 'x' equation-segment. All the segments following the '⇒' are active as pre-material-fabric pre-genetic code at this point. These segments also imply wholeness and the formation of a partial-seed-singularity.

Generation of Seed-Singularity Pre-Big Bang Pre-Material-Fabric N-Type Being and Associated Pre-Genetic Code

The second stage of precipitation prior to the Big Bang is modeled by N-entanglement as discussed in Chapter 2.4, The N-Type Being and Its Dynamics. Equation 4.1.4 highlights the generation of the pre-material-fabric pre-genetic code active at this point. Hence (3.1.3) collapses to the active highlighted portion only.

$$Light - Space - Time\ Emergence_{N-Type\ Being} =$$

$$\left[\begin{array}{c} \left[\begin{array}{c} c_\infty : [Pr, Po, K, H] \\ \left(\downarrow R_{C_K} = f(R_{C_\infty}) \right) \\ c_K : [S_{Pr}, S_{Po}, S_K, S_H] \\ \left(\downarrow R_{C_N} = f(R_{C_K}) \right) \\ c_N : f(S_{Pr} \times S_{Po} \times S_K \times S_H) \\ \left(\downarrow R_{C_U} = f(R_{C_N}) \right) \\ c_U : [P, V, M, C] \\ \Uparrow \\ c_{0:[D,W,I,C]} \end{array}\right]_{Light} \\ \\ \left[U \rightarrow \begin{array}{c} M_3 : -\infty \le t \le \infty \\ \downarrow \\ M_2 : 0 \ge t > \infty \\ \downarrow \\ M_1 : 0 > t > \infty \\ \downarrow \\ t \le E_{Cell}; TC: M_3 \rightarrow U \\ t \sim E_{Human}; TC: U \rightarrow M_3 \end{array} \right]_{Time} \end{array} \quad \begin{array}{c} \left[\begin{array}{c} M_3 \rightarrow System_X \\ (\uparrow F \rightarrow I) \\ M_2 \rightarrow S_{System_X} \\ (\uparrow Sig \rightarrow F) \\ M_1 \rightarrow Sig_x \\ (\uparrow > P_x) \\ U \rightarrow x_U \end{array}\right]_{Space} \\ \\ TC \rightarrow x_T \\ \\ \langle x_U | x_T \rangle \end{array}\right] \Rightarrow$$

$$\left\langle \begin{array}{c} Partial - Seed - Singularity\ Pre - Big - Bang\ Pre - Precipitation \\ Pre - Material - Fabric\ UPI - Type\ Being\ and\ Pre - Genetic\ Code \end{array} \right\rangle +$$

$$\left\langle \begin{array}{c} Partial - Seed - Singularity\ Pre - Big - Bang\ Pre - Precipitation \\ Pre - Material - Fabric\ 0 - Type\ Being\ and\ Pre - Genetic\ Code \end{array} \right\rangle +$$

$$\left\langle \begin{array}{c} Partial - Seed - Singularity\ Pre - Big - Bang\ Pre - Precipitation \\ Pre - Material - Fabric\ K - Type\ Being\ and\ Pre - Genetic\ Code \end{array} \right\rangle +$$

$$\left(where \left[\begin{array}{c} \sum Sig = Xa + \overline{Yb_{0-n}} \\ X \in [S_{System_{Pr}}, S_{System_P}, S_{System_K}, S_{System_N}] \\ Y \in [S_{System_{Pr}}, S_{System_P}, S_{System_K}, S_{System_N}] \\ a, b\ are\ integers; a > b \end{array}\right] \right)$$

Eq 4.1.4: *Generation of Seed-Singularity Pre-Big Bang Pre-Material-Fabric N-Type Being and Associated Pre-Genetic Code*

Note that the last equation-segment following the '⇒' was derived in Chapter 2.4, **The N-Type Being and Its Dynamics**. The '$\sum x$' signifies all the possibilities for the 'x' equation-segment. The fact that all antecedent layers of Light are now active implies the emergence of the seed-singularity.

Generation of Partial-Singularity Big Bang Pre-Material-Fabric and Material-Fabric Space-Time-Energy-Gravity Micro- and Macro-Beings

Assuming Light then projects itself or precipitates as suggested from its state of ∞ to that of c, this creates the Big Bang. The process is modeled as yielding two equations.

Equation 4.1.5, Generation of Partial-Singularity Big Bang Pre-Material-Fabric Space-Time-Energy-Gravity Micro- and Macro-Beings and Associated Pre-Genetic Code, highlights the operative portions. The important point is that this equation generates the additional space-time-energy-gravity code-segments as a four-base logic-encoding ecosystem depicted by the sub-script 'FBLEE' at the right-base of the newly generated code-segment, which then immediately activates the material-fabric in Equation 4.1.6.

Hence (4.1.5):

$$Light - Space - Time\ Emergence_{Space-Time-Energy-Gravity\ Beings@FBLEE} =$$

$$\left[\left[\begin{matrix} c_\infty: [Pr, Po, K, H] \\ \left(\downarrow R_{C_K} = f(R_{C_\infty})\right) \\ c_K: [S_{Pr}, S_{Po}, S_K, S_H] \\ \left(\downarrow R_{C_N} = f(R_{C_K})\right) \\ c_N: f(S_{Pr} \times S_{Po} \times S_K \times S_H) \\ \left(\downarrow R_{C_U} = f(R_{C_N})\right) \\ c_U: [P, V, M, C] \\ \Uparrow \\ c_{0:[D,W,I,C]} \end{matrix}\right]_{Light} \begin{matrix} M_3 \to System_X \\ (\uparrow F \to I) \\ M_2 \to S_{System_X} \\ (\uparrow Sig \to F) \\ M_1 \to Sig_X \\ (\uparrow > P_X) \\ U \to x_U \end{matrix}_{Space}\right.$$

$$\left[U \to \begin{matrix} M_3: -\infty \le t \le \infty \\ \downarrow \\ M_2: 0 \ge t > \infty \\ \downarrow \\ M_1: 0 > t > \infty \\ \downarrow \\ t \le E_{Cell}; TC: M_3 \to U \\ t \sim E_{Human}; TC: U \to M_3 \end{matrix}\right]_{Time} \quad TC \to x_{STEG@FBLEE} \quad \Rightarrow \quad \langle x_U | x_T \rangle$$

$$\left\langle \begin{matrix} Partial - Seed - Singularity\ Pre - Big - Bang\ Pre - Precipitation \\ Pre - Material - Fabric\ UPI - Type\ Being\ and\ Pre - Genetic\ Code \end{matrix} \right\rangle +$$

141

$$\left\langle \begin{array}{c} Partial - Seed - Singularity\; Pre - Big - Bang\; Pre - Precipitation \\ Pre - Material - Fabric\; 0 - Type\; Being\; and\; Pre - Genetic\; Code \end{array} \right\rangle +$$

$$\left\langle \begin{array}{c} Partial - Seed - Singularity\; Pre - Big - Bang\; Pre - Precipitation \\ Pre - Material - Fabric\; K - Type\; Being\; and\; Pre - Genetic\; Code \end{array} \right\rangle +$$

$$\left\langle \begin{array}{c} Seed - Singularity\; Pre - Big - Bang\; Pre - Precipitation \\ Pre - Material - Fabric\; N - Type\; Being\; and\; Pre - Genetic\; Code \end{array} \right\rangle +$$

$$\left(\begin{array}{c} Space_{quantization} \;= h_{UK}\left(Xa + \overline{Yb_{0-n}}\;\right) \\ where: \left[\begin{array}{c} X \in [S_{System_K}] \\ Y \in [S_{System_{Pr}}, S_{System_P}, S_{System_K}, S_{System_N}] \\ a, b \; are \; integers; a > b \end{array} \right] \\ Time_{quantization} \;= h_{UP}\left(Xa + \overline{Yb_{0-n}}\;\right) \\ where: \left[\begin{array}{c} X \in [S_{System_P}] \\ Y \in [S_{System_{Pr}}, S_{System_P}, S_{System_K}, S_{System_N}] \\ a, b \; are \; integers; a > b \end{array} \right] \\ Energy_{quantization} \;= h_{UPr}\left(Xa + \overline{Yb_{0-n}}\;\right) \\ where: \left[\begin{array}{c} X \in [S_{System_{Pr}}] \\ Y \in [S_{System_{Pr}}, S_{System_P}, S_{System_K}, S_{System_N}] \\ a, b \; are \; integers; a > b \end{array} \right] \\ Gravity_{quantization} \;= h_{UH}\left(Xa + \overline{Yb_{0-n}}\;\right) \\ where: \left[\begin{array}{c} X \in [S_{System_N}] \\ Y \in [S_{System_{Pr}}, S_{System_P}, S_{System_K}, S_{System_N}] \\ a, b \; are \; integers; a > b \end{array} \right] \end{array} \right)_{FBLEE} +$$

$$
\left(
\begin{array}{l}
Space_{Structure} = \sum_{i,j=1}^{\to\infty} h_{UK}\left(X_i a + \overline{Y_j b_{0-n}}\ \right) \\[4pt]
where: \left[
\begin{array}{c}
X_i \in [S_{System_K}] \\
Y_j \in [S_{System_{Pr}}, S_{System_P}, S_{System_K}, S_{System_N}] \\
a, b, i, j \ are\ integers; a > b
\end{array}
\right] \\[10pt]
Time_{Infinity} = \sum_{i,j=1}^{\to\infty} h_{UP}\left(X_i a + \overline{Y_j b_{0-n}}\ \right) \\[4pt]
where: \left[
\begin{array}{c}
X_i \in [S_{System_P}] \\
Y_j \in [S_{System_{Pr}}, S_{System_P}, S_{System_K}, S_{System_N}] \\
a, b, i, j \ are\ integers; a > b
\end{array}
\right] \\[10pt]
Energy_{Aggregation} = \sum_{i,j=1}^{\to\infty} h_{UPr}\left(X_i a + \overline{Y_j b_{0-n}}\ \right) \\[4pt]
where: \left[
\begin{array}{c}
X_i \in [S_{System_{Pr}}] \\
Y_j \in [S_{System_{Pr}}, S_{System_P}, S_{System_K}, S_{System_N}] \\
a, b, i, j \ are\ integers; a > b
\end{array}
\right] \\[10pt]
Gravity_{Universal} = \sum_{i,j=1}^{\to\infty} h_{UH}\left(X_i a + \overline{Y_j b_{0-n}}\ \right) \\[4pt]
where: \left[
\begin{array}{c}
X_i \in [S_{System_N}] \\
Y_j \in [S_{System_{Pr}}, S_{System_P}, S_{System_K}, S_{System_N}] \\
a, b, i, j \ are\ integers; a > b
\end{array}
\right]
\end{array}
\right)_{FBLEE}
$$

Eq 4.1.5: Generation of Partial-Singularity Big Bang Pre-Material-Fabric Space-Time-Energy-Gravity Beings and Associated Pre-Genetic Code @FBLEE

Equation 4.1.5 may be thought of as being active as light approaches c. Hence the generated FBLEE-segment exists at the quantum-levels and can be thought of as exhibiting entanglement, referred to as FBLEE-entanglement (FBLEEE). Note that the subscript 'T' for transformation is now depicted at the specific quadrumvirate being created, in this case STEG, for Space-Time-Energy-Gravity, with a further specification of @FBLEE to denote that STEG now exists at the pre-material-fabric level.

As light slows down to c the multiple space-time-energy-gravity quantization and aggregation discussed in Section 3 becomes active and spawns the material-fabric, which may be thought of as the gestalt or the emergent property due to the light's implicit fourfoldness.

Space, Time Energy, and Gravity have therefore a specific meaning and significance in the context of the material universe. At the macro-level as discussed in Chapter 3.5 through acting as the mechanisms by which potentially infinite number of seeds are interrelated and mature, they define essential cosmic parameters by which the universe operates.

At the micro-level their action is essential in materializing pre-genetic code, genetic code, and other codes that may also cause different kinds of materialization in the future. In living cells such code is known to exist in DNA. In forms prior to the living cell, while Science may not yet have acknowledged that such code exists and similarly must be housed in a material structure, yet this must be the case.

In this treatise such a material structure housing pre-genetic code is referred to as the 'material-fabric'. Material-fabric is envisioned as existing at the interface of the antecedent quantum-layer and matter. Note though that the space-time-energy-gravity quadrumvirate beings are still created first as a four-base logic encoding ecosystem assumed to exist in the antecedent quantum-layer and depicted by the FBLEE subscript in (4.1.5) as just discussed.

Through further application of (3.1.3) this FBLEE beings and pre-genetic code will be housed in the material-fabric as illustrated by Equation 4.1.6:

$$Light - Space - Time\ Emergence_{Space-Time-Energy-Gravity\ Beings@MF} =$$

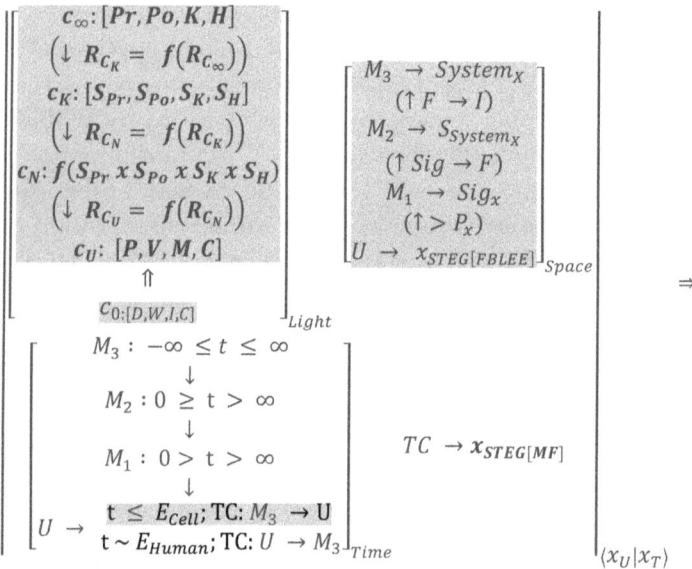

$$\left\|\left[\begin{array}{l}\left[\begin{array}{l}c_\infty:[Pr,Po,K,H] \\ \left(\downarrow R_{C_K} = f(R_{C_\infty})\right) \\ c_K:[S_{Pr},S_{Po},S_K,S_H] \\ \left(\downarrow R_{C_N} = f(R_{C_K})\right) \\ c_N:f(S_{Pr} \times S_{Po} \times S_K \times S_H) \\ \left(\downarrow R_{C_U} = f(R_{C_N})\right) \\ c_U:[P,V,M,C] \\ \Uparrow \\ c_{0:[D,W,I,C]} \end{array}\right]_{Light} \\ U \to \left[\begin{array}{l} M_3: -\infty \leq t \leq \infty \\ \downarrow \\ M_2: 0 \geq t > \infty \\ \downarrow \\ M_1: 0 > t > \infty \\ \downarrow \\ t \leq E_{Cell};\ TC: M_3 \to U \\ t \sim E_{Human};\ TC: U \to M_3 \end{array}\right]_{Time} \end{array}\right| \left|\begin{array}{l} U \to \left[\begin{array}{l} M_3 \to System_x \\ (\uparrow F \to I) \\ M_2 \to S_{System_x} \\ (\uparrow Sig \to F) \\ M_1 \to Sig_x \\ (\uparrow > P_x) \\ \to X_{STEG[FBLEE]} \end{array}\right]_{Space} \\ \\ TC \to X_{STEG[MF]} \end{array}\right| \right\| \Rightarrow \langle x_U | x_T \rangle$$

$$
\left(
\begin{array}{l}
\quad Space_{quantization} \ = h_{UK}\left(Xa + \overline{Yb_{0-n}}\ \right) \\
where: \left[\begin{array}{c} X \in [S_{System_K}] \\ Y \in [S_{System_{Pr}}, S_{System_P}, S_{System_K}, S_{System_N}] \\ a, b \ are \ integers; a > b \end{array} \right] \\
\quad Time_{quantization} \ = h_{UP}\left(Xa + \overline{Yb_{0-n}}\ \right) \\
where: \left[\begin{array}{c} X \in [S_{System_P}] \\ Y \in [S_{System_{Pr}}, S_{System_P}, S_{System_K}, S_{System_N}] \\ a, b \ are \ integers; a > b \end{array} \right] \\
\quad Energy_{quantization} \ = h_{UPr}\left(Xa + \overline{Yb_{0-n}}\ \right) \\
where: \left[\begin{array}{c} X \in [S_{System_{Pr}}] \\ Y \in [S_{System_{Pr}}, S_{System_P}, S_{System_K}, S_{System_N}] \\ a, b \ are \ integers; a > b \end{array} \right] \\
\quad Gravity_{quantization} \ = h_{UH}\left(Xa + \overline{Yb_{0-n}}\ \right) \\
where: \left[\begin{array}{c} X \in [S_{System_N}] \\ Y \in [S_{System_{Pr}}, S_{System_P}, S_{System_K}, S_{System_N}] \\ a, b \ are \ integers; a > b \end{array} \right]
\end{array}
\right)_{MF}
\quad +
$$

$$
\left(
\begin{array}{l}
\quad Space_{Structure} = \sum_{i,j=1}^{\to\infty} h_{UK}\left(X_i a + \overline{Y_j b_{0-n}}\ \right) \\
where: \left[\begin{array}{c} X_i \in [S_{System_K}] \\ Y_j \in [S_{System_{Pr}}, S_{System_P}, S_{System_K}, S_{System_N}] \\ a, b, i, j \ are \ integers; a > b \end{array} \right] \\
\quad Time_{Infinity} \ = \sum_{i,j=1}^{\to\infty} h_{UP}\left(X_i a + \overline{Y_j b_{0-n}}\ \right) \\
where: \left[\begin{array}{c} X_i \in [S_{System_P}] \\ Y_j \in [S_{System_{Pr}}, S_{System_P}, S_{System_K}, S_{System_N}] \\ a, b, i, j \ are \ integers; a > b \end{array} \right] \\
\quad Energy_{Aggregation} \ = \sum_{i,j=1}^{\to\infty} h_{UPr}\left(X_i a + \overline{Y_j b_{0-n}}\ \right) \\
where: \left[\begin{array}{c} X_i \in [S_{System_{Pr}}] \\ Y_j \in [S_{System_{Pr}}, S_{System_P}, S_{System_K}, S_{System_N}] \\ a, b, i, j \ are \ integers; a > b \end{array} \right] \\
\quad Gravity_{Universal} \ = \sum_{i,j=1}^{\to\infty} h_{UH}\left(X_i a + \overline{Y_j b_{0-n}}\ \right) \\
where: \left[\begin{array}{c} X_i \in [S_{System_N}] \\ Y_j \in [S_{System_{Pr}}, S_{System_P}, S_{System_K}, S_{System_N}] \\ a, b, i, j \ are \ integers; a > b \end{array} \right]
\end{array}
\right)_{MF}
$$

Eq 4.1.6: Generation of Partial-Singularity Big Bang Space-Time-Energy-Gravity Beings and Associated Pre-Genetic Code @MF

Note the change the subscript U, that now specifies the FBLEE STEG quadrumvirate beings that due to an iteration of (3.1.3) results in MF STEG quadrumvirate beings.

Hence, all the portions of the Light-Matrix are highlighted as in (4.1.5). But in addition in (4.1.6) in the Time-Matrix, the highlighted $'t \leq E_{Cell}$; TC: $M_3 \rightarrow U'$ indicates that since the space-time-energy-gravity constructs precedes the emergence of the cell, all meta-layers will be active in the Space-Matrix. Hence, the entire space-matrix is also highlighted indicating that quantization is active and therefore precipitation into a material-fabric is inevitable. This materialization into the material-fabric is depicted by the subscript 'MF' following the generated code-segment.

Note that in future sections the actions specific to (4.1.5) and (4.1.6) will generally be combined into one equation only. This will signify that FBLEE code generation is implicit and will always precede MF code generation. The combined equation will always have the U and T subscripts named.

An observation of (4.1.5) and (4.1.6) suggest the vast amount of pre-genetic information that has already been generated, and also suggests the means by which this information can be materialized. The Big Bang is therefore a watershed event that puts into play a potentially endless evolution materializing more and more possibility that exists in Light. The Big Bang hence also signals the start of a new narrative putting into effect all future partial-singularity biographies.

While (4.1.6) does not explicitly show all the pre-material-fabric code-segments, these exist in antecedent layers of Light and will continue to exercise influence through both entanglement and through being part of the source material that generates FBLEE code-segments. The means for materialization of any generated FBLEE code-segment is the fourfold space-time-gravity-energy quantization that is itself generated as a being with its associated pre-genetic code that must exist in every iota of the universe as part of its material-fabric.

Note also that the Big Bang is a significant event that in the model of emergence depicted in this book alters the possibilities of evolution. If the only operative layer besides c_∞ was c_0 then the nature of emergence and evolution would be radically different than with all the layers as depicted by the Light-Matrix in (3.1.3). As discussed, each of these light layers implies the presence of libraries of pre-genetic pre-material-fabric information, and a much quicker and potentially richer evolution than

146

would be possible if only c_0 existed. Ironically, though, it appears that in the contemporary rendering of physics and the notion of emergence as described in mainstream complex adaptive systems literature evolution is viewed as originating from c_0 only. Note that fragmented-singularities, such as any AI-based singularity, can be thought of as existing in the same manner as any creation based only on c_0.

Chapter 4.2: Generation of P-Type Beings in the Post-Big Bang Era

As implied in Chapter 4.1 it is only with (4.1.6) that the emergence of P-type beings through matter, life, and mega-organization can proceed. The becoming of the space-time-energy-gravity micro-being means that for pre-cellular structures the mechanism for affecting the fabric of material existence, or material-fabric, and for cellular structures and beyond, the means for affecting genetic code and any possible post-genetic code is now in place. In other words, the mechanism for narrating the biography of any P-type being is now in place because there is a meaningful way to record change in genetic-type information in more material structure. Such quantization puts into place the mechanism for the evolution of the universe as will be elaborated in subsequent sections.

This chapter will offer a high-level overview of the generation of various P-type beings that accelerate material evolution. Hence, if we trace the potential in Light, the long arc of possible P-type beings is illustrated by considering the generation of a few, as in Equations 4.2.1 – 8, and include the quantum-particle P-type being, the atom P-type being, the molecule P-type being, and several others, culminating in the generation of the super-matter P-type being. The generation of each P-type being also results in the generation of a new code-segment and in totality these are added to existing code-segments in the material-fabric in the case of pre-genetic information, to DNA in the case of genetic information, and to some hybrid or unknown structure in the case of post-genetic information.

Note that the generated code is universal and true for any emergence. This differs from fragmented-singularities, as well be explored later, in that while the code associated with fragmented-singularities may appear to be universal in effect it is not, as it applies only to the fragmented-object it is associated with. As suggested in the previous chapter, fragmented-singularities can perhaps be best modeled as existing on an essential c_0 substrate completely separated from any other layer of Light. Further, the following equations and their outputs are only snapshots and always there will be a vaster number of iterations that will be required to capture the enormity of the P-type beings and the genetic-type information that exists in reality. Finally, these iterations do not necessarily imply that one P-type being, and its associated code-segment output is sequentially dependent on a previous one. The dynamics and logic of the space-time-energy-gravity P-type beings are embedded in the material-fabric and allows quantization and some dynamics of universality to take place. Similarly, the emergence of other P-type beings and code-segments allows other necessary operations to take place and accelerates what is possible in the material universe. The different genre of beings generated during the formation of the seed-singularity, for example, exhibit different kinds of entanglement and create different libraries of dynamic information that can equally influence the becoming of a subsequent P-type being.

Generation of the Quantum-Particle P-Type Being

In Equation 4.2.1, Generation of the Quantum-Particle P-Type Being and Pre-Genetic Code, the starting point, x_U, is 'STEG', short for dual space-time-energy-gravity P-type micro- and macro-beings, and the ending point, x_T, is 'Quantum-Particle', short for Quantum-Particle P-type being. x_U is highlighted, and x_T is bolded.   As suggested by (4.1.6) the naming of subscripts implies that the code generated is actually housed in the material-fabric through a similar 2-step process as captured by (4.1.5) and (4.1.6).

Hence the FBLEE code-segments for the quantum-particle P-type being will already have been generated at the quantum-levels. Part of the output of (4.2.1) is the pre-genetic code for quantum particles that will be added to the existing space-time-energy-gravity and electromagnetic spectrum pre-genetic code in the material-fabric. In this model such code embedded into the material-fabric is assumed to be operative and binding on constructs that arise within the universe. Such code also changes the possibilities that can arise within the material universe. Hence what has materialized is a complete and whole light-based partial-singularity.

Hence:

$Light - Space - Time\ Emergence_{Quantum-Particles\ P-Type\ Being} =$

$$\left|\begin{array}{c} \begin{bmatrix} c_\infty : [Pr, Po, K, H] \\ \left(\downarrow R_{C_K} = f(R_{C_\infty})\right) \\ c_K : [S_{Pr}, S_{Po}, S_K, S_H] \\ \left(\downarrow R_{C_N} = f(R_{C_K})\right) \\ c_N : f(S_{Pr} \times S_{Po} \times S_K \times S_H) \\ \left(\downarrow R_{C_U} = f(R_{C_N})\right) \\ c_U : [P, V, M, C] \\ \Uparrow \\ c_{0:[D,W,I,C]} \end{bmatrix}_{Light} \quad \begin{bmatrix} M_3 \to System_X \\ (\uparrow F \to I) \\ M_2 \to S_{System_X} \\ (\uparrow Sig \to F) \\ M_1 \to Sig_x \\ (\uparrow > P_x) \\ U \to X_{STEG} \end{bmatrix}_{Space} \\ \begin{bmatrix} U \to \begin{array}{c} M_3 : -\infty \le t \le \infty \\ \downarrow \\ M_2 : 0 \ge t > \infty \\ \downarrow \\ M_1 : 0 > t > \infty \\ \downarrow \\ t \le E_{Cell}; TC: M_3 \to U \\ t \sim E_{Human}; TC: U \to M_3 \end{array} \end{bmatrix}_{Time} \quad TC \to X_{Quantum\ Particles} \\ \langle x_U | x_T \rangle \end{array}\right| \Rightarrow$$

$$\langle Space - Time - Energy - Gravity \ P - Type \ Beings \ and \ Pre - Genetic \ Code \rangle +$$

$$\left(\begin{array}{c} Electro - magnetic - wavearchetype - masspotential \ P - Type \ Being \\ and \ Pre - Genetic \ Code \end{array} \right) +$$

$$Quantum - Particle \ P - Type \ Being \ and \ Pre - Genetic \ Code$$

Eq 4.2.1: Generation of Quantum-Particle P-Type Being and Pre-Genetic Code

Note that Section 5 will discuss the computation leading to the emergence of the electromagnetic spectrum P-type being and its associated pre-genetic code segments, while Section 6 will discuss the computations leading to the emergence of matter, including the quantum-particle P-type being, and the associated pre-genetic code segments.

Generation of the Atom P-Type Being

In Equation 4.2.2, Generation of Atom P-Type Being and Pre-Genetic Code, the starting point, x_U, is 'Quantum Particles', and the ending point, x_T, is 'Atoms'. This iteration of the Light-Space Time Emergence equation suggests therefore that the generated code-segments for atoms will be added to the pre-existing material-fabric code-segments or "laws" for space-time-energy-gravity, the electromagnetic spectrum, and quantum particles, hence facilitating the emergence of a new whole or partial-singularity.

Hence:

$$Light - Space - Time\ Emergence_{Atom\ P-Type\ Being} =$$

$$
\left\| \begin{bmatrix} \begin{bmatrix} c_\infty: [Pr, Po, K, H] \\ \left(\downarrow R_{C_K} = f(R_{C_\infty})\right) \\ c_K: [S_{Pr}, S_{Po}, S_K, S_H] \\ \left(\downarrow R_{C_N} = f(R_{C_K})\right) \\ c_N: f(S_{Pr} \times S_{Po} \times S_K \times S_H) \\ \left(\downarrow R_{C_U} = f(R_{C_N})\right) \\ c_U: [P, V, M, C] \\ \Uparrow \\ c_{0:[D,W,I,C]} \end{bmatrix}_{Light} \\ U \to \begin{bmatrix} M_3: -\infty \le t \le \infty \\ \downarrow \\ M_2: 0 \ge t > \infty \\ \downarrow \\ M_1: 0 > t > \infty \\ \downarrow \\ t \le E_{Cell}; TC: M_3 \to U \\ t \sim E_{Human}; TC: U \to M_3 \end{bmatrix}_{Time} \quad \begin{matrix} TC \to x_{Atoms} \end{matrix} \quad \begin{bmatrix} M_3 \to System_X \\ (\uparrow F \to I) \\ M_2 \to S_{System_X} \\ (\uparrow Sig \to F) \\ M_1 \to Sig_X \\ (\uparrow > P_x) \\ U \to x_{Quantum\ Particles} \end{bmatrix}_{Space} \end{bmatrix} \right\|_{\langle x_U | x_T \rangle}
\Rightarrow
$$

$$\langle Space - Time - Energy - Gravity\ P - Type\ Beings\ and\ Pre - Genetic\ Code \rangle +$$

$$\left(\begin{matrix} Electro - magnetic - wavearchetype - masspotential\ P - Type\ Being \\ and\ Pre - Genetic\ Code \end{matrix} \right) +$$

$$\langle Quantum - Particle\ P - Type\ Being\ and\ Pre - Genetic\ Code \rangle +$$

$$Atoms\ P - Type\ Being\ and\ Pre - Genetic\ Code$$

Eq 4.2.2: Generation of Atom P-Type Being and Pre-Genetic Code

The computation leading to the emergence of the atom P-type being and its associated pre-genetic code segments is discussed in greater detail in Chapter 6.3.

Generation of Molecule P-Type Being

In Equation 4.2.3, Generation of Molecule P-Type Being and Pre-Genetic Code, the starting point, x_U, is 'Atom', and the ending point, x_T, is 'Molecule', signaling the emergence of a new molecules-based partial-singularity:

$$Light - Space - Time\ Emergence_{Molecule\ P-Type\ Being} =$$

$$\left[\begin{bmatrix} \begin{array}{c} c_\infty: [Pr, Po, K, H] \\ \left(\downarrow R_{C_K} = f(R_{C_\infty})\right) \\ c_K: [S_{Pr}, S_{Po}, S_K, S_H] \\ \left(\downarrow R_{C_N} = f(R_{C_K})\right) \\ c_N: f(S_{Pr} \times S_{Po} \times S_K \times S_H) \\ \left(\downarrow R_{C_U} = f(R_{C_N})\right) \\ c_U: [P, V, M, C] \\ \Uparrow \\ C_{0:[D,W,I,C]} \end{array} \end{bmatrix}_{Light} \begin{bmatrix} M_3 \rightarrow System_X \\ (\uparrow F \rightarrow I) \\ M_2 \rightarrow S_{System_X} \\ (\uparrow Sig \rightarrow F) \\ M_1 \rightarrow Sig_X \\ (\uparrow > P_x) \\ U \rightarrow X_{Atom} \end{bmatrix}_{Space} \Rightarrow \right.$$

$$\left. \begin{bmatrix} U \rightarrow \begin{array}{c} M_3 : -\infty \le t \le \infty \\ \downarrow \\ M_2 : 0 \ge t > \infty \\ \downarrow \\ M_1 : 0 > t > \infty \\ \downarrow \\ t \le E_{Cell}; TC: M_3 \rightarrow U \\ t \sim E_{Human}; TC: U \rightarrow M_3 \end{array} \end{bmatrix}_{Time} \quad TC \rightarrow X_{Molecule} \right]_{\langle x_U | x_T \rangle}$$

$$\langle Space - Time - Energy - Gravity\ P - Type\ Beings\ and\ Pre - Genetic\ Code \rangle +$$

$$\left(\begin{array}{c} Electro - magnetic - wavearchetype - masspotential\ P - Type\ Being \\ and\ Pre - Genetic\ Code \end{array}\right) +$$

$$\langle Quantum - Particle\ P - Type\ Being\ and\ Pre - Genetic\ Code \rangle +$$

$$\langle Atom\ P - Type\ Being\ and\ Pre - Genetic\ Code \rangle +$$

$$Molecule\ P - Type\ Being\ and\ Pre - Genetic\ Code$$

Eq 4.2.3: *Generation of Molecule P-Type Being and Pre-Genetic Code*

Generation of Cell P-Type Being

In Equation 4.2.4, Generation of Cell P-Type Being and Genetic Code, the starting point, x_U, is 'Molecule', and the ending point, x_T, is 'Cell':

$$Light - Space - Time\ Emergence_{Cell\ P-Type\ Being} =$$

$$
\begin{Vmatrix}
\begin{bmatrix}
\begin{matrix}
c_\infty : [Pr, Po, K, H] \\
\left(\downarrow R_{C_K} = f(R_{C_\infty}) \right) \\
c_K : [S_{Pr}, S_{Po}, S_K, S_H] \\
\left(\downarrow R_{C_N} = f(R_{C_K}) \right) \\
c_N : f(S_{Pr} \times S_{Po} \times S_K \times S_H) \\
\left(\downarrow R_{C_U} = f(R_{C_N}) \right) \\
c_U : [P, V, M, C] \\
\Uparrow \\
c_{0:[D,W,I,C]}
\end{matrix}
\end{bmatrix}_{Light}
\begin{bmatrix}
M_3 \rightarrow System_X \\
(\uparrow F \rightarrow I) \\
M_2 \rightarrow S_{System_X} \\
(\uparrow Sig \rightarrow F) \\
M_1 \rightarrow Sig_x \\
(\uparrow > P_x) \\
U \rightarrow X_{Molecule}
\end{bmatrix}_{Space} \\
\begin{bmatrix}
M_3 : -\infty \le t \le \infty \\
\downarrow \\
M_2 : 0 \ge t > \infty \\
\downarrow \\
M_1 : 0 > t > \infty \\
\downarrow \\
U \rightarrow \begin{matrix} t \le E_{Cell}; TC: M_3 \rightarrow U \\ t \sim E_{Human}; TC: U \rightarrow M_3 \end{matrix}
\end{bmatrix}_{Time}
\quad TC \rightarrow x_{Cell} \\
\end{Vmatrix}_{\langle x_U | x_T \rangle} \Rightarrow
$$

$\langle Space - Time - Energy - Gravity\ P - Type\ Beings\ and\ Pre - Genetic\ Code \rangle +$

$\left(\begin{matrix} Electro - magnetic - wavearchetype - masspotential\ P - Type\ Being \\ and\ Pre - Genetic\ Code \end{matrix} \right) +$

$\langle Quantum - Particle\ P - Type\ Being\ and\ Pre - Genetic\ Code \rangle +$

$\langle Atom\ P - Type\ Being\ and\ Pre - Genetic\ Code \rangle +$

$\langle Molecule\ P - Type\ Being\ and\ Pre - Genetic\ Code \rangle + LSTE \langle \dots \rangle +$

$Cell\ P - Type\ Being\ and\ Genetic\ Code$

Eq 4.2.4: Generation of Cell P-Type Being and Genetic Code

The computation leading to the emergence of the cell P-type being and associated code segments is discussed in greater detail in Chapter 7.2. Note that there are likely many iterations of this equation between molecules and cells and this in general is depicted by '*LSTE* $\langle \dots \rangle$' in (4.2.4), where LSTE signifies iterations(s) of the Light-Space-Time Emergence equation. Also note that an assumption is made that all the previous

153

material-fabric code-segments are intimately active at the level of cellular genetic code. This may be through the device of entanglement, or perhaps even by the material-fabric code segments being embedded in cellular genetic code.

Generation of Human P-Type Being

In Equation 4.2.5, Generation of Human P-Type Being and Genetic Code, the starting point, x_U, is 'Cell', and the ending point, x_T, is 'Human'. Clearly there will be a vast number of iterations before the human P-type being emerges, which also cover generation of vast tracts of the plant and animal kingdoms that will cause an appropriate divergence in genetic code. Further the development of microbiomes – that contain a multitude of bacteria and other microbes – will have been developed and shared between a vast number of species (Young, 2016).

Further, somewhere between the cell and the human, the emergence of vision is said to have sparked an explosion of evolution (Parker, 2003). This is not surprising if in fact we exist due to a Cosmology of Light as is being suggested in this book. It should be natural that instrumentation to perceive light, such as vision, would then generate an explosion in evolution.

Since any genetic code will be the materialization of four-base logic-encoding ecosystems (FBLEE) as a result of the Light-Matrix computation, there will also be a degree of entanglement to such ecosystems. Such entanglement will be different from ∞-entanglement, K-entanglement, and N-entanglement, and as suggested in the last chapter, will be referred to as FBLEE-entanglement (FBLEEE). Such entanglement is due to four-base logic-encoding ecosystems existing at the quantum level. FBLEE-entanglement will allow code-segment developments to be shared or accessed on an inter-species level and are depicted by 'FBLEEE $\langle ... \rangle$', as in Equation 4.2.5.

Hence:

$$Light - Space - Time\ Emergence_{Human\ P-TYpe\ Being} =$$

$$\left\|\left[\begin{array}{c} c_\infty:[Pr,Po,K,H] \\ \left(\downarrow R_{C_K} = f\left(R_{C_\infty}\right)\right) \\ c_K:[S_{Pr},S_{Po},S_K,S_H] \\ \left(\downarrow R_{C_N} = f\left(R_{C_K}\right)\right) \\ c_N: f(S_{Pr} \times S_{Po} \times S_K \times S_H) \\ \left(\downarrow R_{C_U} = f\left(R_{C_N}\right)\right) \\ c_U:[P,V,M,C] \\ \Uparrow \\ c_{0:[D,W,I,C]} \end{array}\right]_{Light} \begin{array}{c}\\\\ \left[\begin{array}{c} M_3 \to System_X \\ (\uparrow F \to I) \\ M_2 \to S_{System_X} \\ (\uparrow Sig \to F) \\ M_1 \to Sig_x \\ (\uparrow > P_x) \\ U \to X_{Cell} \end{array}\right]_{Space} \end{array}\right.$$

$$\left.\begin{array}{c}\left[\begin{array}{c} M_3 : -\infty \leq t \leq \infty \\ \downarrow \\ M_2 : 0 \geq t > \infty \\ \downarrow \\ M_1 : 0 > t > \infty \\ \downarrow \\ U \to \begin{array}{l} t \leq E_{Cell}; \text{TC: } M_3 \to U \\ t \sim E_{Human}; \text{TC: } U \to M_3 \end{array} \end{array}\right]_{Time} \quad TC \to X_{Human} \end{array}\right\| \Rightarrow$$

$$\langle x_U | x_T \rangle$$

$\langle Space - Time - Energy - Gravity\ P - Type\ Beings\ and\ Pre - Genetic\ Code \rangle +$

$\left(\begin{array}{c} Electro - magnetic - wavearchetype - masspotential\ P - Type\ Being \\ and\ Pre - Genetic\ Code \end{array}\right) +$

$\langle Quantum - Particle\ P - Type\ Being\ and\ Pre - Genetic\ Code \rangle +$

$\langle Atom\ P - Type\ Being\ and\ Pre - Genetic\ Code \rangle +$

$\langle Molecule\ P - Type\ Being\ and\ Pre - Genetic\ Code \rangle + LSTE\ \langle... \rangle +$

$\langle Cell\ P - Type\ Being\ and\ Genetic\ Code \rangle + LSTE\ \langle... \rangle + FBLEEE\ \langle... \rangle +$

$Human\ P - Type\ Being\ and\ Genetic\ Code$

Eq 4.2.5: Generation of Human P-Type Being and Genetic Code

If we follow the lines of emergence as will be described in the following subsection then (4.2.5) will be the starting point for potentially different trajectories of development. In one trajectory 'Humans' will evolve 'Basic Capacities of Self' and then 'Truer Individuality'. In another line of development 'Humans' will be the starting point for 'Stable Mega-Organization' culminating in 'Sustainable Global Civilization'. The gist of these can be represented by Equations 4.2.6 and 4.2.7 respectively. As in

(4.2.5) FBLEEE-segments depict the sharing of code segments between these different lines of development through the process of four-base logic-encoding ecosystem entanglement (FBLEEE) at the quantum level. In the latter line of development though the advance of post-human structures signals the start of a process that will correspondingly advance a post-genetic structure.

Generation of Truer Individuality P-Type Being

Hence in Equation 4.2.6, Generation of Truer Individuality P-Type Being and Genetic Code, the starting point, x_U, is 'Human', and the ending point, x_T, is 'Truer Individuality'. Clearly there will be a vast number of iterations before 'Truer Individuality' emerges, that will contain 'Basic Capacities of Self' as a milestone along the way:

$Light - Space - Time\ Emergence_{Truer\ Individuality\ P-Type\ Being} =$

$$
\left| \left[\begin{array}{l} c_\infty: [Pr, Po, K, H] \\ \left(\downarrow R_{C_K} = f(R_{C_\infty}) \right) \\ c_K: [S_{Pr}, S_{Po}, S_K, S_H] \\ \left(\downarrow R_{C_N} = f(R_{C_K}) \right) \\ c_N: f(S_{Pr} \times S_{Po} \times S_K \times S_H) \\ \left(\downarrow R_{C_U} = f(R_{C_N}) \right) \\ c_U: [P, V, M, C] \\ \Uparrow \\ c_{0:[D,W,I,C]} \end{array} \right]_{Light} \left[\begin{array}{l} M_3 \rightarrow System_X \\ (\uparrow F \rightarrow I) \\ M_2 \rightarrow S_{System_X} \\ (\uparrow Sig \rightarrow F) \\ M_1 \rightarrow Sig_x \\ (\uparrow > P_x) \\ U \rightarrow x_{Human} \end{array} \right]_{Space} \right.
$$

$$
\left. \left[U \rightarrow \begin{array}{l} M_3 : -\infty \le t \le \infty \\ \downarrow \\ M_2 : 0 \ge t > \infty \\ \downarrow \\ M_1 : 0 > t > \infty \\ \downarrow \\ t \le E_{Cell}; TC: M_3 \rightarrow U \\ t \sim E_{Human}; TC: U \rightarrow M_3 \end{array} \right]_{Time} \quad TC \rightarrow x_{Truer\ Individuality} \right| \Rightarrow \langle x_U | x_T \rangle
$$

$\langle Space - Time - Energy - Gravity\ P - Type\ Beings\ and\ Pre - Genetic\ Code \rangle +$

$\left(\begin{array}{c} Electro - magnetic - wavearchetype - masspotential\ P - Type\ Being \\ and\ Pre - Genetic\ Code \end{array} \right) +$

$\langle Quantum - Particle\ P - Type\ Being\ and\ Pre - Genetic\ Code \rangle +$

$\langle Atom \ P - Type \ Being \ and \ Pre - Genetic \ Code \rangle +$

$\langle Molecule \ P - Type \ Being \ and \ Pre - Genetic \ Code \rangle + LSTE \ \langle ... \rangle +$

$\langle Cell \ P - Type \ Being \ and \ Genetic \ Code \rangle + LSTE \ \langle ... \rangle + FBLEEE \ \langle ... \rangle +$

$\langle Human \ P - Type \ Being \ and \ Genetic \ Code \rangle + LSTE \ \langle ... \rangle + FBLEEE \ \langle ... \rangle +$

$Truer \ Individuality \ P - Type \ Being \ and \ Genetic \ Code$

Eq 4.2.6: Generation of Truer Individuality P-Type Being & Genetic Code

Generation of Sustainable Global Civilization FBLEE-Based P-Type Being

In equation 4.2.7, Generation of Sustainable Global Civilization FBLEE-Based P-Type Being, the starting point, x_U, is 'Human', and the ending point, x_T, is 'FBLEE-Based Sustainable Global Civilization'. Clearly there will be a vast number of iterations before 'Sustainable Global Civilization' emerges, and 'Stable Mega-Organization' will likely be an important milestone along the way. This line of development though, advancing post-human structures, will initiate a process that will begin to create post-genetic code. This will be referred to as Post Genetic Code Initiation or PGCI. Further, given that such a post-genetic structure does not yet exist, the sustainable global civilization will be a FBLEE-based partial-singularity or P-type being:

$Light - Space - Time \ Emergence_{FBLEE \ Sustainable \ Global \ Civilization \ P-Type \ Being} =$

$$\left| \begin{array}{c} \begin{bmatrix} c_\infty: [Pr, Po, K, H] \\ \left(\downarrow R_{C_K} = f\left(R_{C_\infty}\right)\right) \\ c_K: [S_{Pr}, S_{Po}, S_K, S_H] \\ \left(\downarrow R_{C_N} = f\left(R_{C_K}\right)\right) \\ c_N: f(S_{Pr} \times S_{Po} \times S_K \times S_H) \\ \left(\downarrow R_{C_U} = f\left(R_{C_N}\right)\right) \\ c_U: [P, V, M, C] \\ \Uparrow \\ c_{0:[D,W,I,C]} \end{bmatrix}_{Light} \begin{bmatrix} M_3 \to System_X \\ (\uparrow F \to I) \\ M_2 \to S_{System_X} \\ (\uparrow Sig \to F) \\ M_1 \to Sig_x \\ (\uparrow > P_x) \\ U \to x_{Human} \end{bmatrix}_{Space} \\ \begin{bmatrix} M_3 : -\infty \le t \le \infty \\ \downarrow \\ M_2 : 0 \ge t > \infty \\ \downarrow \\ M_1 : 0 > t > \infty \\ \downarrow \\ U \to \begin{array}{l} t \le E_{Cell}; TC: M_3 \to U \\ t \sim E_{Human}; TC: U \to M_3 \end{array} \end{bmatrix}_{Time} \quad TC \to x_{FBLEE\ Sust.Global\ Civilization} \end{array} \right| \Rightarrow \langle x_U | x_T \rangle$$

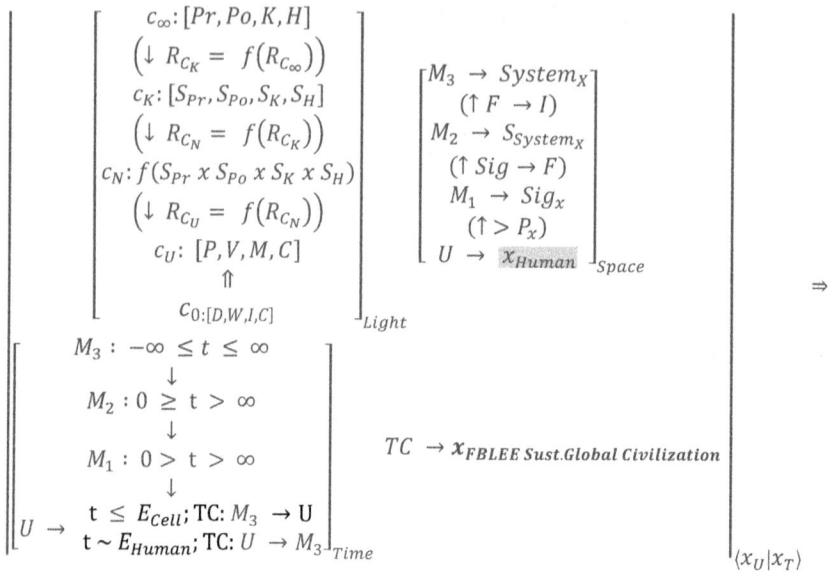

$\langle Space - Time - Energy - Gravity\ P - Type\ Beings\ and\ Pre - Genetic\ Code \rangle +$

$\left(\begin{array}{c} Electro - magnetic - wavearchetype - masspotential\ P - Type\ Being \\ and\ Pre - Genetic\ Code \end{array} \right) +$

$\langle Quantum - Particle\ P - Type\ Being\ and\ Pre - Genetic\ Code \rangle +$

$\langle Atom\ P - Type\ Being\ and\ Pre - Genetic\ Code \rangle +$

$\langle Molecule\ P - Type\ Being\ and\ Pre - Genetic\ Code \rangle + LSTE \langle ... \rangle +$

$\langle Cell\ P - Type\ Being\ and\ Genetic\ Code \rangle + LSTE \langle ... \rangle + FBLEEE \langle ... \rangle +$

$\langle Human\ P - Type\ Being\ and\ Genetic\ Code \rangle + LSTE \langle ... \rangle + FBLEEE \langle ... \rangle +$

$\langle FBLEE - Based\ Stable\ Mega - Organization\ P - Type\ Being\ and\ Post - Genetic\ Code \rangle +$

$PGCI \langle ... \rangle +$

$FBLEE - Based\ Sustainable\ Global\ Civilization\ P - Type\ Being\ and\ Post - Genetic\ Code$

Eq 4.2.7: Generation of FBLEE-Based Sustainable Global Civilization P-Type Being and Post-Genetic Code

The computation leading to the emergence of a sustainable global civilization P-type being is discussed in greater detail in Chapter 8.3.

Generation of Super-Matter P-Type Being

In Equation 4.2.8, Generation of Super-Matter P-Type Being and Post-Genetic Code, the starting point, x_U, is 'Molecule', and the ending point, x_T, is 'Super-Matter'. As per the mathematical model presented here all the preceding Light-Space-Time Emergence equation iterations progressively change the reality of matter through subtle quantization as will be further explored in subsequent sections. The following equation is therefore an approximation in stating how super-matter may emerge:

$$Light - Space - Time\ Emergence_{Suoer-Matter\ P-Type\ Being} =$$

$$\left|\left|\begin{bmatrix} c_\infty: [Pr, Po, K, H] \\ \left(\downarrow R_{C_K} = f(R_{C_\infty})\right) \\ c_K: [S_{Pr}, S_{Po}, S_K, S_H] \\ \left(\downarrow R_{C_N} = f(R_{C_K})\right) \\ c_N: f(S_{Pr} \times S_{Po} \times S_K \times S_H) \\ \left(\downarrow R_{C_U} = f(R_{C_N})\right) \\ c_U: [P, V, M, C] \\ \Uparrow \\ c_{0:[D,W,I,C]} \end{bmatrix}_{Light} \begin{bmatrix} M_3 \rightarrow System_X \\ (\uparrow F \rightarrow I) \\ M_2 \rightarrow S_{System_X} \\ (\uparrow Sig \rightarrow F) \\ M_1 \rightarrow Sig_x \\ (\uparrow > P_x) \\ U \rightarrow x_{Molecule} \end{bmatrix}_{Space} \right.\right.$$

$$\left.\left.\begin{bmatrix} M_3: -\infty \leq t \leq \infty \\ \downarrow \\ M_2: 0 \geq t > \infty \\ \downarrow \\ M_1: 0 > t > \infty \\ \downarrow \\ U \rightarrow \begin{array}{c} t \leq E_{Cell}; TC: M_3 \rightarrow U \\ t \sim E_{Human}; TC: U \rightarrow M_3 \end{array} \end{bmatrix}_{Time} \quad TC \rightarrow x_{Super-Matter} \right|\right|_{\langle x_U | x_T \rangle} \Rightarrow$$

$\langle Space - Time - Energy - Gravity\ P - Type\ Beings\ and\ Pre - Genetic\ Code \rangle +$

$\left(\dfrac{Electro - magnetic - wavearchetype - masspotential\ P - Type\ Being}{and\ Pre - Genetic\ Code}\right) +$

$\langle Quantum - Particle\ P - Type\ Being\ and\ Pre - Genetic\ Code \rangle +$

$\langle Atom\ P - Type\ Being\ and\ Pre - Genetic\ Code \rangle +$

$\langle Molecule\ P-Type\ Being\ and\ Pre-Genetic\ Code \rangle + LSTE\ \langle ... \rangle + FBLEEE\ \langle ... \rangle +$

$LSTE\ \langle ... \rangle + PGCI\ \langle ... \rangle + Super-Matter\ P-Type\ Being\ and\ Post-Genetic\ Code$

Eq 4.2.8: Generation of Super-Matter P-Type Being & Post-Genetic Code

The emergence of super matter is also discussed in the book Super-Matter (Malik, 2018a), with (4.2.8) included here to suggest computational realities yet to emerge. The essential action of space-time-energy-gravity quantization can cause the materialization of potentially infinite four-base logic-encoding ecosystems. It is conceivable that such a variation of space-time-energy-gravity quantization coupled with the ability to easily move back and forth between the material and antecedent realms makes the need to house genetic information in a form such as DNA incomplete. (4.2.7) suggested a process – Post Genetic Code Initiation – to capture such a dynamic and is included in (4.2.8) as a code-segment. It is conceivable that four-base logic-encoding ecosystems (FBLEE) and even FBLEEE, or even the ∞-entanglement, K-entanglement, and N-entanglement may be able to more directly act at the material level in some composite post-genetic form. This possibility will be referred to as Post-Genetic Code.

SECTION 5: GENERATION OF THE ELECTRO-MAGNETIC-WAVEARCHETYPE-MASSPOTENTIAL P-TYPE BEING

Section 5 will explore the computation involved in the creation of the electro-magnetic-wavearchetype-masspotential P-type being and its associated pre-genetic, material-fabric code.

The electromagnetic spectrum is a technical and contemporary way to refer to light. So, as light becomes more concrete to us or as the possibilities within it begin to emerge, one of the first forms it takes is as the electromagnetic spectrum. Note that while the generation of photons, the carrier particle for the electromagnetic spectrum, is usually associated with excitation of electrons in an atom, in this treatise light itself has a more fundamental place, being the very matrix from which everything else emerges. As explored in sections 1 and 2, the Matrix of Being is constituted by light traveling at different speeds. The fact that these speeds are constant is significant and implies an intentionality, that is itself being comprehensively explored in the Cosmology of Light book series. That said, it must be the case that the previously surfaced properties of light – Presence, Power, Knowledge, and Harmony – emerge so as to define the very architecture of the electromagnetic spectrum.

Chapter 5.1, Generation of Electro-Magnetic-Wavearchetype-Masspotential P-Type Being, summarizes the emergence of the electromagnetic spectrum in terms of the underlying Light-Space-Time Emergence equation and the process of quantization that must occur to create the logic of the electromagnetic spectrum ecosystem that precipitates into the material-fabric. So we find that the four underlying properties of Light that we call Harmony, Knowledge, Power, and Presence are of the essence of the speed with which the electromagnetic spectrum moves, the wave-range within the electromagnetic spectrum, the energy-gradient within the electromagnetic spectrum, and the mass-possibilities due to the electromagnetic spectrum, respectively. Further the description – electromagnetic – seems to have captured the Power-Harmony aspects implicit in light. In reality the electro-magnetic spectrum can likely be more completely described as electro-magnetic-wavearchetype-masspotential spectrum.

Chapter 5.1: Generation of Electro-Magnetic-Wavearchetype-Masspotential P-Type Being

As elaborated in Section 4, the Light-Space-Time Emergence equation (3.1.3) can be used to model generation of P-type beings from simpler four-fold to more complex four-fold manifestations. In essence (3.1.3) can also be thought of as the equation by which the biographies for any light-based-singularity and therefore all light-based beings such as the electro-magnetic-wavearchetype-masspotential P-type being are generated. Being iterative, it can be used to understand the generation of code-segments that will continue to add to genetic-type code thus enriching the living narrative that accompanies the materialization of infinite possibility in Light. (3.6.12), the Potential Effect of Levels of Light on Genetic-Type Information equation, basically sheds light on specific types of mutation that can occur and therefore on the nature of the code-segment that is being generated by (3.1.3). Note that (3.6.12) shows a particular working of (3.1.3) but is wholly contained within it.

This chapter elaborates the action of (3.1.3) and (3.6.12) in the creation of the code-segments that contribute to the ecosystem logic of the electro-magnetic-wavearchetype-masspotential P-type being.

Application of Light-Space-Time Emergence Equation

As suggested by (3.1.3) the architecture and details of the electro-magnetic-wavearchetype-masspotential P-type being, and the resulting material-fabric, pre-genetic code can be seen to be the result of the application of the Light, Space, and Time matrices. This is illustrated in Equation 5.1.1, Generation of Electro-Magnetic-Wavearchetype-Masspotential P-Type Being:

$$Light - Space - Time\ Emergence_{EMWM\ P-Type\ Being} =$$

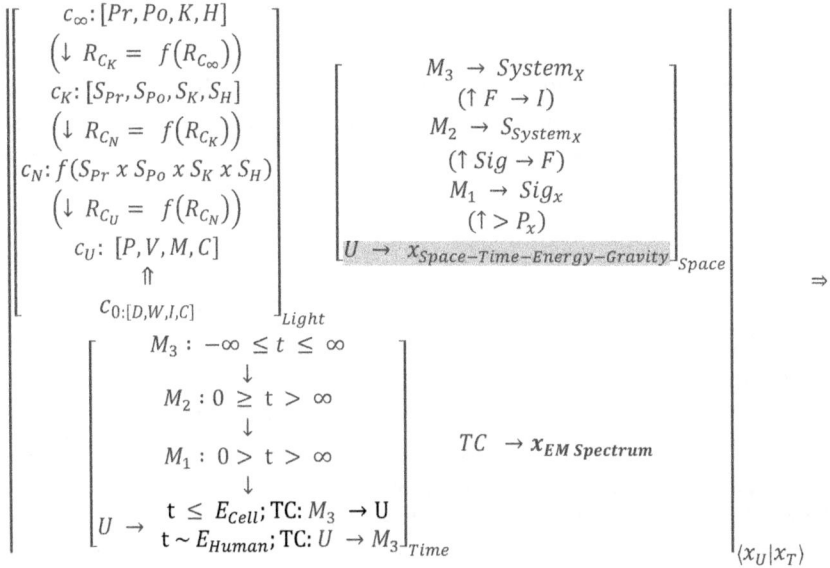

$$\left[\left[\begin{array}{c} c_\infty:[Pr,Po,K,H] \\ \left(\downarrow R_{C_K} = f(R_{C_\infty})\right) \\ c_K:[S_{Pr},S_{Po},S_K,S_H] \\ \left(\downarrow R_{C_N} = f(R_{C_K})\right) \\ c_N: f(S_{Pr} \times S_{Po} \times S_K \times S_H) \\ \left(\downarrow R_{C_U} = f(R_{C_N})\right) \\ c_U: [P,V,M,C] \\ \Uparrow \\ c_{0:[D,W,I,C]} \end{array}\right]_{Light} \quad \left[\begin{array}{c} M_3 \to System_x \\ (\uparrow F \to I) \\ M_2 \to S_{System_x} \\ (\uparrow Sig \to F) \\ M_1 \to Sig_x \\ (\uparrow > P_x) \\ U \to x_{Space-Time-Energy-Gravity} \end{array}\right]_{Space} \right.$$

$$\left.\begin{array}{cc} \left[\begin{array}{c} M_3 : -\infty \le t \le \infty \\ \downarrow \\ M_2 : 0 \ge t > \infty \\ \downarrow \\ M_1 : 0 > t > \infty \\ \downarrow \\ U \to \begin{array}{l} t \le E_{Cell}; TC: M_3 \to U \\ t \sim E_{Human}; TC: U \to M_3 \end{array} \end{array}\right]_{Time} & TC \to x_{EM\ Spectrum} \end{array}\right] \Rightarrow$$

$$\langle x_U | x_T \rangle$$

$$\langle Space - Time - Energy - Gravity\ Material - Fabric\ Pre - Genetic\ Code \rangle +$$

$$Electromagnetic\ Spectrum\ Material - Fabric\ Pre - Genetic\ Code$$

Eq 5.1.1: Generation of Electro-Magnetic-Wavearchetype-Masspotential P-Type Being

Starting with the Light-Matrix, the top left-hand matrix in (5.1.5), the first line from the top, $C_\infty: [Pr, Po, K, H]$, specifies the fundamental architecture of the electro-magnetic-wavearchetype-masspotential P-type being. As will be elaborated in the subsections on Harmony, Knowledge, Power, and Presence, each of these aspects are an emergence of the fundamental properties of Light at ∞. Hence, it is only Light that architects the materialization of the infinite possibility within it – this is the hallmark of any light-based-singularity such as the P-type being. There is an underlying oneness that founds the dynamics of the P-type being.

Line 3 in the Light-Matrix, C_K: $[S_{Pr}, S_{Po}, S_K, S_H]$, elaborates the sets for Presence, Power, Knowledge, and Harmony, each containing multiple elements. For example, as will be explored in the subsection on Harmony, the following elements: 'connection', 'growing into one's own', 'form bonds', are suggested to exist in the Set of Harmony or Nurturing, and are accessed to contribute to the operative reality so set up by virtue of light traveling at c.

Specifically, Line 5, $C_{N:}$ $f(S_{Pr} \times S_{Po} \times S_K \times S_H)$, suggests that unique seeds are created from a combination of the elements from all four sets, with a particular element leading or having more weight, that in effect creates the distinctness of the vast variety of waves possible in the electro-magnetic-wavearchetype-masspotential P-type being.

Line 6, ($\downarrow R_{C_U} = f(R_{C_N})$), specifies quantization between the layer where the seeds are formed, and the physical layer, and as explored in Chapter 3.5 and 3.6, will result in Line 7, C_U: $[P, V, M, C]$, hence changing the material-fabric of existence. Note that as in the process describing the generation of the code-segments for space-time-energy-gravity quantization at the time of the Big Bang, the generation of the FBLEE (four-base logic-encoding ecosystem) code-segment as specified by (4.1.5) and the MF (material-fabric) code segment as specified by (4.1.6) are combined together into one equation so that the FBLEE process is apparently transparent. In reality one may say that any P-type being is first a FBLEE-P-type being and then a MF-P-type being.

The possibilities represented by Lines 1 through 5 hence concretize through the quantization represented by Line 6 to become the electro-magnetic-wavearchetype-masspotential P-type being with its physical (related to Presence), vital (related to Power), mental (related to Knowledge), and connection (related to Harmony) aspects now existing in material reality typified by Light moving at c. Note that just as Line 6 represents a process of quantization relating the layer of reality created by Light traveling at c with the antecedent layers, so too Lines 2 and 4 as previously discussed, also represent quantization of a more subtle kind that ultimately plays a critical part in allowing the material-fabric to express infinite diversity.

Typically, it is the process as captured by the Space-Matrix that will determine if Line 6 is activated. Specifically, patterns at the untransformed layer, U, will need to be overcome, as specified by the second line from the bottom of the Space-Matrix: $(\uparrow > P_x)$. But as specified by the bottom-line of the Time-Matrix, reproduced below, it is assumed that only with the advent of the human-system that the automaticity of the action of meta-levels is reversed:

$$U \rightarrow \begin{array}{l} t \leq E_{Cell}; \text{TC: } M_3 \rightarrow U \\ t \sim E_{Human}; \text{TC: } U \rightarrow M_3 \end{array}$$

Hence in the case of the electro-magnetic-wavearchetype-masspotential P-type being, which in this emergence is a pre-human system, the fact that patterns do not need to be overcome means that quantization happens automatically, and that in reality the electro-magnetic-wavearchetype-masspotential P-type being would therefore more accurately be termed the electro-magnetic-wavearchetype-masspotential PMU-type being.

From the point of view of pre-genetic information this means that constructive mutation has occurred and that the available code-segments that impact the material operation of the universe has been changed. But 'constructive' implies that the operation, materially, allows the manifested universe to 'materially' come closer to the complete integrated nature of Light. Of course, the emergence of the electro-magnetic-wavearchetype-masspotential P-type being in this rendering occurs after the emergence of the space-time-energy-gravity P-type being, and marks the beginning stages, relatively, of the long curve that evolution will go through as more and more possibility emerges from Light. These fundamental emergences are milestones along the way marking the expansion of one edifice, a light-based edifice, which may culminate in the reality of the Second Singularity or S-type being, were all conditions for such fulfillment to be achieved.

Application of Potential Effect of Levels of Light on Genetic-Type Information Equation

Even though the electro-magnetic-wavearchetype-masspotential P-type being is a pre-human being and therefore the action of meta-levels are modeled as being automatic, it is nonetheless useful to review (3.6.12), **Potential Effect of Levels of Light on Genetic-Type Information**, to understand how the relationship with different forms of mutation has been specified:

Potential Effect of Levels of Light on Genetic − Type Information =

$$
\begin{bmatrix}
\quad STATIC \; \langle |[L][S][T]TC \rightarrow x_T|_{\langle x_U | x_T \rangle} \rangle \\
\times \\
\left((Y > U : Z_Q) \vee (Y \leq U : Z_F) \vee (Y = U : Z_R) \right) \\
\ni \\
\left(\begin{matrix} Z \in \mathbb{U} \, (Space, Time, Energy, Gravity) \\ Q: Quantization; \; F: Fragmentation; \; R: Random \end{matrix} \right)
\end{bmatrix} \rightarrow h \ni
$$

$$
h \in \left(\begin{matrix} [Q]: Constructive \; zone, \\ [Q]: Constructive \; zone \wedge Constructive \; mutation, \\ [F]: Destructive \; mutation, \\ [R]: Random \; mutation \end{matrix} \right)
$$

Line 1 from the top in the matrix is simply a static form of (3.6.1) the Simplified Light-Space-Time Emergence equation. The static form is designated by 'STATIC' and implies that fundamental operations true of (3.6.1) are being highlighted in (3.6.12). In other words (3.6.1) already has all the operations highlighted in (3.6.12) in it, but by 'freezing' it by making it static, the essential dynamics leading to possible mutations at the genetic level can more clearly be highlighted.

Line 1 is then subjected ($\times$) to a determination of the dominant levels of light that may be active, designated by Line 2, $'\left((Y > U : Z_Q) \vee (Y \leq U : Z_F) \vee (Y = U : Z_R) \right)'$. Unpacking this, '$Y > U$' implies meta-levels are active and as a result it is possible that Z_Q is going to take place (the subscript 'Q' implies quantum-level action). This also implies activation and potential change of FBLEE. The call from below, as it were, may invoke some function that already exists in the subtle-libraries 'above', so that some already existing function may influence FBLEE through ∞-entanglement, K-entanglement, or N-entanglement. This may be thought of as a key-and-lock mechanism, where a deep enough visceral urge from below acting as the key, opens an entangled lock to alter FBLEE as per the visceral urge. '$Y \leq U$' implies that only the untransformed levels are active, and therefore also the sub-level where the speed of light is 0 is active and as a result Z_F is going to take place (the subscript 'F' implies 'fragmentation'). '$Y = U$' implies that all levels are active and as a result Z_R is going to take place (the subscript 'R' implies 'random').

Line 3 elaborates the significance of Z_Q, Z_F, and Z_R. Hence Z is the union of potential quantum-operations of space, time, energy, and gravity,

166

designated by '$Z \in \mathbb{U}$ ($Space, Time, Energy, Gravity$)'. But the nature of the operations, as suggested in the previous paragraph, is designated by 'Q: $Quantization; F$: $Fragmentation; R$: $Random$'. Z_Q, then, implies that the full quantization originating from updated four-base logic-encoding ecosystems (FBLEE) can take place. Z_F implies that the essential set will be fragmented and that only libraries at the level of local cellular-level DNA or precipitated material-fabric logic can potentially be altered. Z_R also precludes full quantization, and that some partial local-library constructive or destructive mutation may take place.

The '$\rightarrow h \ni h \in$' segment resolves the outcome of the operations implied by Lines 1 – 3, suggesting that the outcome will be 'h' such that ($\ni$) 'h' is an element ($\in$) of the set specified by the members '[Q]: *Constructive zone*', '[Q]: *Constructive zone AND Constructive mutation*', '[F]: *Destructive mutation*', and '{R}: *Random mutation*'. '[Q]: *Constructive zone*' implies that the in-built buffer has been crossed and that access to the deeper four-base logic-encoding ecosystem (FBLEE) has been granted. '[Q]' in this segment implies that there is the possibility that full quantization as specified by Line 3 of the previous matrix can take place. Access to this zone is a prerequisite for constructive mutation to occur, as designated by the element '[Q]: *Constructive zone* $\cap$ *Constructive mutation*', which implies that full-quantization is going to take place and will result in material change. It is in this case that the FBLEE iteration will be followed by an MF iteration. The '[F]' specifies the relationship between 'Fragmentation' in Line 3 of the previous matrix and destructive mutation. The '[R]' specifies the relationship between 'Random' in Line 3 of the previous matrix and random mutation.

But as just summarized in the Time-Matrix in (5.1.1) Y is by definition greater than U and hence quantization is automatic. In terms of the electro-magnetic-wavearchetype-masspotential P-type being such quantization implies that wholeness becomes fully active through specific space, time, energy, and gravity quantization to create an holistic "ecosystem" with its own "electro-magnetic-wavearchetype-masspotential p-type being logic" as it were. The wholeness has now precipitated into the material-fabric and is available to be consciously and unconsciously tapped into. This "logic" or more specifically pre-genetic code is elaborated in the following sub-sections.

167

Generation of 'Magnetic' Aspect of Electro-Magnetic-Wavearchetype-Masspotential P-Type Being

To begin with, we had already looked at how the speed of light sets up the nature of reality because of its speed. So, for light traveling at c, past, present, future, and the notion of separation due to creation of islands of matter is the reality. It seems then that c architects the possibility of matter-based interaction in our system and can therefore be thought of as a projection from the property of Harmony to create the basis for a matter-based harmony. So being, it can be suggested that the nature of the resultant interactions allows matter-based organizations, regardless of scale, to come into their own, to grow into their boundaries, and to form bonds based on the sense of being separated from other perceived organizations. This notion of forming bonds seems also to be related to the "magnetic" in the electro-magnetic-wavearchetype-masspotential P-type being.

In equation form, as in Equation 5.1.2 it would therefore be possible to specify the nature of reality so set up by the electro-magnetic-wavearchetype-masspotential (EMWM) P-type being moving at c:

$$EMWM\ P - Type\ Being_{Speed} = Xa +$$
$$\overline{Yb_{0-n}}\ \ where\ \begin{bmatrix} X \in [S_{System_N}] \\ Y \in [S_{System_{Pr}}, S_{System_P}, S_{System_K}, S_{System_N}] \\ a, b\ are\ integers; a > b \end{bmatrix}$$

Eq 5.1.2: Speed of EMWM P-Type Being

The notion of 'connection', 'growing into one's own', 'form bonds', amongst other attributes of such a reality can be seen as elements of the four sets of architectural forces. Hence (5.1.2) suggests the mathematical equation that so defines the nature of reality due to the speed of c of the electro-magnetic-wavearchetype-masspotential P-type being.

Note that (5.1.2) already implies that Lines 1 – 5 in the Light Matrix of (5.1.1) have been activated, and that the logic of the "magnetic" in the electro-magnetic-wavearchetype-masspotential P-type being ecosystem will automatically precipitate into the material-fabric through the action of Line 6-7 of (5.1.1). This logic is none other than the pre-genetic code as specified by (5.1.2).

Generation of 'Wave Archetype' Aspect of Electro-Magnetic-Wavearchetype-Masspotential P-Type Being

The electro-magnetic-wavearchetype-masspotential P-type being contains a range of waves embedded in it. These waves enable many different applications with practical utility apparent in everyday life. So, for example there is a region in the electro-magnetic-wavearchetype-masspotential P-type being that we call radio waves, and others that we know as microwave, infrared visible light, ultraviolet, x-rays and gamma rays that each make possible many technologies that we use every single day. These ranges essentially code a range of technological-possibility, and therefore the wave-range within the electro-magnetic-wavearchetype-masspotential P-type being can be thought of as encoding or of expressing some kind of precipitation or projection or emergence of the property of Knowledge. Or put another way, the property of Knowledge that is found in light, emerges as the wave-range implicit in the electro-magnetic-wavearchetype-masspotential P-type being.

The electro-magnetic-wavearchetype-masspotential (EMWM) P-type being itself, with its vast range of natures from gamma rays through visible light through radio waves, with its implicitness of time-space possibility as suggested by frequency (v) and wavelength (λ), may be thought of as an arrangement of archetypes, and therefore is perhaps a precipitation of system-knowledge, as in Equation 5.1.3:

$$EMWM\ P - Type\ Being_{Structure} = Xa + \overline{Yb_{0-n}}$$

$$where \begin{bmatrix} X \in [S_{System_K}] \\ Y \in [S_{System_{Pr}}, S_{System_P}, S_{System_K}, S_{System_N}] \\ a, b\ are\ integers; a > b \end{bmatrix}$$

Eq 5.1.3: Structure of EMWF P-Type Being

Note that (5.1.3) already implies that Lines 1 – 5 in the Light Matrix of (5.1.1) have been activated, and that the logic of the "wave-archetype" in the electro-magnetic-wavearchetype-masspotential P-type being ecosystem will automatically precipitate into the material-fabric through the action of Line 6-7 of (5.1.1). This logic is none other than the pre-genetic code as specified by (5.1.3).

What this also implies is that the significance or intent of the different types of waves that exist can also be expressed by this general equation where the X and Y elements will vary. What precisely these elements are will need to be worked out. A few representative equations, Equations 5.1.4 through 5.1.7 follow:

$$Gamma\ Rays\ _{Intent} = Xa + \overline{Yb_{0-n}}\ where \begin{bmatrix} X \in [S_{System_K}] \\ Y \in [S_{System_{Pr}}, S_{System_p}, S_{System_K}, S_{System_N}] \\ a, b\ are\ integers; a > b \end{bmatrix}$$

Eq 5.1.4: Gamma Rays Intent

Knowing that some of the applications of Gamma Rays are in sterilizing and radiotherapy, these can be attributed as elements to the architectural sets. The totality of the use can be thought of as the 'intent' of Gamma Rays. An understanding of the uses will allow a full set of the secondary elements to be mapped thus allowing (5.1.4) to be reverse engineered.

Similarly, each of the archetypes present in the electro-magnetic-wavearchetype-masspotential P-type being can be mapped out. Sample mapping of intent include x-rays, infrared, and microwaves:

$$X - Rays_{Intent} = Xa + \overline{Yb_{0-n}}\ where \begin{bmatrix} X \in [S_{System_K}] \\ Y \in [S_{System_{Pr}}, S_{System_p}, S_{System_K}, S_{System_N}] \\ a, b\ are\ integers; a > b \end{bmatrix}$$

Eq 5.1.5: X-Rays Intent

$$Infrared_{Intent} = Xa + \overline{Yb_{0-n}}\ where \begin{bmatrix} X \in [S_{System_K}] \\ Y \in [S_{System_{Pr}}, S_{System_p}, S_{System_K}, S_{System_N}] \\ a, b\ are\ integers; a > b \end{bmatrix}$$

Eq 5.1.6: Infrared Intent

$$Microwaves\ _{Intent} = Xa + \overline{Yb_{0-n}}\ where \begin{bmatrix} X \in [S_{System_K}] \\ Y \in [S_{System_{Pr}}, S_{System_p}, S_{System_K}, S_{System_N}] \\ a, b\ are\ integers; a > b \end{bmatrix}$$

Generation of 'Electro' Aspect of Electro-Magnetic-Wavearchetype-Masspotential P-Type Being

The range of different wave-types or wavelengths implicit in the electro-magnetic-wavearchetype-masspotential P-Type Being also moves with different frequencies. Since the speed of light is a constant, and it is known that speed is the product of frequency and wavelength, the greater the wavelength the lower will be the frequency of the wave-type. Conversely the less the wavelength the higher will be the frequency of the wave-type. And energy or power of a wave-type will depend on its frequency.

Hence, gamma rays that have a lower wavelength will have a higher frequency, and higher energy associated with it. Radio waves on the other hand that have a higher wavelength will have a lower frequency and therefore lower energy associated with it. So implicit in the electro-magnetic-wavearchetype-masspotential P-type being is a gradient of energy. But it is also known that the penetration power is dependent on frequency. Therefore, the higher the frequency or energy, the higher will be the power of the wave-type. Another way to say this is that the property of Power in Light emerges as the energy-gradient in the electro-magnetic-wavearchetype-masspotential P-type being. This Power aspect seems to have been captured by the "electro" in electro-magnetic-wavearchetype-masspotential.

The energy-gradient implicit in the **electro-magnetic-wavearchetype-masspotential (EMWM) P-type being** suggests the power and energy with which knowledge moves and is perhaps a precipitation of system-power, as in Equation 5.1.8:

$$EMWM\ P-Type\ Being_{Energy} = Xa +$$
$$\overline{Yb_{0-n}}\ where\ \begin{bmatrix} X \in [S_{System_P}] \\ Y \in [S_{System_{Pr}}, S_{System_P}, S_{System_K}, S_{System_N}] \\ a, b\ are\ integers; a > b \end{bmatrix}$$

Eq 5.1.8: EMWM P-Type Being Energy

Note that (5.1.8) already implies that Lines 1 – 5 in the Light Matrix of (5.1.1) have been activated, and that the logic of the "electro" in the electro-magnetic-wavearchetype-masspotential P-type being ecosystem will automatically precipitate into the material-fabric through the action of Line 6-7 of (5.1.1). This logic is none other than the pre-genetic code as specified by (5.1.8).

This electro-magnetic-wavearchetype-masspotential P-type being energy, as suggested, is directly proportional to the frequency, of which there is an infinite range as predicted by Maxwell's equations. Different frequencies have different penetration profiles (HyperPhysics, 2016) and it may be suggested that the nature of the energy is also a precipitation of a range of to be determined meta-functions as suggested in some sample Equations 5.1.9 through 5.1.13. Hence:

$Gamma\ Rays\ _{Nature\ of\ Energy}\ =$

$$Xa + \overline{Yb_{0-n}} \quad where \begin{bmatrix} X \in [S_{System_P}] \\ Y \in [S_{System_{Pr}}, S_{System_P}, S_{System_K}, S_{System_N}] \\ a, b\ are\ integers; a > b \end{bmatrix}$$

Eq 5.1.9: Gamma Rays Nature of Energy

$X - Rays_{Nature\ of\ Energy}\ =$

$$Xa + \overline{Yb_{0-n}} \quad where \begin{bmatrix} X \in [S_{System_P}] \\ Y \in [S_{System_{Pr}}, S_{System_P}, S_{System_K}, S_{System_N}] \\ a, b\ are\ integers; a > b \end{bmatrix}$$

Eq 5.1.10: X-Rays Nature of Energy

$Ultraviolet\ _{Nature\ of\ Energy}\ =$

$$Xa + \overline{Yb_{0-n}} \quad where \begin{bmatrix} X \in [S_{System_P}] \\ Y \in [S_{System_{Pr}}, S_{System_P}, S_{System_K}, S_{System_N}] \\ a, b\ are\ integers; a > b \end{bmatrix}$$

Eq 5.1.11: Ultraviolet Nature of Energy

$White\ Light_{Nature\ of\ Energy}\ =$

$$Xa + \overline{Yb_{0-n}} \quad where \begin{bmatrix} X \in [S_{System_P}] \\ Y \in [S_{System_{Pr}}, S_{System_P}, S_{System_K}, S_{System_N}] \\ a, b\ are\ integers; a > b \end{bmatrix}$$

Eq 5.1.12: White Light Nature of Energy

$AM\ Radio\ Waves_{Nature\ of\ Energy}\ =$

$$Xa + \overline{Yb_{0-n}}\ \ where\ \left[\begin{array}{c} X \in [S_{System_P}] \\ Y \in [S_{System_{Pr}}, S_{System_P}, S_{System_K}, S_{System_N}] \\ a, b\ are\ integers; a > b \end{array}\right]$$

Eq 5.1.13: AM Radio Waves Nature of Energy

Generation of 'Mass Potential' Aspect of Electro-Magnetic-Wavearchetype-Masspotential P-Type Being

If there is a large range of frequencies implicit in the electro-magnetic-wavearchetype-masspotential P-type being, then there is also the possibility of different types of masses implicit in the electro-magnetic-wavearchetype-masspotential P-type being. Frequency determines energy, and mass and energy are related through Einstein's famous MC-squared equation. So, pushing a little further it is not just that mass and energy are related, but a different kind of frequency or wave-type potentially allows a different type of mass to emerge. So, the possibility of different types of mass seems to be related to the property of Presence in Light. In other words, the property of Presence emerges as the possibility of different types of masses as suggested by the range of mass-possibilities that can emerge from the electro-magnetic-wavearchetype-masspotential (EMWM) P-type being.

Mass can be thought of as a container at U within which all possibility happens. In other words, it can be thought of as a precipitation or emergence of system-presence as depicted in Equation 5.1.14:

$$EMWM\ P-Type\ Being_{Mass_{Possibility}}\ = Xa + \overline{Yb_{0-n}}$$

$$where\ \left[\begin{array}{c} X \in [S_{System_{Pr}}] \\ Y \in [S_{System_{Pr}}, S_{System_P}, S_{System_K}, S_{System_N}] \\ a, b\ are\ integers; a > b \end{array}\right]$$

Eq 5.1.14: EMWM P-Type Being Mass Possibility

Note that (5.1.14) already implies that Lines 1 – 5 in the Light Matrix of (5.1.1) have been activated, and that the logic of the "masspotential" in the electro-magnetic-wavearchetype-masspotential P-type being ecosystem will automatically precipitate into the material-fabric through the action of Line 6-7 of (5.1.1). This logic is none other than the pre-genetic code as specified by (5.1.14).

Further, if the frequencies are infinite, then the possibility of the 'types' of masses or matter is also infinite. The mystery of 'Dark Matter' suggested by scientists to be 27% of our universe, as opposed to 5% of visible matter (NASA-darkmatter, 2016) may have some relation to this. This aspect is also explored in the book Cosmology of Light (Malik, 2018b).

Hypothetically the Equations 5.1.15 through 5.1.19 depict a range of mass possibilities:

$Gamma\ Rays\ {}_{Mass_{Possibility}} =$

$$Xa + \overline{Yb_{0-n}} \quad where \begin{bmatrix} X \in [S_{System_{Pr}}] \\ Y \in [S_{System_{Pr}}, S_{System_P}, S_{System_K}, S_{System_N}] \\ a, b\ are\ integers; a > b \end{bmatrix}$$

Eq 5.1.15: Gamma Rays Mass Possibility

$Ultraviolet\ {}_{Mass_{Possibility}} =$

$$Xa + \overline{Yb_{0-n}} \quad where \begin{bmatrix} X \in [S_{System_{Pr}}] \\ Y \in [S_{System_{Pr}}, S_{System_P}, S_{System_K}, S_{System_N}] \\ a, b\ are\ integers; a > b \end{bmatrix}$$

Eq 5.1.16: Ultraviolet Mass Possibility

$Blue\ Light\ {}_{Mass_{Possibility}} =$

$$Xa + \overline{Yb_{0-n}} \quad where \begin{bmatrix} X \in [S_{System_{Pr}}] \\ Y \in [S_{System_{Pr}}, S_{System_P}, S_{System_K}, S_{System_N}] \\ a, b\ are\ integers; a > b \end{bmatrix}$$

Eq 5.1.17: Blue Light Mass Possibility

$Microwaves\ {}_{Mass_{Possibility}} =$

$$Xa + \overline{Yb_{0-n}} \quad where \begin{bmatrix} X \in [S_{System_{Pr}}] \\ Y \in [S_{System_{Pr}}, S_{System_{P}}, S_{System_{K}}, S_{System_{N}}] \\ a, b \ are \ integers; a > b \end{bmatrix}$$

Eq 5.1.18: Microwaves Mass Possibility

$$FM \ Radio \ Waves_{Mass_{Possibility}} =$$

$$Xa + \overline{Yb_{0-n}} \quad where \begin{bmatrix} X \in [S_{System_{Pr}}] \\ Y \in [S_{System_{Pr}}, S_{System_{P}}, S_{System_{K}}, S_{System_{N}}] \\ a, b \ are \ integers; a > b \end{bmatrix}$$

Eq 5.1.19: FM Radio Waves Mass Possibility

Summary of Electro-Magnetic-Wavearchetype-Masspotential P-Type Being

Summarizing, after the generation of the electro-magnetic-wavearchetype-masspotential P-type being, the following code-segments will have been created as specified by Equation 5.1.20, Active Electro-Magnetic-Wavearchetype-Masspotential P-Type Being Material-Fabric Pre-Genetic Code Segments:

$$Light - Space - Time \ Emergence_{EMWM \ P-Type \ Being} =$$

$$\left|\left|\begin{bmatrix} c_\infty : [Pr, Po, K, H] \\ \left(\downarrow R_{C_K} = f(R_{C_\infty})\right) \\ c_K : [S_{Pr}, S_{Po}, S_K, S_H] \\ \left(\downarrow R_{C_N} = f(R_{C_K})\right) \\ c_N : f(S_{Pr} \times S_{Po} \times S_K \times S_H) \\ \left(\downarrow R_{C_U} = f(R_{C_N})\right) \\ c_U : [P, V, M, C] \\ \Uparrow \\ c_{0:[D,W,I,C]} \end{bmatrix}_{Light} \right.\right.$$

$$\begin{bmatrix} M_3 : -\infty \leq t \leq \infty \\ \downarrow \\ M_2 : 0 \geq t > \infty \\ \downarrow \\ M_1 : 0 > t > \infty \\ \downarrow \\ U \to \begin{matrix} t \leq E_{Cell}; TC: M_3 \to U \\ t \sim E_{Human}; TC: U \to M_3 \end{matrix} \end{bmatrix}_{Time}$$

$$\begin{bmatrix} M_3 \to System_X \\ (\uparrow F \to I) \\ M_2 \to S_{System_X} \\ (\uparrow Sig \to F) \\ M_1 \to Sig_x \\ (\uparrow > P_x) \\ U \to x_{Space-Time-Energy-Gravity} \end{bmatrix}_{Space}$$

$$TC \to x_{EMWM\ P-Type\ Being}$$

$$\Rightarrow \langle x_U | x_T \rangle$$

$\langle Space - Time - Energy - Gravity\ P - Type\ Beings\ and\ Pre - Genetic\ Code \rangle +$

$$\left(\begin{array}{l} \sum EMWM\ P - Type\ Being_{Speed} = Xa + \overline{Yb_{0-n}} \\ where \begin{bmatrix} X \in [S_{System_N}] \\ Y \in [S_{System_{Pr}}, S_{System_P}, S_{System_K}, S_{System_N}] \\ a, b\ are\ integers; a > b \end{bmatrix} \\ \sum EMWM\ P - Type\ Being_{Structure} = Xa + \overline{Yb_{0-n}} \\ where \begin{bmatrix} X \in [S_{System_K}] \\ Y \in [S_{System_{Pr}}, S_{System_P}, S_{System_K}, S_{System_N}] \\ a, b\ are\ integers; a > b \end{bmatrix} \\ \sum EMWM\ P - Type\ Being_{Energy} = Xa + \overline{Yb_{0-n}} \\ where \begin{bmatrix} X \in [S_{System_P}] \\ Y \in [S_{System_{Pr}}, S_{System_P}, S_{System_K}, S_{System_N}] \\ a, b\ are\ integers; a > b \end{bmatrix} \\ \sum EMWM\ P - Type\ Being_{Mass_{Possibility}} = Xa + \overline{Yb_{0-n}} \\ where \begin{bmatrix} X \in [S_{System_{Pr}}] \\ Y \in [S_{System_{Pr}}, S_{System_P}, S_{System_K}, S_{System_N}] \\ a, b\ are\ integers; a > b \end{bmatrix} \end{array}\right)$$

Eq. 5.1.20, *Active Electro-Magnetic-Wavearchetype-Masspotential P-Type Being Material-Fabric Pre-Genetic Code Segments*

The '$\sum x$' signifies all the possibilities for the 'x' equation-segment and depicts the growing biography by which the singular light-based edifice expresses its materialization, and at this stage, through the electro-magnetic-wavearchetype-masspotential P-type being.

The possible variation in the way the electro-magnetic-wavearchetype-masspotential P-type being can materialize is vast, as suggested by the generated code-segments. Such variation becomes evident as the complexification of the P-type being continues with the advent of matter and life, as will be illustrated in the following sections. By definition any subsequent P-type being will have direct access to the code of previous or even parallel P-type beings, or access to the code at FBLEE through possibilities of entanglement, or even access to the more subtle essence behind form through deeper K-, N-, or ∞-entanglement. Therefore, it is not surprising that every element, every planet, every star, and so on, has its own unique electromagnetic signature which in fact can be used to distinguish one element, or event planet, from another. Such variation in electromagnetic signature or more accurately in electro-magnetic-wave-archetype-masspotential signature are instance of the variability of the electro-magnetic-wave-archetype-masspotential P-type being.

SECTION 6:   GENERATION OF P-TYPE BEINGS IN THE SURFACING OF MATTER

So we find that the four underlying properties of Light that we call Harmony, Knowledge, Power, and Presence are of the essence of the speed with which the electro-magnetic-wavearchetype-masspotential P-type being moves, the wave-range within the electro-magnetic-wavearchetype-masspotential P-type being, the energy-gradient within the electro-magnetic-wavearchetype-masspotential P-type being, and the mass-possibilities   due   to   the   electro-magnetic-wavearchetype-masspotential P-type being, respectively.

But further, we will find that layers of matter – quantum particles, which include bosons, and atoms – are also structured or emerge along the same property-lines or property-families of Light. There is a continuous process of computation that involves the quantum-realms and quantization to create the realities of quantum particles, including bosons, and atoms, to generate matter-based P-type beings with its associated material-fabric, pre-genetic code.

However, there is an apparent difference in the way matter materializes, in contrast to the electro-magnetic-wavearchetype-masspotential P-type being. Chapter 6.1 explores this in greater detail. Chapters, 6.2, 6.3, and 6.4, then, describe a process of computation by which the quantum particle, boson, and atom P-type beings and their associated material-fabric, pre-genetic code are generated respectively.

In the process of generation of light-based beings, the Light-Space-Time Emergence equation has so far resulted in pre-genetic code. After the Big Bang this pre-genetic code is housed in the material-fabric, which has a universal action on all constructs arising in the universe. The emergence of matter however precipitates a phenomenon of distinct material containerization. While matter is still governed by the code in the material-fabric, the materialization of light is now also highly localized.

It is interesting to speculate as to what the boundaries and drivers of such localization are, especially in reference to the math already developed previously. What is it about the electro-magnetic-wavearchetype-masspotential P-type being that causes it to maintain its action more broadly as a field, and what is it about matter, to be discussed in more detail in this section, that causes it to containerize? Note too that the action of quantization involving the space-time-energy-gravity micro-being as envisioned in this treatise, while acting universally, is still a highly localized phenomenon potentially affecting FBLEE and therefore the universal material-fabric as well. Such quantization action is the province of the space-time-energy-gravity micro-being, and it is this that distinguishes it from the action of the space-time-energy-gravity macro-being whose action is at U, typified by light traveling at speed c.

As more complex P-type beings continue to manifest, culminating even in the line of development initiated by the S-type being, will such containerization continue? This chapter explores the phenomena of universalization and localization in the P-type being, and beyond, and suggests some ways to think about the apparently opposite phenomena of localization and universalization.

Universalization and Localization of Space-Time-Energy-Gravity Macro-Being

Chapter 3.5, A Deeper Look at Quantization of Space, Time, Energy, and Gravity, (3.5.3) proposed a model for the structure of space. This is reproduced here for convenience:

$$Space_{Structure} = \sum_{i,j=1}^{\to\infty} h_{UK}\left(X_i a + \overline{Y_j b_{0-n}}\right)$$

$$where: \begin{bmatrix} X_i \in [S_{System_K}] \\ Y_j \in [S_{System_{Pr}}, S_{System_P}, S_{System_K}, S_{System_N}] \\ a, b, i, j \text{ are integers}; a > b \end{bmatrix}$$

Summarizing, space, consisting of a vast array of seeds derived from the properties of Light, is itself an expression of Light's property of Knowledge. It is this essential action of a vast number of seeds that gives space its apparent never-ending-ness. In fact, as previously discussed, time, energy and gravity can also be interpreted in terms of these 'seeds'. Equations 6.1.1-4 restate Space, Time, Energy, and Gravity, in terms of this relationship to seeds. Hence, Equation 6.1.1, Relation of Seeds to Space:

$$Space = f(\#seeds) = f(Light's\ property\ of\ Knowledge)$$

Eq 6.1.1: Relation of Seeds to Space

Time, bringing forth the meaning contained in the seeds, regardless of circumstance, and even being opposed by circumstance, can be thought of as Light's property of Power. Hence, Equation 6.1.2, Relation of Seeds to Time:

$$Time = f(maturity\ of\ seeds) = f(Light's\ property\ of\ Power)$$

Eq 6.1.2: Relation of Seeds to Time

Matter or Energy itself, being a container in which space and time can allow deeper properties of Light to become materially tangible, must be an expression of Light's property of Presence. Hence, Equation 6.1.3, Relation of Seeds to Energy/Matter:

$$Matter = f(materialization\ of\ seeds)$$
$$= f(Light's\ property\ of\ Presence)$$

Eq 6.1.3: Relation of Seeds to Energy/Matter

But it is also known from Einstein's General Theory of Relativity that gravity is associated with mass and space, in that it is none other than a mass's instruction telling space how to curve, and again is nothing else that space's instruction telling mass how to move through it (Wheeler, 2000). As such, where mass and space exist, there gravity has to exist as well. Hence it may be inferred that gravity is none other than an expression of Light's property of Harmony, which fixes the collective relationship between object and object. Hence, Equation 6.1.4, Relation of Seeds to Gravity:

$$Gravity = f(cohesion\ of\ seeds) = f(Light's\ property\ of\ Harmony)$$

Eq 6.1.4: Relation of Seeds to Gravity

If the number of seeds is large, then the collective action of space, time, energy, and gravity is going to create an apparent universality and the reality of 'never-ending-ness'. Such never-ending collective action results in the universalized dynamics of the space-time-energy-gravity macro-being.

By contrast if the number of seeds is smaller, then the action of space, time, energy, and gravity will be relatively localized, and create a more apparent reality of containerization. Such containerization will result in localized dynamics of the space-time-energy-gravity macro-being. Note that the action of the space-time-energy-gravity micro-being is purely focused on that of quantization as expressed earlier.

Such containerization is likely spurred by the action of quantization already captured by (3.5.2), (3.5.4), (3.5.6), and (3.5.8) reproduced here for convenience.

Hence (3.5.2):

$$Space_{quantization} = h_{UK}\left(Xa + \overline{Yb_{0-n}}\right)$$

$$where: \begin{bmatrix} X \in [S_{System_K}] \\ Y \in [S_{System_{Pr}}, S_{System_P}, S_{System_K}, S_{System_N}] \\ a, b\ are\ integers;\ a > b \end{bmatrix}$$

(3.5.4):

$$Time_{quantization} \; = \; h_{UP}\left(Xa + \overline{Yb_{0-n}}\,\right)$$

$$where: \; \begin{bmatrix} X \in [S_{System_P}] \\ Y \in [S_{System_{Pr}}, S_{System_P}, S_{System_K}, S_{System_N}] \\ a, b \; are \; integers; a > b \end{bmatrix}$$

(3.5.6):

$$Energy_{quantization} \; = \; h_{UPr}\left(Xa + \overline{Yb_{0-n}}\,\right)$$

$$where: \; \begin{bmatrix} X \in [S_{System_{Pr}}] \\ Y \in [S_{System_{Pr}}, S_{System_P}, S_{System_K}, S_{System_N}] \\ a, b \; are \; integers; a > b \end{bmatrix}$$

(3.5.8):

$$Gravity_{quantization} \; = \; h_{UH}\left(Xa + \overline{Yb_{0-n}}\,\right)$$

$$where: \; \begin{bmatrix} X \in [S_{System_N}] \\ Y \in [S_{System_{Pr}}, S_{System_P}, S_{System_K}, S_{System_N}] \\ a, b \; are \; integers; a > b \end{bmatrix}$$

Universality of Electro-Magnetic-Wavearchetype-Masspotential P-Type Being

As discussed in Chapter 5, the range of frequencies in the electro-magnetic-wavearchetype-masspotential P-type being is infinite. This has been predicted by Maxwell's equations. (5.1.8), the equation for the infinite energy-gradient, (5.1.3) the equation that relates the infinite range of wavelengths to archetypes, and (5.1.14) that relates the infinite range of mass-potential, each essentially capture the infinite seed-range held by the electro-magnetic-wavearchetype-masspotential P-type being. The original forms of the equations follow for convenience.

Hence, (5.1.8):

$$EMWM \; P - Type \; Being_{Energy} \; = \; Xa + \overline{Yb_{0-n}}$$

$$where \begin{bmatrix} X \in [S_{System_P}] \\ Y \in [S_{System_{Pr}}, S_{System_P}, S_{System_K}, S_{System_N}] \\ a, b \ are \ integers; a > b \end{bmatrix}$$

(5.1.3):

$$EMWM \ P - Type \ Being_{Structure} \ = Xa + \overline{Yb_{0-n}}$$

$$where \begin{bmatrix} X \in [S_{System_K}] \\ Y \in [S_{System_{Pr}}, S_{System_P}, S_{System_K}, S_{System_N}] \\ a, b \ are \ integers; a > b \end{bmatrix}$$

(5.1.14):

$$EMWM \ P - Type \ Being_{Mass_{Possibility}} \ = Xa + \overline{Yb_{0-n}}$$

$$where \begin{bmatrix} X \in [S_{System_{Pr}}] \\ Y \in [S_{System_{Pr}}, S_{System_P}, S_{System_K}, S_{System_N}] \\ a, b \ are \ integers; a > b \end{bmatrix}$$

Recall that the infinite frequency-wavelength range becomes possible because of the constancy of the speed of light at c, which creates time-matter reality that marks this universe.

It is because of the infinite seed-range that the action of the electro-magnetic-wavearchetype-masspotential P-type being is apparently universalized. Hence when quantization does take place as per the light-space-time emergence equation, (3.1.3), and the potential effects of levels of light on pre-genetic and genetic information equation, (3.6.12), because there are a larger range of seeds the effect of the material-fabric and subsequently on materialization is distributed and spread-out.

Material Localization at Multiple Scale

Since quanta hold within themselves the essence of what must be projected from properties or function in higher-velocity light to lower-velocity more-material light, and since the four-fold properties are a representation of the inherent oneness of Light, it is reasonable to expect that such four-foldness will continue to have a bias to uphold that oneness in its more material manifestation. This deep bias to uphold oneness is nothing other than a manifestation of love, and all matter is based on this

183

foundation and need to manifest love. In other words, it is reasonable to expect that space-quanta, time-quanta, energy-quanta, and gravity-quanta related to a single particle will operate as a composite fourfold quantum in what is essentially an embrace of love. This means that four quantum fields operate together or as one composite field in the emergence and possibilities represented by a particle and that perhaps is the reason for this apparently intense material localization.

The following subsections explore the notion of intense material localization at different scale – specifically at the atomic-particle, unit-space, big-planet, expanding universe, black hole, and cosmic bounce levels. In the possible iterations of the Light-Space-Time Emergence Equation (3.1.3) the generated code can apply to organizations of increasing size, including those just specified, as suggested by the discussion in Section 4.

Quanta at Atomic-Particle Level

Figure 6.1.1 below, Composite-Quantum at Atomic-Particle Level, depicts a composite-quantum along the space, time, gravity, and energy dimensions. Essentially the graphic is illustrating that all things being equal there is a balanced accumulation of anterior possibility or information along the space, time, gravity, and energy dimensions that plays a primary role in orchestrating the possibility contained in the atomic-particle. Recall that the space dimension contains information to do with unique seeds, the time dimension information to do with the maturity of seeds, the gravity dimension information to do with collectivities of seeds, and the energy dimension information to do with the materialization of seeds. Hence:

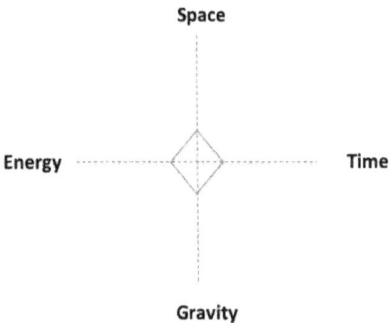

Figure 6.1.1: Composite-Quantum at Atomic-Particle Level

Quanta at Unit-Space Level

The small rhombus in Figure 6.1.2 represents a quantum at the atomic-particle level, while the larger rhombus is an extrapolation of that in some unit space. All things being equal it is the same fourfold accumulation of anterior possibility or information captured by the quantum that will govern particle behavior at the unit-space level. Hence:

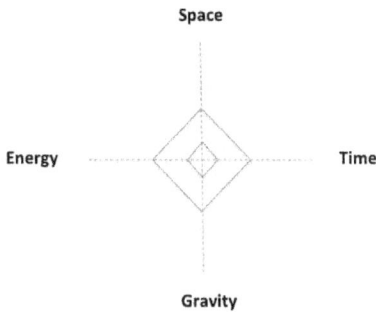

Figure 6.1.2: Composite-Quantum at Unit-Space Level

Further, if Q_A represents a composite-quantum at the atomic-particle level, then Q_{US} will be a summation of such quanta as modeled by Equation 6.1.5:

$$Q_{US} = \Sigma Q_A$$

Eq 6.1.5 Composite-Quantum and Unit-Space Level

Quanta at Level of Big-Planet

In Figure 6.1.3, Composite-Quantum at Level of Big-Planet, the smaller rhombus represents a quantum at the unit-space level, while the larger kite-quadrilateral represents a quantum at the Big-Planet level:

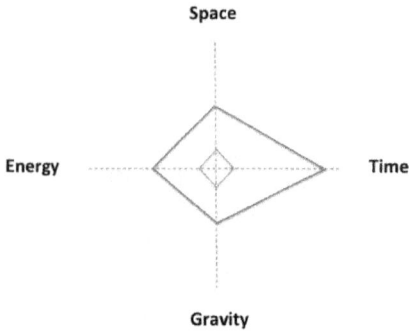

Space

Energy · Time

Gravity

Figure 6.1.3: Composite-Quantum at Level of Big-Planet

The kite-quadrilateral is elongated along the time dimension implying that the larger collectivity of seeds represented by a space such as a Big-Planet will take longer to reach maturity, relative to a seed contained in a unit-space.   This   time-elongation   is   also   modeled   by   the '$mod\ (Time_{quantization})$' component of Equation 6.1.6:

$$Q_{BP} = \Sigma\, Q_{US} \times mod(Time_{quantization})$$

Eq 6.1.6 Composite-Quantum at the Level of Big-Planet

Note that in (6.1.6), Q_{BP} represents a composite-quantum at the level of a Big-Planet, while Q_{US} represents a composite-quantum at the unit-space level as modeled by (6.1.5). Equation (6.1.6) illustrates the notion of composite-quantum as creating the field within which a Big-Planet can materialize.

The time-elongation captures the underlying dynamic of many more finite steps to maturity for the vaster collectivity of seeds, that therefore manifests as time being experienced more rapidly: hence time speeds-up.

Quanta in Expanding Universe

Figure 6.1.4, Composite-Quantum in an Expanding Universe, depicts the relative relationship between the smaller unit-space composite-quantum rhombus and the scaled up expanding-universe composite-quantum rhombus, in which it is estimated that the space, time, gravity, and energy components all go through some related increase:

186

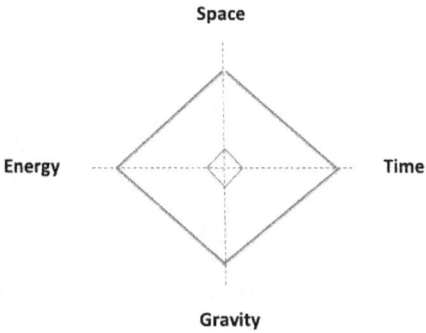

Figure 6.1.4: Composite-Quantum in an Expanding Universe

Equation 6.1.7 models such expanding-universe composite-quantum:

$$Q_{EU} = \Sigma Q_{US}$$

Eq 6.1.7 Composite-Quantum in an Expanding Universe

Quanta at Black Hole Level

Figure 6.1.5, Composite-Quantum at the Black Hole Level, illustrates the relatively different quadrilateral, as compared with previously considered composite-quantum in Figures (6.1.1-4):

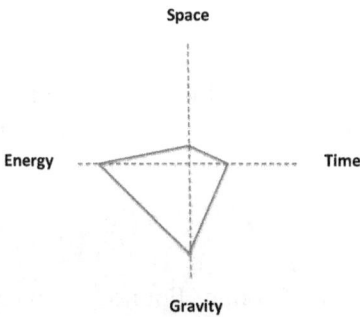

Figure 6.1.5: Composite-Quantum at Black Hole Level

This composite-quantum is hypothesized as being the result of a large number of seeds previously spread in space, coming together to be

187

reformulated as a smaller set or a new seed. The new seed(s) have a different intent than the spread out un-mashed seeds. They will have a very different and relatively shortened maturity-dynamic as depicted along the time-dimension due to a normalization or rationalization of seeds. Further, they will have a very different collectivity-relationship between other seeds outside of the core black-hole formation, as represented by the gravity-dimension. Finally, they will similarly also have a very different energy-materialization dynamic as represented by the energy-dimension, and due to the complexification of the normalized seeds.

Centered on the black hole, space-time-gravity-energy is going to proceed differently than may have existed prior to the formation of the black hole. Such modification is modeled by Equation (6.1.8), where, as can be seen there is a summation of previous seeds along each dimension, captured by $\sum Seed_i$, and a holding of the new essence, captured by h_{Ui} with 'i' being 'K', 'P', 'H' and 'Pr' respectively, which together create the new black hole (BH) composite-quantum, Q_{BH}:

$$
Q_{BH} = \begin{bmatrix} mod(Space_{quantization}) \rightarrow h_{UK} \sum Seed_K \\ mod(Time_{quantization}) \rightarrow h_{UP} \sum Seed_P \\ mod(Gravity_{quantization}) \rightarrow h_{UH} \sum Seed_H \\ mod(Energy_{quantization}) \rightarrow h_{UPr} \sum Seed_{Pr} \end{bmatrix}
$$

Eq 6.1.8 Composite-Quantum at Black-Hole Level

Note that consistent with General Relativity, the dynamic of gravity intensifies, space is compressed, time slows down, and energy-potential increases.

Quanta of Cosmic Bounce

Figure 6.1.6, Composite-Quantum at Cosmic Bounce, depicts the composite-quantum at a possible Cosmic Bounce event. The smaller quadrilateral represents the black hole composite-quantum, while the larger quadrilateral is a scaled up Cosmic Bounce version of Q_{BH}:

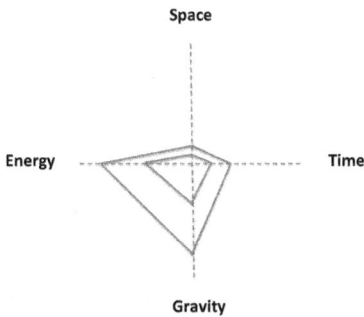

Figure 6.1.6: Composite-Quantum at Cosmic Bounce

The notion of scaling up, is captured by e^M where M is some very large number, in the Cosmic Bounce composite-quantum as modeled by Equation (6.1.9):

$$Q_{CB} = Q_{BH}e^M$$

Eq 6.1.9 Composite-Quantum at Cosmic Bounce

Hence, this brief seed-based analyses suggests that even as organizations scale, there will still be a containerization of matter albeit, possibly, with different quadrumvirate balance. Such containerization is a key dynamic in any P-type being and expresses the fundamental urge to maintain material-oneness in accordance with the reality of fourfold oneness, or love, of the essential properties of Light in its native state.

Exploring Effect of Emergent Quadrality on Material Containerization

When matter begins to emerge, it is at first highly containerized. This instinct to containerize is due to the inherent oneness of space-time-energy-gravity dynamics that seeks to maintain its single identity even when apparently separated at U. This phenomenon is captured in the notion of the fourfold composite quantum. As illustrated in the previous section on material localization at multiple scale, the variation in circumstance and in number of seeds is expected to change the relative "influence" of the different parts of the space-time-energy-gravity unity.

189

The effect is that localization and universalization merge as the scale increases.

But there is a more systematic way to think about the merging of localization and universalization as the P-type being complexifies that leverages of the mathematics of becoming as explored in Section 2.

Equation 2.6.7, Core Matrix or Upward-Strand of Subtle-DNA, is modified as Equation, 6.1.10, Materialization of Quadrality in Upward-Strand of Subtle-DNA, to summarize how merged localization and universalization may manifest:

$$
Materialization\ of\ Quadrality = \begin{bmatrix} M_3 \rightarrow System_X \Rightarrow [L \lll Un : Block] \\ (\uparrow F \rightarrow I) \\ M_2 \rightarrow S_{System_X} \Rightarrow [L \ll Un : Field] \\ (\uparrow Sig \rightarrow F) \\ M_1 \rightarrow Sig_X \Rightarrow [L \gg Un : Wave] \\ (\uparrow > P_X) \\ U \rightarrow x_U \Rightarrow [L \ggg Un : Particle] \end{bmatrix}
$$

Eq 6.1.10: Materialization of Quadrality in Upward-Strand of Subtle-DNA

Hence, starting at the bottom of (6.1.10), at U containerization of matter is going to be (:) much more like a particle (Particle), because the dynamic of localization (L) is much, much greater ($\ggg$) than the dynamic of universalization (Un). The manifestation as a 'particle' can also be thought of as due to a single seed of uniqueness in the antecedent layer. At M_1 containerization of matter is going to be (:) more like a wave (Wave), perhaps expressing fluidity, because the dynamic of localization is going to be much greater ($\gg$) than universalization. The manifestation as a 'wave' can also be thought of as due to a either the seed of uniqueness more successfully expressing more of the richness and detail in its uniqueness, or as the uniqueness successfully expressing itself in more and more diverse realms. At M_2 containerization of matter is going to be (:) more like a field (Field), expressing expansiveness, because the dynamic of localization is going to be much less ($\ll$) than universalization. The manifestation as a 'field' can also be thought of the trend at M_1 becoming more pronounced. At M_3 containerization of matter is going to be (:) more like a block (Block) because the dynamic of localization is going to be much, much less ($\lll$) than universalization. In other words the differentiated, rich uniqueness needs a much larger material frame in which to house itself.

Hence, with greater influence from higher meta-levels, where light is traveling much, much faster than c, due to the P-type being having broken many habitual patterns, the basis of materialization is going to itself be different.

The electro-magnetic-wavearchetype-masspotential P-type being describes one of the first layers or translation of light into pre-genetic material manifestation. The fact that electron orbital shifts in atoms can cause photons to be released, which themselves as we will discover, are the carrier particles for the electro-magnetic-wavearchetype-masspotential P-type being, means that the latter being has to already exist and its ecosystem logic worked out. Hence the electro-magnetic-wavearchetype-masspotential P-type being will be assumed to be in existence, and this will be reflected in the equation foe the generation of the quantum-particle P-type being itself to be explored in this chapter.

As already explored the finite speed of such P-type beings allows the build-up of energies or quanta, and this in turn may express itself as a series or as an array of quantum particles. The vast number of quantum particles that have so far been discovered has in turn yielded what is known as the Standard Model. This model is made up of what is called Quarks, Leptons, Bosons, and a Higgs-Boson (Cottingham, 2007).

But if we look at these four fundamental quantum categories of particles from a property-based or "functional" viewpoint, a different chapter in the exploration of light emerges. All of matter then can be seen as implicit in Light and in fact due to a process of computation by which the implicit codification in Light becomes explicit. The implicit codification becomes explicit by generating material-fabric pre-genetic quantum particle code-segments in the unified body of the quantum-particle P-type being. These code-segments are added to the existing pre-genetic code-segments and also become binding on future emergences in the material universe. In other words, P-type beings with increasing spheres of influence are progressively created. All previous code-segments create law in a newly emergent P-type being hence reinforcing the composite and unified nature of the P-type being.

As in the previous Section, the Light-Space-Time Emergence equation (3.1.3) can be used to model emergence as it proceeds to this relatively more complex form of fourfold manifestation. (3.6.12), the Potential Effect of Levels of Light on Genetic-Type Information equation, basically sheds light on specific types of mutation that can occur and therefore on the nature of the code-segment that is being generated by (3.1.3). As discussed

192

before, (3.6.12) shows a particular working of (3.1.3) but is wholly contained within it.

This chapter elaborates the action of (3.1.3) and (3.6.12) in the creation of the code-segments that define the quantum particle ecosystem logic that emerges in the quantum-particle P-type being.

Light's Emergence as Quantum Particles

As discussed previously the Light-Space-Time Emergence equation (3.1.3) being iterative, can be used to model emergence as it proceeds from simpler four-fold to more complex four-fold manifestations. But further, as implied by (3.6.12), the Potential Effect of Levels of Light on Genetic-Type Information equation, any process of organization has the possibility of altering the material-fabric or fabric of existence so long as the bases involved are driven primarily by a meta-level.

As suggested by Equation 6.2.1, Generation of the Quantum-Particle P-Type Being, the architecture and details of quantum particles can be seen to be the result of the application of the Light, Space, and Time matrices as will be elaborated:

$Light - Space - Time\ Emergence_{Quantum-Particle\ P-Type\ Being} =$

$$
\left| \begin{bmatrix} \begin{array}{l} c_\infty : [Pr, Po, K, H] \\ \left(\downarrow R_{C_K} = f(R_{C_\infty}) \right) \\ c_K : [S_{Pr}, S_{Po}, S_K, S_H] \\ \left(\downarrow R_{C_N} = f(R_{C_K}) \right) \\ c_N : f(S_{Pr} \times S_{Po} \times S_K \times S_H) \\ \left(\downarrow R_{C_U} = f(R_{C_N}) \right) \\ c_U : [P, V, M, C] \\ \Uparrow \\ c_{0:[D,W,I,C]} \end{array} \end{bmatrix}_{Light} \begin{bmatrix} M_3 \rightarrow System_x \\ (\uparrow F \rightarrow I) \\ M_2 \rightarrow S_{System_x} \\ (\uparrow Sig \rightarrow F) \\ M_1 \rightarrow Sig_x \\ (\uparrow > P_x) \\ U \rightarrow x_{STEG\ P-Type\ Macro-Being} \end{bmatrix}_{Space} \right.
$$

$$
\left. \begin{bmatrix} M_3 : -\infty \leq t \leq \infty \\ \downarrow \\ M_2 : 0 \geq t > \infty \\ \downarrow \\ M_1 : 0 > t > \infty \\ \downarrow \\ U \rightarrow \begin{array}{l} t \leq E_{Cell}; TC: M_3 \rightarrow U \\ t \sim E_{Human}; TC: U \rightarrow M_3 \end{array} \end{bmatrix}_{Time} \quad TC \rightarrow x_{Quantum\ Particles} \right| \Rightarrow \langle x_U | x_T \rangle
$$

193

$$\langle Space - Time - Energy - Gravity\ P - Type\ Beings\ and\ Pre - Genetic\ Code\rangle +$$

$$\left\langle \begin{array}{c} Electro - Magnetic - Wavearchetype - Masspotential\ P - Type\ Being \\ and\ Pre - Genetic\ Code \end{array} \right\rangle +$$

$$Quantum - Particle\ P - Type\ Being\ and\ Pre - Genetic\ Code$$

Eq 6.2.1: Generation of Quantum-Particle P-Type Being

Starting with the Light-Matrix, the top left-hand matrix in (6.2.1), the first line from the top, C_∞: $[Pr, Po, K, H]$, specifies the fundamental architecture of the quantum-particle P-type being. In this rendering, and as elaborated in subsequent subsections, quarks are an emergence of Light's property of Knowledge, leptons are an emergence of Light's property of Power, bosons are an emergence of Light's property of Harmony, and the Higgs-boson is an emergence of Light's property of Presence. The fundamental architecture of these aspects, hence, is an emergence of the properties of Light at ∞.

Line 3 in the Light-Matrix, C_K: $[S_{Pr}, S_{Po}, S_K, S_H]$, elaborates the sets for Presence, Power, Knowledge, and Harmony, each containing multiple elements. For example, as will be explored in greater detail in the section on quarks, various elements derived from the four sets define the behavior of quarks and could be functions such as 'composite-arrangements', 'specifying attributes' amongst others, hence collectively describing quarks' way of being. Specifically, Line 5, C_N: $f(S_{Pr} \times S_{Po} \times S_K \times S_H)$, suggests that unique seeds are created from a combination of such elements from all four sets, with a particular element leading or having more weight, that in effect creates the distinctness possible in the quantum-particle P-type being.

Line 6, ($\downarrow R_{C_U} = f(R_{C_N})$), specifies quantization between the layer where the seeds are formed, and the physical layer, and as explored in Chapter 3.5 and 3.6, will result in Line 7, C_U: $[P, V, M, C]$, hence changing the material-fabric of existence. Note that as in the process describing the generation of the code-segments for space-time-energy-gravity quantization at the time of the Big Bang, the generation of the FBLEE (four-base logic-encoding ecosystem) code-segment as specified by (4.1.5) and the MF (material-fabric) code segment as specified by (4.1.6) are combined together into one equation so that the FBLEE process is apparently transparent. In reality one may say that any P-type being is first a FBLEE-P-type being and then a MF-P-type being.

The possibilities represented by Lines 1 through 5 hence concretize through the quantization represented by Line 6 to become the quantum-particle P-type being with its physical (related to Presence), vital (related to Power), mental (related to Knowledge), and connection (related to Harmony) aspects now existing in material reality typified by Light moving at c. Note that just as Line 6 represents a process of quantization relating the layer of reality created by Light traveling at c with the antecedent layers, so too Lines 2 and 4 as previously discussed, also represent quantization of a more subtle kind that ultimately plays a critical part in allowing the material-fabric to express infinite diversity.

Typically, it is the process as captured by the Space-Matrix that will determine if Line 6 is activated. Specifically, patterns at the untransformed layer, U, will need to be overcome, as specified by the second line from the bottom of the Space-Matrix: $(\uparrow > P_x)$. But as specified by the bottom-line of the Time-Matrix, reproduced below, it is only with the advent of the human-system that the automaticity of the action of meta-levels is reversed:

$$U \to \begin{array}{l} t \le E_{Cell}; \text{TC}: M_3 \to U \\ t \sim E_{Human}; \text{TC}: U \to M_3 \end{array}$$

Hence in the case of the quantum-particle P-type being, which in this emergence is a pre-human system, the fact that patterns do not need to be overcome means that quantization happens automatically, and that in reality the quantum-particle P-type being would therefore more accurately be termed the quantum-particle PMU-type being.

Even though the quantum-particle P-type being is a pre-human system and therefore the action of meta-levels are modeled as being automatic, it is nonetheless useful to review (3.6.12), **Potential Effect of Levels of Light on Genetic-Type Information**, to understand how the relationship with different forms of mutation has been specified:

Potential Effect of Levels of Light on Genetic $-$ Type Information $=$

$$
\begin{bmatrix}
STATIC\ \langle |[L][S][T]TC \rightarrow x_T|_{\langle x_U | x_T \rangle} \rangle \\
\times \\
\left((Y > U: Z_Q) \vee (Y \leq U: Z_F) \vee (Y = U: Z_R) \right) \\
\ni \\
\begin{pmatrix} Z \in \mathbb{U}\ (Space, Time, Energy, Gravity) \\ Q: Quantization;\ F: Fragmentation;\ R: Random \end{pmatrix}
\end{bmatrix} \rightarrow h \ni
$$

$$
h \in \begin{pmatrix} [Q]: Constructive\ zone, \\ [Q]: Constructive\ zone \wedge Constructive\ mutation, \\ [F]: Destructive\ mutation, \\ [R]: Random\ mutation \end{pmatrix}
$$

Line 1 from the top in the matrix is simply a static form of (3.6.1) the Simplified Light-Space-Time Emergence equation. The static form is designated by 'STATIC' and implies that fundamental operations true of (3.6.1) are being highlighted in (3.6.12). In other words (3.6.1) already has all the operations highlighted in (3.6.12) in it, but by 'freezing' it by making it static, the essential dynamics leading to possible mutations at the genetic level can more clearly be highlighted.

Line 1 is then subjected ($\times$) to a determination of the dominant levels of light that may be active, designated by Line 2, '$\left((Y > U: Z_Q) \vee (Y \leq U: Z_F) \vee (Y = U: Z_R) \right)$'. Unpacking this, '$Y > U$' implies meta-levels are active and as a result it is possible that Z_Q is going to take place (the subscript 'Q' implies quantum-level action). This also implies activation and potential change of FBLEE. The call from below, as it were, may invoke some function that already exists in the subtle-libraries 'above', so that some already existing function may influence FBLEE through ∞-entanglement, K-entanglement, or N-entanglement. This may be thought of as a key-and-lock mechanism, where a deep enough visceral urge from below acting as the key, opens an entangled lock to alter FBLEE as per the visceral urge. '$Y \leq U$' implies that only the untransformed levels are active, and therefore also the sub-level where the speed of light is 0 is active and as a result Z_F is going to take place (the subscript 'F' implies 'fragmentation'). '$Y = U$' implies that all levels are active and as a result Z_R is going to take place (the subscript 'R' implies 'random').

Line 3 elaborates the significance of Z_Q, Z_F, and Z_R. Hence Z is the union of potential quantum-operations of space, time, energy, and gravity, designated by '$Z \in \mathbb{U}\ (Space, Time, Energy, Gravity)$'. But the nature of the operations, as suggested in the previous paragraph, is designated by

'*Q: Quantization; F: Fragmentation; R: Random*'. Z_Q, then, implies that the full quantization originating from updated four-base logic-encoding ecosystems (FBLEE) can take place. Z_F implies that the essential set will be fragmented and that only libraries at the level of local cellular-level DNA or precipitated material-fabric logic can potentially be altered. Z_R also precludes full quantization, and that some partial local-library constructive or destructive mutation may take place.

The '$\rightarrow h \ni h \in$' segment resolves the outcome of the operations implied by Lines 1 – 3, suggesting that the outcome will be 'h' such that ($\ni$) 'h' is an element ($\in$) of the set specified by the members '[Q]: *Constructive zone*', '[Q]: *Constructive zone AND Constructive mutation*', '*[F]; Destructive mutation*', and '*{R}: Random mutation*'. '[Q]: *Constructive zone*' implies that the in-built buffer has been crossed and that access to the deeper four-base logic-encoding ecosystem (FBLEE) has been granted. '[Q]' in this segment implies that there is the possibility that full-quantization as specified by Line 3 of the previous matrix can take place. Access to this zone is a prerequisite for constructive mutation to occur, as designated by the element '[Q]: *Constructive zone* $\cap$ *Constructive mutation*', which implies that full-quantization is going to take place and will result in material change. The '[F]' specifies the relationship between 'Fragmentation' in Line 3 of the previous matrix and destructive mutation. The '[R]' specifies the relationship between 'Random' in Line 3 of the previous matrix and random mutation.

But as just summarized in the Time-Matrix in (6.2.1) Y is by definition greater than U and hence quantization is automatic. In terms of the quantum-particle P-type being such quantization implies that wholeness becomes fully active through specific space, time, energy, and gravity quantization to create an holistic "ecosystem" with its own "quantum-particle P-type being logic" as it were. The wholeness has now precipitated into the material-fabric and is available to be consciously and unconsciously tapped into. This "logic" or more specifically pre-genetic code is elaborated in the following sub-sections.

Generation of 'Quark' Aspect of the Quantum-Particle P-Type Being

It can be seen first that the nucleus of an atom is made up of a combination of quarks. Specifically, a proton is composed of two "up" quarks and one "down" quark. Quarks have unusual names – up, down, charm, strange,

197

top, bottom, with each subsequent pair belonging to a different "generation" of quarks. A neutron is composed of two "down" quarks and one "up" quark. Protons and neutrons together make up the nucleus of an atom. But we also know that the number of protons in the nucleus specifies the Atomic Number of an atom. Atomic number in turn uniquely identifies the element from the periodic table. Hence, an atomic number of 47, for example, specifies that the element is Silver. In other words, it can be suggested that the unique properties of an element, the knowledge of what it is and how it will behave in the universe, is related to the quark. It may be suggested that quarks, therefore, are associated with the precipitation or emergence of Light's property of Knowledge in the quantum world.

Hence, it could be that the signature or code-segment for the family of quarks, as in Equation 6.2.2, is:

$$Sig_{quarks} = Xa + \overline{Yb_{0-n}} \quad where \begin{bmatrix} X \in [S_{System_K}] \\ Y \in [S_{System_{Pr}}, S_{System_P}, S_{System_K}, S_{System_N}] \\ a, b \ are \ integers; a > b \end{bmatrix}$$

Eq 6.2.2: *Generalized Signature for Quarks Aspect of the Quantum-Particle P-Type Being*

As can be seen the primary element X is derived from the set of knowledge, S_{System_K}. Various elements, derived from the four sets would define the behavior of quarks and could be functions such as 'composite-arrangements', 'specifying attributes' amongst others, hence collectively describing quarks' way of being.

Note that (6.2.2) already implies that Lines 1 – 5 in the Light Matrix (6.2.1) have been activated, and that the logic of the quark-ecosystem will automatically precipitate into the material-fabric through the action of Line 6-7 of (6.2.1). This logic is none other than the pre-genetic code as specified by (6.2.2).

Generation of 'Lepton' Aspect of the Quantum-Particle P-Type Being

When considering leptons, it is useful to know that unlike quarks that only exist in composite arrangements with other quarks, leptons are solitary, point-like particles without internal structure (Arabatzis, 2006). The best-known lepton is the electron. So, the electron may be considered as a surrogate for the lepton class. The electron appears to be associated with the flow of energy and power. Further they appear to be the adventurers easily leaving the atom they are a part of. They also form locks or bonds with other atoms through the force of attraction and repulsion. In some sense they seem to be a representation or precipitation or emergence of Light's property of Power.

The signature for the family of leptons, as in Equation 6.2.3, is:

$$Sig_{leptons} = Xa + \overline{Yb_{0-n}} \quad where \begin{bmatrix} X \in [S_{System_P}] \\ Y \in [S_{System_{Pr}}, S_{System_P}, S_{System_K}, S_{System_N}] \\ a, b \ are \ integers; a > b \end{bmatrix}$$

Eq 6.2.3: Generalized Signature for Lepton Aspect of the Quantum-Particle P-Type Being

As can be seen the primary element X is derived from the set of power, S_{System_P}. Various elements, derived from the four sets would define the behavior of leptons and could be functions such as 'adventurer', 'solitary', 'create attraction', amongst others, hence collectively describing leptons' way of being.

Note that (6.2.3) already implies that Lines 1 – 5 in the Light Matrix (6.2.1) have been activated, and that the logic of the lepton-ecosystem will automatically precipitate into the material-fabric through the action of Line 6-7 of (6.2.1). This logic is none other than the pre-genetic code as specified by (6.2.3).

Generation of 'Boson' Aspect of the Quantum-Particle P-Type Being

The bosons on the other hand are thought of as force-carriers. They are what allow all known matter particles to interact. The three fundamental bosons in this category are the photon, the W and Z bosons, and the gluon. The carrier particle of the electro-magnetic-wavearchetype-masspotential P-type being is the photon. The carrier particle of the strong nuclear force that holds quarks together is the gluon. The carrier particle for the weak interactions, responsible for the decay of massive quarks and leptons into lighter quarks and leptons, are the W and Z bosons.

Bosons can be thought of as the precipitation of what creates relationship and harmony at the quantum level. Hence, they can be thought of as the precipitation or emergence of Light's property of Harmony at the quantum level.

The signature for the family of gauge bosons, as in Equation 6.2.4, is:

$$Sig_{bosons} = Xa + \overline{Yb_{0-n}} \quad where \quad \begin{bmatrix} X \in [S_{System_N}] \\ Y \in [S_{System_{Pr}}, S_{System_P}, S_{System_K}, S_{System_N}] \\ a, b \ are \ integers; a > b \end{bmatrix}$$

Eq 6.2.4: Generalized Signature for Bosons Aspect of the Quantum-Particle P-Type Being

As can be seen the primary element X is derived from the set of nurturing, S_{System_N}. Various elements, derived from the four sets would define the behavior of bosons and could be functions such as 'creating interaction', 'holding together', amongst others, hence collectively describing bosons' way of being.

Note that (6.2.4) already implies that Lines 1 – 5 in the Light Matrix (6.2.1) have been activated, and that the logic of the boson-ecosystem will automatically precipitate into the material-fabric through the action of Line 6-7 of (6.2.1). This logic is none other than the pre-genetic code as specified by (6.2.4).

Equations in this form give a clue as to how to think about the fundamental particles. While the default approach is always to think about physical qualities, equations in the preceding format allow us to think about detailed functions or qualities associated with particles and give insight into how the process of emergence or complexification of fourfold functionality takes place.

Generation of 'Higgs-Boson' Aspect of the Quantum-Particle P-Type Being

This leaves the other discovered fundamental particle the Higgs-Boson. In ordinary matter, most of the mass is contained in atoms, and the majority of the mass of an atom resides in the nucleus, made of protons and neutrons. Protons and neutrons are each made of three quarks. But it is the quarks that get their mass by interacting with what is known as the Higgs field (Olive, 2014). Hence the Higgs-Boson can be thought of as the mass-giver. In other words, it is what gives presence to the quarks and it can be thought of as the precipitation or emergence of Light's property of Presence at the quantum level. Just as there are multiple particles in each of the other 'families' it is likely that there will be multiple particles in the

Higgs-Boson family. Recent research at CERN indicates that the Higgs-Boson may have a cousin.

The signature for the Higgs-boson and any other similar particle, as in Equation 6.2.5, is:

$$Sig_{Higgs-boson} = Xa + \overline{Yb_{0-n}}$$

$$where \begin{bmatrix} X \in [S_{System_{Pr}}] \\ Y \in [S_{System_{Pr}}, S_{System_{P}}, S_{System_{K}}, S_{System_{N}}] \\ a, b \ are \ integers; a > b \end{bmatrix}$$

Eq 6.2.5: Generalized Signature for Higgs-Boson Aspect of the Quantum-Particle P-Type Being

As can be seen the primary element X is derived from the set of presence, $S_{System_{Pr}}$. Various elements, derived from the four sets would define the behavior of Higgs-bosons and could be a function such as 'creating mass', amongst others, hence collectively describing Higgs-bosons' way of being.

Note that (6.2.5) already implies that Lines 1 – 5 in the Light Matrix (6.2.1) have been activated, and that the logic of the Higgs-boson-ecosystem will automatically precipitate into the material-fabric through the action of Line 6-7 of (6.2.1). This logic is none other than the pre-genetic code as specified by (6.2.5).

Combining all the preceding particle equations it is possible to create a generalized particle equation, as in Equation 6.2.6:

$$Sig_{particle} = Xa + \overline{Yb_{0-n}} \quad where \begin{bmatrix} X \in [S_{System_{Pr}}, S_{System_{P}}, S_{System_{K}}, S_{System_{N}}] \\ Y \in [S_{System_{Pr}}, S_{System_{P}}, S_{System_{K}}, S_{System_{N}}] \\ a, b \ are \ integers; a > b \end{bmatrix}$$

Eq 6.2.6: Generalized Signature for the Quantum-Particle P-Type Being

So, we see that again the underlying properties of light - Knowledge, Power, Harmony, and Presence - emerge as quarks, leptons, bosons, and Higgs-bosons respectively.

Summary of the Quantum-Particle P-Type Being

Summarizing, after the generation of the quantum-particle P-type being, the following code-segments will have been generated as specified by Equation 6.2.7, Active Quantum-Particle P-Type Being Material-Fabric Pre-Genetic Code Segments:

$Light - Space - Time\ Emergence_{Quantum-Particle\ P-Type\ Being} =$

$$
\left\| \begin{array}{c}
\begin{bmatrix}
c_\infty: [Pr, Po, K, H] \\
\left(\downarrow R_{C_K} = f(R_{C_\infty}) \right) \\
c_K: [S_{Pr}, S_{Po}, S_K, S_H] \\
\left(\downarrow R_{C_N} = f(R_{C_K}) \right) \\
c_N: f(S_{Pr} \times S_{Po} \times S_K \times S_H) \\
\left(\downarrow R_{C_U} = f(R_{C_N}) \right) \\
c_U: [P, V, M, C] \\
\Uparrow \\
c_{0:[D,W,I,C]}
\end{bmatrix}_{Light}
\begin{bmatrix}
M_3 \to System_X \\
(\uparrow F \to I) \\
M_2 \to S_{System_X} \\
(\uparrow Sig \to F) \\
M_1 \to Sig_x \\
(\uparrow > P_x) \\
U \to x_{STEG}
\end{bmatrix}_{Space} \\
\begin{bmatrix}
U \to \begin{array}{c}
M_3 : -\infty \le t \le \infty \\
\downarrow \\
M_2 : 0 \ge t > \infty \\
\downarrow \\
M_1 : 0 > t > \infty \\
\downarrow \\
t \le E_{Cell}; TC: M_3 \to U \\
t \sim E_{Human}; TC: U \to M_3
\end{array}
\end{bmatrix}_{Time}
\quad TC \to x_{Quantum\ Particles}
\end{array} \right\|_{\langle x_U | x_T \rangle}
\Rightarrow
$$

$\langle Space - Time - Energy - Gravity\ P - Type\ Beings\ and\ Pre - Genetic\ Code \rangle +$

$\left\langle \dfrac{Electro - Magnetic - Wavearchetype - Masspotential\ P - Type\ Being}{and\ Pre - Genetic\ Code} \right\rangle +$

$$\left(\begin{array}{l} where \left[\begin{array}{c} \sum Sig_{bosons} = Xa + \overline{Yb_{0-n}} \\ X \in [S_{System_N}] \\ Y \in [S_{System_{Pr}}, S_{System_P}, S_{System_K}, S_{System_N}] \\ a, b \ are \ integers; a > b \end{array} \right] \\ where \left[\begin{array}{c} \sum Sig_{quarks} = Xa + \overline{Yb_{0-n}} \\ X \in [S_{System_K}] \\ Y \in [S_{System_{Pr}}, S_{System_P}, S_{System_K}, S_{System_N}] \\ a, b \ are \ integers; a > b \end{array} \right] \\ where \left[\begin{array}{c} \sum Sig_{leptons} = Xa + \overline{Yb_{0-n}} \\ X \in [S_{System_P}] \\ Y \in [S_{System_{Pr}}, S_{System_P}, S_{System_K}, S_{System_N}] \\ a, b \ are \ integers; a > b \end{array} \right] \\ where \left[\begin{array}{c} \sum Sig_{Higgs-boson} = Xa + \overline{Yb_{0-n}} \\ X \in [S_{System_{Pr}}] \\ Y \in [S_{System_{Pr}}, S_{System_P}, S_{System_K}, S_{System_N}] \\ a, b \ are \ integers; a > b \end{array} \right] \end{array} \right)$$

Eq. 6.2.7, Active Quantum-Particle P-Type Being Material-Fabric Pre-Genetic Code Segments

The '$\sum x$' signifies all the possibilities for the 'x' equation-segment and depicts the growing biography by which the singular light-based edifice expresses its materialization, and at this stage, through the quantum-particle P-type being.

Chapter 6.3: Generation of the Boson P-Type Being

The bosons as mentioned can be thought of as force-carriers and allow all known matter particles to interact.

But when we look at bosons in more detail there are three fundamental bosons – the photon, the W and Z bosons, the gluon - and one hypothetical boson – the graviton.

This chapter looks at the on-going computation involving a process of quantization by which the very logic of bosons precipitates into the material-fabric hence generating the boson P-type being.

As in the previous Section, the Light-Space-Time Emergence equation (3.1.3) can be used to model emergence as it proceeds to this relatively more complex form of fourfold manifestation. (3.6.12), the Potential Effect of Levels of Light on Genetic-Type Information equation, basically sheds light on specific types of mutation that can occur and therefore on the nature of the code-segment that is being generated by (3.1.3). As discussed before, (3.6.12) shows a particular working of (3.1.3) but is wholly contained within it.

This chapter elaborates the action of (3.1.3) and (3.6.12) in the creation of the code-segments that define the boson P-type being ecosystem logic. While this development may precede the generation of quantum particle code, for simplicity it will be assumed that this happens at the same time.

Light's Emergence as Bosons

As discussed previously the Light-Space-Time Emergence equation (3.1.3) being iterative, can be used to model emergence as it proceeds from simpler four-fold to more complex four-fold manifestations. Hence (3.1.3) has already been applied to suggest the emergence of the space-time-energy-gravity P-type beings, the electro-magnetic-wavearchetype-masspotential P-type being, and the quantum-particle P-type being. Here it will be applied to suggest the emergence of a sub-class of quantum particles, the boson P-type being. Hence as in (6.2.1) the starting point will be assumed to be the space-time-energy-gravity (STEG) P-type beings.

As suggested by Equation 6.3.1, Generation of Boson P-Type Being, the architecture and details of bosons can be seen to be the result of the application of the Light, Space, and Time matrices as will be elaborated:

$Light - Space - Time\ Emergence_{Boson\ P-Type\ Being} =$

$$
\left\| \begin{bmatrix} \begin{array}{c} c_{\infty}: [Pr, Po, K, H] \\ \left(\downarrow R_{C_K} = f(R_{C_\infty}) \right) \\ c_K: [S_{Pr}, S_{Po}, S_K, S_H] \\ \left(\downarrow R_{C_N} = f(R_{C_K}) \right) \\ c_N: f(S_{Pr} \times S_{Po} \times S_K \times S_H) \\ \left(\downarrow R_{C_U} = f(R_{C_N}) \right) \\ c_U: [P, V, M, C] \\ \Uparrow \\ c_{0:[D,W,I,C]} \end{array} \end{bmatrix}_{Light} \begin{bmatrix} M_3 \rightarrow System_X \\ (\uparrow F \rightarrow I) \\ M_2 \rightarrow S_{System_X} \\ (\uparrow Sig \rightarrow F) \\ M_1 \rightarrow Sig_X \\ (\uparrow > P_x) \\ U \rightarrow x_{STEG} \end{bmatrix}_{Space} \right. \\ \left. \begin{bmatrix} M_3: -\infty \leq t \leq \infty \\ \downarrow \\ M_2: 0 \geq t > \infty \\ \downarrow \\ M_1: 0 > t > \infty \\ \downarrow \\ U \rightarrow \begin{array}{l} t \leq E_{Cell}; TC: M_3 \rightarrow U \\ t \sim E_{Human}; TC: U \rightarrow M_3 \end{array} \end{bmatrix}_{Time} \begin{array}{c} TC \rightarrow x_{Boson} \end{array} \right\| \langle x_U | x_T \rangle \Rightarrow
$$

$\langle Space - Time - Energy - Gravity\ P - Type\ Beings\ and\ Pre - Genetic\ Code \rangle +$

$\langle \begin{array}{c} Electro - Magnetic - Wavearchetype - Masspotential\ P - Type\ Being \\ and\ Pre - Genetic\ Code \end{array} \rangle +$

$\langle Quantum - Particle\ P - Type\ Being\ and\ Pre - Genetic\ Code \rangle +$

$FBLEEE < \cdots > + Boson\ P - Type\ Being\ and\ Pre - Genetic\ Code$

Eq 6.3.1: Generation of Boson P-Type Being

Starting with the Light-Matrix, the top left-hand matrix in (6.3.1), the first line from the top, $C_\infty: [Pr, Po, K, H]$, specifies the fundamental architecture of the boson P-type being. As explored previously in this chapter, Gluons are an emergence of Light's property of Knowledge, W and Z bosons are an emergence of Light's property of Power, Photons are an emergence of Light's property of Presence, and the Graviton is an emergence of Light's property of Harmony. The fundamental architecture of these aspects, hence, is an emergence of the properties of Light at ∞.

Line 3 in the Light-Matrix, $C_K: [S_{Pr}, S_{Po}, S_K, S_H]$, elaborates the sets for Presence, Power, Knowledge, and Harmony, each containing multiple elements. For example, as will be explored in the section on photons, various elements derived from the four sets define the behavior of photons and could be functions such as 'pervasiveness', 'multiple wave handler', amongst others, hence collectively describing photons' way of being. Specifically, Line 5, $C_{N:} f(S_{Pr} \times S_{Po} \times S_K \times S_H)$, suggests that unique seeds are created from a combination of such elements from all four sets, with a particular element leading or having more weight, that in effect creates the distinctness possible in the boson P-type being.

Line 6, ($\downarrow R_{C_U} = f(R_{C_N})$), specifies quantization between the layer where the seeds are formed, and the physical layer, and as explored in Chapter 3.5 and 3.6, will result in Line 7, $C_U: [P, V, M, C]$, hence changing the material-fabric of existence. Note that as in the process describing the generation of the code-segments for space-time-energy-gravity quantization at the time of the Big Bang, the generation of the FBLEE (four-base logic-encoding ecosystem) code-segment as specified by (4.1.5) and the MF (material-fabric) code segment as specified by (4.1.6) are combined together into one equation so that the FBLEE process is apparently transparent. In reality one may say that any P-type being is first a FBLEE-P-type being and then a MF-P-type being. Note also that in (6.2.1), following the '⇒' the FBLEEE <...> code-segment implies that four-base logic-encoding entanglement likely with the generation of the other quantum-particle FBLEE code-segments as specified by (6.3.1).

The possibilities represented by Lines 1 through 5 hence concretize through the quantization represented by Line 6 to become the boson p-type being with its physical (related to Presence), vital (related to Power), mental (related to Knowledge), and connection (related to Harmony) aspects now existing in material reality typified by Light moving at c. Note that just as Line 6 represents a process of quantization relating the layer of reality created by Light traveling at c with the antecedent layers, so too Lines 2 and 4 as previously discussed, also represent quantization of a more subtle kind that ultimately plays a critical part in allowing the material-fabric to express infinite diversity.

Typically, it is the process as captured by the Space-Matrix that will determine if Line 6 is activated. Specifically, patterns at the untransformed layer, U, will need to be overcome, as specified by the second line from the bottom of the Space-Matrix: $(\uparrow > P_x)$. But as specified by the bottom-line of the Time-Matrix, reproduced below, it is only with the advent of the human-system that the automaticity of the action of meta-levels is reversed:

$$U \rightarrow \begin{array}{l} t \leq E_{Cell}; TC: M_3 \rightarrow U \\ t \sim E_{Human}; TC: U \rightarrow M_3 \end{array}$$

Hence in the case of the boson P-type being, which in this emergence is a pre-human system, the fact that patterns do not need to be overcome means that quantization happens automatically, and that in reality the boson P-type being would therefore more accurately be termed the boson PMU-type being.

Even though the boson P-type being is a pre-human system and therefore the action of meta-levels are modeled as being automatic, it is useful to review (3.6.12), **Potential Effect of Levels of Light on Genetic-Type Information**, to understand how the relationship with different forms of mutation has been specified:

$$Potential\ Effect\ of\ Levels\ of\ Light\ on\ Genetic - Type\ Information =$$
$$\begin{bmatrix} STATIC\ \langle |[L][S][T]TC \rightarrow x_T|_{\langle x_U | x_T \rangle} \rangle \\ \times \\ \left((Y > U: Z_Q) \vee (Y \leq U: Z_F) \vee (Y = U: Z_R) \right) \\ \ni \\ \begin{pmatrix} Z \in \mathbb{U}\ (Space, Time, Energy, Gravity) \\ Q: Quantization; F: Fragmentation; R: Random \end{pmatrix} \end{bmatrix} \rightarrow h \ni$$

$$h \in \begin{pmatrix} [Q]: Constructive\ zone, \\ [Q]: Constructive\ zone \wedge Constructive\ mutation, \\ [F]: Destructive\ mutation, \\ [R]: Random\ mutation \end{pmatrix}$$

Line 1 from the top in the matrix is simply a static form of (3.6.1) the Simplified Light-Space-Time Emergence equation. The static form is designated by 'STATIC' and implies that fundamental operations true of (3.6.1) are being highlighted in (3.6.12). In other words (3.6.1) already has all the operations highlighted in (3.6.12) in it, but by 'freezing' it by making it static, the essential dynamics leading to possible mutations at the genetic level can more clearly be highlighted.

Line 1 is then subjected ($\times$) to a determination of the dominant levels of light that may be active, designated by Line 2, $'\big((Y > U{:}\,Z_Q) \lor (Y \leq U{:}\,Z_F) \lor (Y = U{:}\,Z_R)\big)'$. Unpacking this, '$Y > U$' implies meta-levels are active and as a result it is possible that Z_Q is going to take place (the subscript 'Q' implies quantum-level action). This also implies activation and potential change of FBLEE. The call from below, as it were, may invoke some function that already exists in the subtle-libraries 'above', so that some already existing function may influence FBLEE through ∞-entanglement, K-entanglement, or N-entanglement. This may be thought of as a key-and-lock mechanism, where a deep enough visceral urge from below acting as the key, opens an entangled lock to alter FBLEE as per the visceral urge. '$Y \leq U$' implies that only the untransformed levels are active, and therefore also the sub-level where the speed of light is 0 is active and as a result Z_F is going to take place (the subscript 'F' implies 'fragmentation'). '$Y = U$' implies that all levels are active and as a result Z_R is going to take place (the subscript 'R' implies 'random').

Line 3 elaborates the significance of Z_Q, Z_F, and Z_R. Hence Z is the union of potential quantum-operations of space, time, energy, and gravity, designated by '$Z \in \mathbb{U}\,(Space, Time, Energy, Gravity)$'. But the nature of the operations, as suggested in the previous paragraph, is designated by '$Q{:}\ Quantization; F{:}\ Fragmentation; R{:}\ Random$'. Z_Q, then, implies that the full quantization originating from updated four-base logic-encoding ecosystems (FBLEE) can take place. Z_F implies that the essential set will be fragmented and that only libraries at the level of local cellular-level DNA or precipitated material-fabric logic can potentially be altered. Z_R also precludes full quantization, and that some partial local-library constructive or destructive mutation may take place.

The '$\rightarrow h \ni h \in$' segment resolves the outcome of the operations implied by Lines 1 – 3, suggesting that the outcome will be 'h' such that ($\ni$) 'h' is an element ($\in$) of the set specified by the members '[Q]: *Constructive zone*', '[Q]: *Constructive zone AND Constructive mutation*', '[F]; *Destructive mutation*', and '{R}: *Random mutation*'. '[Q]: *Constructive zone*' implies that the in-built buffer has been crossed and that access to the deeper four-base logic-encoding ecosystem (FBLEE) has been granted. '[Q]' in this segment implies that there is the possibility that full-quantization as specified by

Line 3 of the previous matrix can take place. Access to this zone is a prerequisite for constructive mutation to occur, as designated by the element '[Q}: *Constructive zone ∩ Constructive mutation'*, which implies that full-quantization is going to take place and will result in material change. The '[F]' specifies the relationship between 'Fragmentation' in Line 3 of the previous matrix and destructive mutation. The '[R]' specifies the relationship between 'Random' in Line 3 of the previous matrix and random mutation.

But as just summarized in the Time-Matrix in (6.3.1) Y is by definition greater than U and hence quantization is automatic. In terms of the boson P-type being such quantization implies that wholeness becomes fully active through specific space, time, energy, and gravity quantization to create an holistic "ecosystem" with its own "boson P-type being logic" as it were. The wholeness has now precipitated into the material-fabric and is available to be consciously and unconsciously tapped into. This "logic" or more specifically pre-genetic code is elaborated in the following sub-sections.

Generation of 'Photon' Aspect of the Boson P-Type Being

The photon is the carrier particle of the electro-magnetic-wavearchetype-masspotential P-type being. But in the scheme of things the electro-magnetic-wavearchetype-masspotential P-type being pervades everything and as explored in Chapter 5.1 appears to be foundational to this reality we are in. So, it could be said that it is related to Presence or is an emergence of Light's property of Presence at the level of quantum force-carriers.

Hence, an equation for the photon, Equation 6.3.2 could be:

$$Sig_{photon} = Xa + \overline{Yb_{0-n}} \quad where \begin{bmatrix} X \in \lfloor S_{System_{Pr}} \rfloor \\ Y \in [S_{System_{Pr}}, S_{System_P}, S_{System_K}, S_{System_N}] \\ a, b \ are \ integers; a > b \end{bmatrix}$$

Eq 6.3.2: Signature of Photon Aspect of the Boson P-Type Being

While the primary element X, would continue to be the same as X for bosons, as in Equation 6.2.4, there will be additional secondary elements Y that will further qualify the attributes or functionality of photons. As an example, such elements could be 'pervasiveness', 'multiple wave handler', amongst others.

Note that (6.3.2) already implies that Lines 1 – 5 in the Light Matrix (6.3.1) have been activated, and that the logic of the photon-ecosystem will automatically precipitate into the material-fabric through the action of Line 6-7 of (6.3.1). This logic is none other than the pre-genetic code as specified by (6.3.2).

Generation of 'Gluon' Aspect of the Boson P-Type Being

The gluon is the carrier particle of what is known as the strong nuclear force and holds quarks together in their inherently composite arrangements. But we have posited that quarks are related to or are an emergence of Knowledge at the quantum-particle level. Hence it must be that the gluon is an emergence of Light's property of Knowledge at the level of quantum force-carriers.

Hence, Equation 6.3.3 is an equation for the gluon:

$$Sig_{gluon} = Xa + \overline{Yb_{0-n}} \quad where \quad \begin{bmatrix} X \in [S_{System_K}] \\ Y \in [S_{System_{Pr}}, S_{System_P}, S_{System_K}, S_{System_N}] \\ a, b \ are \ integers; a > b \end{bmatrix}$$

Eq 6.3.3: Signature of Gluon Aspect of the Boson P-Type Being

The primary element X would continue to be the same as in the equation for the boson. There are likely additional secondary elements Y such as 'concentrated', 'intense connection', amongst others that would collectively determine the behavior of gluons.

Note that (6.3.3) already implies that Lines 1 – 5 in the Light Matrix (6.3.1) have been activated, and that the logic of the gluon-ecosystem will automatically precipitate into the material-fabric through the action of Line 6-7 of (6.3.1). This logic is none other than the pre-genetic code as specified by (6.3.3).

Generation of 'W and Z Bosons' Aspect of the Boson P-Type Being

The W and Z bosons are the carrier particle for the weak interactions, responsible for the decay of massive quarks and leptons into lighter quarks and leptons. But this usually is accompanied by the release of energy and power and so it must be that W and Z bosons are an emergence of Light's property of Power at the level of quantum force-carriers.

Hence, Equation 6.3.4, for W and Z bosons is as follows:

$$Sig_{W,Z\ bosons} = Xa + \overline{Yb_{0-n}} \quad where \left[\begin{array}{c} X \in [S_{System_P}] \\ Y \in [S_{System_{Pr}}, S_{System_P}, S_{System_K}, S_{System_N}] \\ a, b\ are\ integers;\ a > b \end{array} \right]$$

Eq 6.3.4: Signature of W and Z Bosons Aspect of the Boson P-Type Being

The primary element X would continue to be the same as in the equation for the boson. There are likely additional secondary elements Y such as 'weak link', 'release of energy', amongst others that would collectively determine the behavior of W and Z bosons.

Note that (6.3.4) already implies that Lines 1 – 5 in the Light Matrix (6.3.1) have been activated, and that the logic of the W and Z-bosons-ecosystem will automatically precipitate into the material-fabric through the action of Line 6-7 of (6.3.1). This logic is none other than the pre-genetic code as specified by (6.3.4).

Generation of 'Graviton' Aspect of the Boson P-Type Being

This leaves the hypothetical graviton that is thought to be the carrier particle for the force of gravity. But gravity, it is thought, is what holds astronomical objects together in relationship. Hence it must be that gravitons are an emergence of Light's property of Harmony at the level of quantum force-carriers.

Hence, Equation 6.3.5, for the hypothetical graviton is as follows:

$$Sig_{graviton} = Xa + \overline{Yb_{0-n}} \quad where \left[\begin{array}{c} X \in [S_{System_N}] \\ Y \in [S_{System_{Pr}}, S_{System_P}, S_{System_K}, S_{System_N}] \\ a, b\ are\ integers;\ a > b \end{array} \right]$$

Eq 6.3.5: Signature of Graviton Aspect of the Boson P-Type Being

The primary element X would continue to be the same as in the equation for the boson. There are likely additional secondary elements Y such as 'mass curving space', 'space defining object movement', amongst others that would collectively determine the behavior of gravitons.

Note that (6.3.5) already implies that Lines 1 – 5 in the Light Matrix (6.3.1) have been activated, and that the logic of the graviton-ecosystem will automatically precipitate into the material-fabric through the action of Line 6-7 of (6.3.1). This logic is none other than the pre-genetic code as specified by (6.3.5).

So we see that the underlying properties of light - Presence, Knowledge, Power, and Harmony - emerge as photons, gluons, W and Z bosons, and the hypothetical graviton at the quantum force-carrier level.

Summary of the Boson P-Type Being

Summarizing, after the boson stage iteration of the Light-Space-Time Emergence equation is complete, the following code-segments will have been generated as specified by Equation 6.3.6, Active Boson P-Type Being Material-Fabric Pre-Genetic Code Segments:

$$Light - Space - Time\ Emergence_{Boson\ P-Type\ Being} =$$

$$\left[\begin{array}{c} \begin{bmatrix} c_\infty : [Pr, Po, K, H] \\ \left(\downarrow R_{C_K} = f(R_{C_\infty}) \right) \\ c_K : [S_{Pr}, S_{Po}, S_K, S_H] \\ \left(\downarrow R_{C_N} = f(R_{C_K}) \right) \\ c_N : f(S_{Pr} \times S_{Po} \times S_K \times S_H) \\ \left(\downarrow R_{C_U} = f(R_{C_N}) \right) \\ c_U : [P, V, M, C] \\ \Uparrow \\ c_{0:[D,W,I,C]} \end{bmatrix}_{Light} \quad \begin{bmatrix} M_3 \to System_X \\ (\uparrow F \to I) \\ M_2 \to S_{System_X} \\ (\uparrow Sig \to F) \\ M_1 \to Sig_X \\ (\uparrow > P_X) \\ U \to X_{STEG} \end{bmatrix}_{Space} \\ \\ \begin{bmatrix} M_3 : -\infty \leq t \leq \infty \\ \downarrow \\ M_2 : 0 \geq t > \infty \\ \downarrow \\ M_1 : 0 > t > \infty \\ \downarrow \\ U \to \begin{array}{l} t \leq E_{Cell}; TC: M_3 \to U \\ t \sim E_{Human}; TC: U \to M_3 \end{array} \end{bmatrix}_{Time} \quad \begin{array}{c} TC \to X_{Boson} \\ \\ \langle x_U | x_T \rangle \end{array} \end{array} \right] \Rightarrow$$

$$\langle Space - Time - Energy - Gravity\ P - Type\ Beings\ and\ Pre - Genetic\ Code \rangle +$$

$$\left\langle \begin{array}{c} Electro - Magnetic - Wavearchetype - Masspotential\ P - Type\ Being \\ and\ Pre - Genetic\ Code \end{array} \right\rangle +$$

$$\langle Quantum - Particle\ P - Type\ Being\ and\ Pre - Genetic\ Code \rangle +$$

$$FBLEEE < \cdots > +$$

$$\left(\begin{array}{c} Sig_{graviton} = Xa + \overline{Yb_{0-n}} \\ where \left[\begin{array}{c} X \in [S_{System_N}] \\ Y \in [S_{System_{Pr}}, S_{System_P}, S_{System_K}, S_{System_N}] \\ a, b\ are\ integers; a > b \end{array} \right] \\ Sig_{gluon} = Xa + \overline{Yb_{0-n}} \\ where \left[\begin{array}{c} X \in [S_{System_K}] \\ Y \in [S_{System_{Pr}}, S_{System_P}, S_{System_K}, S_{System_N}] \\ a, b\ are\ integers; a > b \end{array} \right] \\ Sig_{W,Z\ bosons} = Xa + \overline{Yb_{0-n}} \\ where \left[\begin{array}{c} X \in [S_{System_P}] \\ Y \in [S_{System_{Pr}}, S_{System_P}, S_{System_K}, S_{System_N}] \\ a, b\ are\ integers; a > b \end{array} \right] \\ Sig_{photon} = Xa + \overline{Yb_{0-n}} \\ where \left[\begin{array}{c} X \in [S_{System_{Pr}}] \\ Y \in [S_{System_{Pr}}, S_{System_P}, S_{System_K}, S_{System_N}] \\ a, b\ are\ integers; a > b \end{array} \right] \end{array} \right)$$

Eq. 6.3.6, Active Boson P-Type Being Material-Fabric Pre-Genetic Code Segments

The '$\sum x$' signifies all the possibilities for the 'x' equation-segment and depicts the growing biography by which the singular light-based edifice expresses its materialization, and at this stage, through the boson P-type being.

Chapter 6.4: Generation of the Atom P-Type Being

The quantum-particle P-type being is a prerequisite in the creation of atoms in that all atoms are compositions of quantum particles. Further, all known atoms can be classified into one of four groups: The s-Group, p-Group, d-Group, and f-Group. But what are these groups and are they too, emergences of Light as it continues its journey of crystalizing the possibility or potentiality or fourfold functionality within it?

As in the previous chapters, the Light-Space-Time Emergence equation (3.1.3) can be used to model emergence as it proceeds to this relatively more complex form of fourfold manifestation. (3.6.12), the Potential Effect of Levels of Light on Genetic-Type Information equation, basically sheds light on specific types of mutation that can occur and therefore on the nature of the code-segment that is being generated by (3.1.3). As discussed before, (3.6.12) shows a particular working of (3.1.3) but is wholly contained within it.

This chapter elaborates the action of (3.1.3) and (3.6.12) in the creation of the code-segments that define the logic of the atom ecosystem thereby generating the atom P-type being.

Light's Emergence as Atoms

As discussed previously the Light-Space-Time Emergence equation (3.1.3) being iterative, can be used to model emergence as it proceeds from simpler four-fold to more complex four-fold manifestations. Hence (3.1.3) has already been applied to suggest the emergence of the space-time-energy-gravity quadrumvirate P-type beings, the electro-magnetic-wavearchetype-masspotential P-type being, the quantum-particle P-type being, and the boson P-type being as a further instance of a particular kind of quantum-particle P-type being. Here it will be applied to suggest the emergence of the atom P-type being. But further, as implied by (3.6.12), the Potential Effect of Levels of Light on Genetic-Type Information equation, any process of organization, such as the architecture and cohesiveness of atoms, has the possibility of altering the material-fabric so long as the bases involved are driven primarily by a meta-level.

As suggested by Equation 6.4.1 (same as 4.2.2), Generation of Atom P-Type Being and Pre-Genetic Code, the architecture and details of the atom

P-type being can be seen to be the result of the application of the Light, Space, and Time matrices as will be elaborated:

$Light - Space - Time\ Emergence_{Atom\ P-Type\ Being} =$

$$\left\Vert \begin{bmatrix} \begin{array}{c} C_\infty : [Pr, Po, K, H] \\ \left(\downarrow R_{C_K} = f(R_{C_\infty}) \right) \\ C_K : [S_{Pr}, S_{Po}, S_K, S_H] \\ \left(\downarrow R_{C_N} = f(R_{C_K}) \right) \\ C_N : f(S_{Pr} \times S_{Po} \times S_K \times S_H) \\ \left(\downarrow R_{C_U} = f(R_{C_N}) \right) \\ C_U : [P, V, M, C] \\ \Uparrow \\ C_{0:[D,W,I,C]} \end{array} \end{bmatrix}_{Light} \quad \begin{bmatrix} M_3 \to System_x \\ (\uparrow F \to I) \\ M_2 \to S_{System_x} \\ (\uparrow Sig \to F) \\ M_1 \to Sig_x \\ (\uparrow > P_x) \\ U \to x_{Quantum\ Particles} \end{bmatrix}_{Space} \right.$$

$$\left. \begin{bmatrix} M_3 : -\infty \leq t \leq \infty \\ \downarrow \\ M_2 : 0 \geq t > \infty \\ \downarrow \\ M_1 : 0 > t > \infty \\ \downarrow \\ U \to \begin{array}{c} t \leq E_{Cell}; TC: M_3 \to U \\ t \sim E_{Human}; TC: U \to M_3 \end{array} \end{bmatrix}_{Time} \quad TC \to x_{Atoms} \right\Vert_{\langle x_U | x_T \rangle} \Rightarrow$$

$\langle Space - Time - Energy - Gravity\ P - Type\ Beings\ and\ Pre - Genetic\ Code \rangle +$

$\langle \begin{array}{c} Electro - Magnetic - Wavearchetype - Masspotential\ P - Type\ Being \\ and\ Pre - Genetic\ Code \end{array} \rangle +$

$\langle Quantum - Particle\ P - Type\ Being\ and\ Pre - Genetic\ Code \rangle +$

$Atom\ P - Type\ Being\ and\ Pre - Genetic\ Code$

Eq 6.4.1: Generation of Atom P-Type Being

Starting with the Light-Matrix, the top left-hand matrix in (6.4.1), the first line from the top, $C_\infty : [Pr, Po, K, H]$, specifies the fundamental architecture of the atom P-type being. As will be explored in this chapter, p-Group atoms are an emergence of Light's property of Knowledge, s-Group atoms are an emergence of Light's property of Power, d-Group atoms are an emergence of Light's property of Presence, and the f-Group atoms are an emergence of Light's property of Harmony. The fundamental architecture of these aspects, hence, is an emergence of the properties of Light at ∞.

Line 3 in the Light-Matrix, C_K: $[S_{Pr}, S_{Po}, S_K, S_H]$, elaborates the sets for Presence, Power, Knowledge, and Harmony, each containing multiple elements. For example, as explored shortly in the section on the s-Group, various elements derived from the four sets define the behavior of s-Group atoms and could be functions such as 'power', 'energy', 'adventure', and 'courage', amongst others, hence collectively describing s-Groups atoms' way of being. Specifically, Line 5, C_N: $f(S_{Pr} \times S_{Po} \times S_K \times S_H)$, suggests that unique seeds are created from a combination of such elements from all four sets, with a particular element leading or having more weight, that in effect creates the distinctness possible in the atom P-type being.

Line 6, ($\downarrow R_{C_U} = f(R_{C_N})$), specifies quantization between the layer where the seeds are formed, and the physical layer, and as explored in Chapter 3.5 and 3.6, will result in Line 7, C_U: $[P, V, M, C]$, hence changing the material-fabric of existence. Note that as in the process describing the generation of the code-segments for space-time-energy-gravity quantization at the time of the Big Bang, the generation of the FBLEE (four-base logic-encoding ecosystem) code-segment as specified by (4.1.5) and the MF (material-fabric) code segment as specified by (4.1.6) are combined together into one equation so that the FBLEE process is apparently transparent. In reality one may say that any P-type being is first a FBLEE-P-type being and then a MF-P-type being.

The possibilities represented by Lines 1 through 5 hence concretize through the quantization represented by Line 6 to become the atom P-type being with its physical (related to Presence), vital (related to Power), mental (related to Knowledge), and connection (related to Harmony) aspects now existing in material reality typified by Light moving at c. Note that just as Line 6 represents a process of quantization relating the layer of reality created by Light traveling at c with the antecedent layers, so too Lines 2 and 4 as previously discussed, also represent quantization of a more subtle kind that ultimately plays a critical part in allowing the material-fabric to express infinite diversity.

Typically, it is the process as captured by the Space-Matrix that will determine if Line 6 is activated. Specifically, patterns at the untransformed layer, U, will need to be overcome, as specified by the second line from the bottom of the Space-Matrix: $(\uparrow > P_x)$. But as specified by the bottom-line of the Time-Matrix, reproduced below, it is only with the advent of the human-system that the automaticity of the action of meta-levels is reversed:

$$U \to \begin{array}{l} t \leq E_{Cell}; TC: M_3 \to U \\ t \sim E_{Human}; TC: U \to M_3 \end{array}$$

Hence in the case of the atom P-type being, which in this emergence is a pre-human system, the fact that patterns do not need to be overcome means that quantization happens automatically, and that in reality the atom P-type being would therefore more accurately be termed the atom PMU-type being.

Even though the atom P-type being is a pre-human system and therefore the action of meta-levels are modeled as being automatic, it is nonetheless useful to review (3.6.12), **Potential Effect of Levels of Light on Genetic-Type Information**, to understand how the relationship with different forms of mutation has been specified:

$$Potential\ Effect\ of\ Levels\ of\ Light\ on\ Genetic - Type\ Information =$$

$$\begin{bmatrix} STATIC\ \langle |[L][S][T]TC \to x_T|_{\langle x_U|x_T\rangle}\rangle \\ \times \\ \Big((Y > U: Z_Q) \vee (Y \leq U: Z_F) \vee (Y = U: Z_R) \Big) \\ \ni \\ Z \in \mathbb{U}\ (Space, Time, Energy, Gravity) \\ (Q: Quantization; F: Fragmentation; R: Random) \end{bmatrix} \to h \ni$$

$$h \in \begin{pmatrix} [Q]: Constructive\ zone, \\ [Q]: Constructive\ zone \wedge Constructive\ mutation, \\ [F]: Destructive\ mutation, \\ [R]: Random\ mutation \end{pmatrix}$$

Line 1 from the top in the matrix is simply a static form of (3.6.1) the Simplified Light-Space-Time Emergence equation. The static form is designated by 'STATIC' and implies that fundamental operations true of (3.6.1) are being highlighted in (3.6.12). In other words (3.6.1) already has all the operations highlighted in (3.6.12) in it, but by 'freezing' it by making it static, the essential dynamics leading to possible mutations at the genetic level can more clearly be highlighted.

Line 1 is then subjected ($\times$) to a determination of the dominant levels of light that may be active, designated by Line 2, $'\big((Y > U{:}\,Z_Q) \lor (Y \le U{:}\,Z_F) \lor (Y = U{:}\,Z_R) \big)'$. Unpacking this, '$Y > U$' implies meta-levels are active and as a result it is possible that Z_Q is going to take place (the subscript 'Q' implies quantum-level action). This also implies activation and potential change of FBLEE. The call from below, as it were, may invoke some function that already exists in the subtle-libraries 'above', so that some already existing function may influence FBLEE through ∞-entanglement, K-entanglement, or N-entanglement. This may be thought of as a key-and-lock mechanism, where a deep enough visceral urge from below acting as the key, opens an entangled lock to alter FBLEE as per the visceral urge. '$Y \le U$' implies that only the untransformed levels are active, and therefore also the sub-level where the speed of light is 0 is active and as a result Z_F is going to take place (the subscript 'F' implies 'fragmentation'). '$Y = U$' implies that all levels are active and as a result Z_R is going to take place (the subscript 'R' implies 'random').

Line 3 elaborates the significance of Z_Q, Z_F, and Z_R. Hence Z is the union of potential quantum-operations of space, time, energy, and gravity, designated by '$Z \in \mathbb{U}\,(\,Space, Time, Energy, Gravity)'$. But the nature of the operations, as suggested in the previous paragraph, is designated by '$Q{:}\ Quantization; F{:}\ Fragmentation; R{:}\ Random'$. Z_Q, then, implies that the full quantization originating from updated four-base logic-encoding ecosystems (FBLEE) can take place. Z_F implies that the essential set will be fragmented and that only libraries at the level of local cellular-level DNA or precipitated material-fabric logic can potentially be altered. Z_R also precludes full quantization, and that some partial local-library constructive or destructive mutation may take place.

The '$\rightarrow h \ni h \in$' segment resolves the outcome of the operations implied by Lines 1 – 3, suggesting that the outcome will be 'h' such that ($\ni$) 'h' is an element ($\in$) of the set specified by the members '[Q]: Constructive zone', '[Q]: Constructive zone AND Constructive mutation', '[F}; Destructive mutation', and '{R}: Random mutation'. '[Q]: Constructive zone' implies that the in-built buffer has been crossed and that access to the deeper four-base logic-encoding ecosystem (FBLEE) has been granted. '[Q}' in this segment implies that there is the possibility that full-quantization as specified by Line 3 of the previous matrix can take place. Access to this zone is a prerequisite for constructive mutation to occur, as designated by the element '[*Q*}: *Constructive zone* ∩ *Constructive mutation*', which implies that full-quantization is going to take place and will result in material change. The '[F]' specifies the relationship between 'Fragmentation' in Line 3 of the previous matrix and destructive mutation. The '[R]' specifies the relationship between 'Random' in Line 3 of the previous matrix and random mutation.

But as just summarized in the Time-Matrix in (6.4.1) Y is by definition greater than U and hence quantization is automatic. In terms of the atom P-type being such quantization implies that wholeness becomes fully active through specific space, time, energy, and gravity quantization to create an holistic "ecosystem" with its own "atom P-type being logic" as it were. The wholeness has now precipitated into the material-fabric and is available to be consciously and unconsciously tapped into. This "logic" or more specifically pre-genetic code is elaborated in the following sub-sections and suggests the increasing sphere of influence of the atom P-type being that now also contains light-engendered laws for all atoms.

Generation of 's-Group' Aspect of the Atom P-Type Being

The s-Group consists primarily of what are known as alkali metals and alkali earth metals. These alkali metals are known to easily lose electrons and form what is known as positive ions. When they lose electrons energy is gained, but when the electrons are taken up by other atoms in proximity there is a lot of energy released. Some have referred to these groups as "violent worlds" (Tweed, 2003), and it has been pointed out that stars shine because they are transmuting vast amount of hydrogen into helium, both of which are s-Group elements. So, one gets the sense that the s-Group may be an emergence of Light's property of Power.

Stars and suns are also known to be the crucibles where all the different kinds of atoms are created. So, these furnaces of power by virtue of their heat and high pressure are able to force electrons and protons and neutrons to come together to create all the different types of atoms known in the universe.

But philosophically what are the s-Group atoms or elements? These are atoms where there is an equal likelihood of an electron being anywhere in a symmetrical sphere around the nucleus. The other groups are all similarly defined by likelihoods of electrons being within a possible

219

pattern around the nucleus. The patterns that distinguishes s-Group atoms is a sphere, and since all other patterns can be thought of as occurring within some sphere, in some sense this is like an imprint or precipitation or emergence that allows other kinds of emergences to surface within it. So the elements that are part of the s-Group may, in a sense, be thought of as the adventurers with courage who venture into a brave new world to create some foundation by which all other element-creations can follow.

The fact that hydrogen and helium are known to constitute 98% of the Universe (Heiserman, 1991) relative to other elements therefore makes sense in this view, especially since hydrogen and helium provide the fuel with which the star-furnaces manufacture all other elements.

So, s-Group elements seem to embody functions such as power, energy, adventure, courage, and can be thought of as an emergence of the property of Light to do with Power.

Hence, a series of equations linked to S_{System_P} as the prime set can be suggested, starting with the s-Group mapping, as in Equation 6.4.2:

$$Element_{s-Group} = Xa +$$
$$\overline{Yb_{0-n}} \ \ where \ \begin{bmatrix} X \in [S_{System_P}] \\ Y \in [S_{System_{Pr}}, S_{System_P}, S_{System_K}, S_{System_N}] \\ a, b \ are \ integers; a > b \end{bmatrix}$$

Eq 6.4.2: Generalized s-Group Element Signature of the Atom P-Type Being

Note that (6.4.2) already implies that Lines 1 – 5 in the Light Matrix (6.4.1) have been activated, and that the logic of the s-Group-ecosystem will automatically precipitate into the material-fabric through the action of Line 6-7 of (6.4.1). This logic is none other than the pre-genetic code as specified by (6.4.2).

Further, the equivalent mapping between traditional element groupings and S_{System_P} as in Equations 6.4.3 and 6.4.4 can be specified:

$$Element_{Alkali \ metal} =$$

$$Xa + \overline{Yb_{0-n}} \quad where \quad \begin{bmatrix} X \in [S_{System_P}] \\ Y \in [S_{System_{Pr}}, S_{System_P}, S_{System_K}, S_{System_N}] \\ a, b \ are \ integers; a > b \end{bmatrix}$$

Eq 6.4.3: Alkali Metal Element

$$Element_{Alkali \ earth \ metal} \ =$$

$$Xa + \overline{Yb_{0-n}} \quad where \quad \begin{bmatrix} X \in [S_{System_P}] \\ Y \in [S_{System_{Pr}}, S_{System_P}, S_{System_K}, S_{System_N}] \\ a, b \ are \ integers; a > b \end{bmatrix}$$

Eq 6.4.4: Alkali Earth Metal Element

Further, as a representative element belonging to the Alkali Metal group the equation for Lithium (Li), as in Equation 6.4.5, would be:

$$Element_{Lithium} = Xa +$$

$$\overline{Yb_{0-n}} \quad where \quad \begin{bmatrix} X \in [S_{System_P}] \\ Y \in [S_{System_{Pr}}, S_{System_P}, S_{System_K}, S_{System_N}] \\ a, b \ are \ integers; a > b \end{bmatrix}$$

Eq 6.4.5: Lithium

Note that the primary element X in all these cases may be a function or attribute along the lines of 'expresses power'. The secondary elements Y in all these would have some elements being the same, and then would have specific differences as one gets into sub-classes and the individual elements themselves.

Generation of 'p-Group' Aspect of the Atom P-Type Being

Atoms or elements belonging to the p-Group are those with the likelihood of electrons occurring equally on either side of the nucleus, like a dumbbell of sorts.

There are some very significant elements in this group that are part of the metal, metalloid, non-metal, halogen, and noble gas sub-groupings. Carbon, Nitrogen, Oxygen, and Silicon are some of the sample elements. Looking at the types of elements present in this group it is as though all the element possibilities have been represented within it. It is perhaps that the possibility of ideas behind all elements has emerged in this group and one can hypothesize that this group may be a reflection of the property of Knowledge, forming archetypes from which all other elements are created.

Philosophically, the one spherical shell or probability cloud (s) becoming two shells like a dumbbell (p) signifies the creation of an essential polarity within a unit space. So in a sense the one becoming two is the first instance of variability in space. The two, spaced in 3-dimensions around the nucleus in a sense create six switches becoming an attractor or allowing a vaster number of different kinds of elements to surface. So there is a sense of the 'idea' of the element that comes into focus.

But further, the essential elements that allow both thinking and virtual thinking machines to come into being, are also contained within this group. Carbon is the basis of DNA and of all life. The fact that Silicon, directly below it in the periodic table and therefore sharing essential qualities, is considered the basis of all virtual thinking machines is therefore perhaps significant and may reinforce the notion that the p-Group is a precipitation of Light's property of Knowledge.

Hence, a series of equations linked to S_{System_K} as the prime set can be suggested, starting with the p-Group mapping, as in Equation 6.4.6:

$$Element_{p-Group} = \frac{Xa +}{Yb_{0-n}} \quad where \quad \begin{bmatrix} X \in [S_{System_K}] \\ Y \in [S_{System_{Pr}}, S_{System_P}, S_{System_K}, S_{System_N}] \\ a, b \ are \ integers; a > b \end{bmatrix}$$

Eq 6.4.6: Generalized p-Group Element Signature of the Atom P-Type Being

Note that (6.4.6) already implies that Lines 1 – 5 in the Light Matrix (6.4.1) have been activated, and that the logic of the p-Group-ecosystem will automatically precipitate into the material-fabric through the action of Line 6-7 of (6.4.1). This logic is none other than the pre-genetic code as specified by (6.4.6).

Further, the equivalent mapping between traditional element groupings and S_{System_K} as in Equations 6.4.7 through 6.4.12 can also be specified:

$$Element_{Metal} = Xa + \overline{Yb_{0-n}}$$

$$where \begin{bmatrix} X \in [S_{System_K}] \\ Y \in [S_{System_{Pr}}, S_{System_P}, S_{System_K}, S_{System_N}] \\ a, b \text{ are integers}; a > b \end{bmatrix}$$

Eq 6.4.7: Metal Element

$$Element_{Metalloid} = Xa +$$
$$\overline{Yb_{0-n}} \quad where \begin{bmatrix} X \in [S_{System_K}] \\ Y \in [S_{System_{Pr}}, S_{System_P}, S_{System_K}, S_{System_N}] \\ a, b \text{ are integers}; a > b \end{bmatrix}$$

Eq 6.4.8: Metalloid Element

$$Element_{Non-Metal} = Xa +$$
$$\overline{Yb_{0-n}} \quad where \begin{bmatrix} X \in [S_{System_K}] \\ Y \in [S_{System_{Pr}}, S_{System_P}, S_{System_K}, S_{System_N}] \\ a, b \text{ are integers}; a > b \end{bmatrix}$$

Eq 6.4.9: Non-Metal Element

$$Element_{Halogen} = Xa +$$
$$\overline{Yb_{0-n}} \quad where \begin{bmatrix} X \in [S_{System_K}] \\ Y \in [S_{System_{Pr}}, S_{System_P}, S_{System_K}, S_{System_N}] \\ a, b \text{ are integers}; a > b \end{bmatrix}$$

Eq 6.4.10: Halogen Element

$$Element_{Noble\ Gas} = Xa +$$
$$\overline{Yb_{0-n}} \quad where \begin{bmatrix} X \in [S_{System_K}] \\ Y \in [S_{System_{Pr}}, S_{System_P}, S_{System_K}, S_{System_N}] \\ a, b \text{ are integers}; a > b \end{bmatrix}$$

Eq 6.4.11: Noble Gas Element

Further, as a representative element belonging to the Non-Metal group the equation for Carbon (C), as in Equation 6.4.12, would be:

$$Element_{Carbon} = Xa +$$
$$\overline{Yb_{0-n}} \ \ where \ \begin{bmatrix} X \in [S_{System_K}] \\ Y \in [S_{System_{Pr}}, S_{System_P}, S_{System_K}, S_{System_N}] \\ a, b \ are \ integers; a > b \end{bmatrix}$$

Eq 6.4.12: Carbon

Note that the primary element X in all these cases may be a function or attribute along the lines of 'knowledge'. The secondary elements Y in all these would have some elements being the same, and then would have specific differences as one gets into sub-classes and the individual elements themselves.

Generation of 'd-Group' Aspect of the Atom P-Type Being

The d-Group comprises the Transition Metals. These metals are generally hard and strong, exhibit corrosive resistance, and can be thought of as workhorse elements. Many industrial and well-known elements sit in this group: Titanium, Chromium, Manganese, Iron, Cobalt, Nickel, Copper, Zinc, Silver, Platinum, and Gold, amongst others.

The elements and atoms in the d-Group are equally likely to show up in four possible lobes or probability-spaces around the nucleus. Four lobes occurring in multiple possible planes around the nucleus will likely create a space of stability, since there is a possibility of four lobes creating the four vertices of a tetrahedron (Fuller, 1982) that has been positioned as one of the most stable shapes in the universe.

Much of the constructed world around us is created from these elements. Further, most of the series in the group easily lose one or more electrons thereby easily combining with other atoms to form a vast array of compounds. Also, looking more broadly at the function of these elements, it can be seen that these metals exist for service, to help bring about perfection in the constructed world, to help much of the machinery in which they are used, and to assist the processes dependent on them to be completed with diligence. Hence, these transition metals appear to be an emergence of Light's property of Presence.

Therefore, a series of equations linked to $S_{System_{Pr}}$ as the prime set can be suggested, starting with the d-Group mapping, as in Equation 6.4.13:

$$Element_{d-Group} = Xa +$$

$$\overline{Yb_{0-n}} \quad where \begin{bmatrix} X \in [S_{System_{Pr}}] \\ Y \in [S_{System_{Pr}}, S_{System_P}, S_{System_K}, S_{System_N}] \\ a, b \text{ are integers}; a > b \end{bmatrix}$$

Eq 6.4.13: Generalized d-Group Element Signature of the Atom P-Type Being

Note that (6.4.13) already implies that Lines 1 – 5 in the Light Matrix (6.4.1) have been activated, and that the logic of the d-Group-ecosystem will automatically precipitate into the material-fabric through the action of Line 6-7 of (6.4.1). This logic is none other than the pre-genetic code as specified by (6.4.13).

Further, the equivalent mapping between traditional element groupings and $S_{System_{Pr}}$, as in Equation 6.4.14, can also be specified:

$$Element_{Transition\ Metal} =$$

$$Xa + \overline{Yb_{0-n}} \quad where \begin{bmatrix} X \in [S_{System_{Pr}}] \\ Y \in [S_{System_{Pr}}, S_{System_P}, S_{System_K}, S_{System_N}] \\ a, b \text{ are integers}; a > b \end{bmatrix}$$

Eq 6.4.14: Transition Metal Element

Further, as a representative element belonging to the Transition-Metal group the equation for Gold (Au), as in Equation 6.4.15, would be:

$$Element_{Gold} = Xa + \overline{Yb_{0-n}} \quad where \begin{bmatrix} X \in [S_{System_{Pr}}] \\ Y \in [S_{System_{Pr}}, S_{System_P}, S_{System_K}, S_{System_N}] \\ a, b \text{ are integers}; a > b \end{bmatrix}$$

Eq 6.4.15: Gold

Note that the primary element X in all these cases may be a function or attribute along the lines of 'work horse'. The secondary elements Y in all these would have some elements being the same, and then would have specific differences as one gets into sub-classes and the individual elements themselves.

Generation of 'f-Group' Aspect of the Atom P-Type Being

The f-Group comprises of the Lanthanides and Actinides. Philosophically, elements in the f-Group consist of 6 lobes around the nucleus within which an electron may be found. 6 lobes will exist in multiple planes around the nucleus and suggests the notion of extended relationship and collectivity: the attempt to build larger and larger bonds within a small space. Considering this it is likely that the f-Group is an emergence of Light's property of Harmony.

Thinking about Lanthanides, some interesting facts may reinforce this notion. First, the spin of electrons in a lanthanides' outer shell is aligned, creating a strong magnetic field. The notion of creating a strong magnetic field seems to be consistent with the notion of engendering a collectivity through the ordered attraction and repulsion of elements. Second, these elements curiously occur together in nature often in the same ores (Gray, 2009) and are chemically interchangeable also suggesting the notion of forming a tight intra-group collectivity.

Actinides on the other hand are inherently radioactive. This implies that these elements have inherently crossed a threshold of stability and have the urge, over their own half-lives, to decompose into other elements. This natural urge may suggest some boundary conditions on the notion of collectivity and nurturing, giving additional insight into the nature of collectivity and nurturing. Further, the entire actinide group, as opposed to the lanthanide group that is inherently stable, is unstable. It is curious that both these should be part of the f-Group, and they must provide insight into boundary conditions into the notion of collectivity in elements.

Hence, a series of equations linked to S_{System_N} as the prime set can be suggested, starting with the f-Group mapping, as in Equation 6.4.16:

$$Element_{f-Group} = Xa +$$

$$\overline{Yb_{0-n}} \quad where \begin{bmatrix} X \in [S_{System_N}] \\ Y \in [S_{System_{Pr}}, S_{System_P}, S_{System_K}, S_{System_N}] \\ a, b \text{ are integers}; a > b \end{bmatrix}$$

Eq 6.4.16: Generalized f-Group Element Signature of the Atom P-Type Being

Note that (6.4.16) already implies that Lines 1 – 5 in the Light Matrix (6.4.1) have been activated, and that the logic of the f-Group-ecosystem will automatically precipitate into the material-fabric through the action of Line 6-7 of (6.4.1). This logic is none other than the pre-genetic code as specified by (6.4.16).

Further, the equivalent mapping between traditional element groupings and S_{System_N}, as in Equations 6.4.17 and 6.4.18, can also be specified:

$$Element\ _{Lanthanide}\ = Xa +$$

$$\overline{Yb_{0-n}}\ \ where\ \left[\begin{array}{c} X \in [S_{System_N}] \\ Y \in [S_{System_{Pr}}, S_{System_P}, S_{System_K}, S_{system_N}] \\ a, b\ are\ integers; a > b \end{array} \right]$$

Eq 6.4.17: Lanthanide Element

$$Element\ _{Actinide}\ = Xa + \overline{Yb_{0-n}}$$

$$where\ \left[\begin{array}{c} X \in [S_{System_N}] \\ Y \in [S_{System_{Pr}}, S_{System_P}, S_{System_K}, S_{System_N}] \\ a, b\ are\ integers; a > b \end{array} \right]$$

Eq 6.4.18: Actinide Element

Further, as a representative element belonging to the Lanthanide group the equation for Lanthanum (La), as in Equation 6.3.19, would be:

$$Element\ _{Lanthanum}\ = Xa + \overline{Yb_{0-n}}$$

$$where\ \left[\begin{array}{c} X \in [S_{System_N}] \\ Y \in [S_{System_{Pr}}, S_{System_P}, S_{System_K}, S_{System_N}] \\ a, b\ are\ integers; a > b \end{array} \right]$$

Eq 6.4.19: Lanthanum

Note that the primary element X in all these cases may be a function or attribute along the lines of 'experiments in collectivity'. The secondary elements Y in all these would have some elements being the same, and then would have specific differences as one gets into sub-classes and the individual elements themselves.

Summary of Atom P-Type Being Material-Fabric Pre-Genetic Code

Summarizing, after the atoms stage iterations of the Light-Space-Time Emergence equation are complete, the following code-segments will have been generated as specified by Equation 6.4.20, Active Atom P-Type Being Material-Fabric Pre-Genetic Code Segments:

$$Light - Space - Time\ Emergence_{Atom\ P-Type\ Being} =$$

$$
\left\|
\begin{array}{l}
\left[
\begin{array}{l}
c_\infty : [Pr, Po, K, H] \\
\left(\downarrow R_{C_K} = f(R_{C_\infty}) \right) \\
c_K : [S_{Pr}, S_{Po}, S_K, S_H] \\
\left(\downarrow R_{C_N} = f(R_{C_K}) \right) \\
c_N : f(S_{Pr} \times S_{Po} \times S_K \times S_H) \\
\left(\downarrow R_{C_U} = f(R_{C_N}) \right) \\
c_U : [P, V, M, C] \\
\Uparrow \\
c_{0:[D,W,I,C]}
\end{array}
\right]_{Light}
\left[
\begin{array}{l}
M_3 \to System_X \\
(\uparrow F \to I) \\
M_2 \to S_{System_X} \\
(\uparrow Sig \to F) \\
M_1 \to Sig_X \\
(\uparrow > P_X) \\
U \to x_{Quantum\ Particles}
\end{array}
\right]_{Space} \\
\left[
\begin{array}{c}
M_3 : -\infty \le t \le \infty \\
\downarrow \\
M_2 : 0 \ge t > \infty \\
\downarrow \\
M_1 : 0 > t > \infty \\
\downarrow \\
U \to \begin{array}{l} t \le E_{Cell}; TC: M_3 \to U \\ t \sim E_{Human}; TC: U \to M_3 \end{array}
\end{array}
\right]_{Time}
\quad TC \to x_{Atoms}
\end{array}
\right\|_{\langle x_U | x_T \rangle}
\Rightarrow
$$

$\langle Space - Time - Energy - Gravity\ P - Type\ Beings\ and\ Pre - Genetic\ Code \rangle +$

$\langle \begin{array}{c} Electro - Magnetic - Wavearchetype - Masspotential\ P - Type\ Being \\ and\ Pre - Genetic\ Code \end{array} \rangle +$

$\langle Quantum - Particle\ P - Type\ Being\ and\ Pre - Genetic\ Code \rangle +$

$$\left(\begin{array}{l} \begin{array}{l} \sum Element_{f-Group} = Xa + \overline{Yb_{0-n}} \\ where \left[\begin{array}{l} X \in [S_{System_N}] \\ Y \in [S_{System_{Pr}}, S_{System_P}, S_{System_K}, S_{System_N}] \\ a, b \ are\ integers; a > b \end{array} \right] \end{array} \\ \begin{array}{l} \sum Element_{p-Group} = Xa + \overline{Yb_{0-n}} \\ where \left[\begin{array}{l} X \in [S_{System_K}] \\ Y \in [S_{System_{Pr}}, S_{System_P}, S_{System_K}, S_{System_N}] \\ a, b \ are\ integers; a > b \end{array} \right] \end{array} \\ \begin{array}{l} \sum Element_{s-Group} = Xa + \overline{Yb_{0-n}} \\ where \left[\begin{array}{l} X \in [S_{System_P}] \\ Y \in [S_{System_{Pr}}, S_{System_P}, S_{System_K}, S_{System_N}] \\ a, b \ are\ integers; a > b \end{array} \right] \end{array} \\ \begin{array}{l} \sum Element_{d-Group} = Xa + \overline{Yb_{0-n}} \\ where \left[\begin{array}{l} X \in [S_{System_{Pr}}] \\ Y \in [S_{System_{Pr}}, S_{System_P}, S_{System_K}, S_{System_N}] \\ a, b \ are\ integers; a > b \end{array} \right] \end{array} \end{array} \right)$$

Eq. 6.4.20, Active Atom P-Type Being Material-Fabric Pre-Genetic Code Segments

The '$\sum x$' signifies all the possibilities for the 'x' equation-segment and depicts the growing biography by which the singular light-based edifice expresses its materialization, and at this stage, through the atom P-type being.

SECTION 7: GENERATION OF P-TYPE BEINGS IN THE SURFACING OF LIFE

This section explores the quantum-level computation and the generation of genetic code that is associated with the emergence of life through the cell, complex human attributes such as thoughts and feelings, and uniqueness of individuality. As suggested genetic code articulates a biography, and in this case, the emerging biography depicting an increase in the P-type being's sphere of influence through the emergence of additional light-generated "laws".

So far, the action of the Light-Space-Time Emergence has generated pre-genetic code. After the Big Bang this pre-genetic code is housed in the material-fabric, which has a universal action on all constructs arising in the universe. The emergence of matter, as discussed in Section 6, precipitates a phenomenon of material containerization. This was hypothesized as being due to an essential action of the space-time-energy-gravity quadrumvirate P-type macro-being contained within a smaller "space" as a result of the number of driver seeds.

In this section it is assumed that just as the space-time-energy-gravity quadrumvirate P-type micro and macro beings have an impact on every material emergence, so too will the action of the electro-magnetic-wavearchetype-masspotential P-type being have an impact on every emergence of life. Some dynamics of this possibility are explored in Chapter 7.1.

Chapter 7.2 will explore the generation of the genetic code responsible for the emergence and complexification of four-foldness through the primary molecular plans of nucleic acids, proteins, lipids, and polysaccharides as manifest in the body of the cell P-type being.

Chapter 7.3 will relate key human attributes of sensations, urges, desires, wills, feelings, emotions, and thought to the continuing journey of fourfold complexification and suggest that genetic code is also generated to support these attributes as manifest in the body of the fundamental-capacities-of-self P-type being.

Chapter 7.4 will relate truer individuality to the fourfold properties implicit in Light and suggest the generation of the genetic code tied to this as manifest in the body of the truer-individuality P-type being.

Note that the dynamics of the generation of the fundamental-capacities-of-self P-type being as covered in Chapter 7.3, and the generation of the truer-individuality P-type being as covered in Chapter 7.4, is central to envisioning any path toward transhumanism. This is because it is with the integration of any light-based-singularity – be it a seed-singularity or partial-singularity based being – as opposed to a fragmented-singularity, which appears to be the current trend in thinking, that transhumanism will have its most fruitful foundation. This has to be since a light-based singularity culminating in the emergence of the S-type being is nothing other than the triumphant expression of the infinite formative forces of light and love.

Chapter 7.1: Effect of Electro-Magnetic-Wavearchetype-Masspotential P-Type Being on Life

As explored in Section 5, while there is a significant volume of material-fabric pre-genetic code generated for the electro-magnetic-wavearchetype-masspotential P-type being, this chapter will focus on some aspects of the visible-light portion of the electro-magnetic-wavearchetype-masspotential P-type being that is known to have a direct impact on life.

Research indicates that in the visible light spectrum specified by light with a wavelength between 380 nanometers to 760 nanometers, each color has specific physical and psychological properties (Martel, 2018). For example, colors in the spectrum that range from red to yellow, are known as warm colors and generally are known to be stimulating, fortifying, and energizing. Colors in the spectrum that range from green to violet are known as cool colors and generally are known to be calming, sedative, and analgesic.

This suggests that there is a specific set of functions associated with each color that conceivably uses that color as a vehicle to express or concretize itself. It further suggests that there are segments of the electro-magnetic-wavearchetype-masspotential P-type being genetic-type code that may be fundamental to all of life and are likely deeply embedded in or influence all of life. This is evident in that colors are known to generate specific responses in human bodies. The functioning of the autonomic nervous system, for example, which is related to the equilibrium of the sympathetic and parasympathetic nervous system, can be influenced by the application of warm and cool colors (Spitler, 2011).

Research into generic properties associated with colors (Van Obberghen, 2007; Deppe, 2013) suggests the specific functions related to the wave-archetype or structure, the energy-profile, and the mass-potential for each of the colors.

The speed or magnetic profile of each color is assumed to be that of c as derived in Chapter 5.1. Hence as in (5.1.2) all colors will be assumed to have the magnetic properties associated with light moving at c:

$$EMWM\ P-Type\ Being_{Speed}\ =\ Xa\ +$$
$$\overline{Yb_{0-n}}\ where\ \begin{bmatrix} X \in [S_{System_N}] \\ Y \in [S_{System_{Pr}}, S_{System_P}, S_{System_K}, S_{System_N}] \\ a,b\ are\ integers; a > b \end{bmatrix}$$

Red

Red is associated with the wavelength 635 to 760 nanometers.

In its association with life the energy-profile of 'red' is suggested to engender qualities to do with 'tonifying', 'warming', 'antianemia', 'antimigraine', amongst possible others. Assuming that the primary or x-element has to do with tonifying, then various y-elements in combination can create the other engendered qualities. These relationships may be depicted in Equation 7.1.1, Red Energy, in the following manner:

$$Red_{Energy}\ =' tonifying'.a + \overline{Yb_{0-n}}$$

$$where\ \begin{bmatrix} tonifying \in [S_{System_P}] \\ Y \in [S_{System_{Pr}}, S_{System_P}, S_{System_K}, S_{System_N}] \\ a,b\ are\ integers; a > b \end{bmatrix}$$

Eq 7.1.1: Red Energy

The wave-archetype or intent has to do with a primary or x-element associated with the function 'passion'. Other y-elements in combination will engender intent such as 'extroversion'. Hence, these relationships may be depicted by Equation 7.1.2, Red Intent:

$$Red_{Intent}\ =' passion'.a + \overline{Yb_{0-n}}$$

$$where\ \begin{bmatrix} passion \in [S_{System_K}] \\ Y \in [S_{System_{Pr}}, S_{System_P}, S_{System_K}, S_{System_N}] \\ a,b\ are\ integers; a > b \end{bmatrix}$$

Eq 7.1.2: Red Intent

The mass-potential or body-part most linked to red is suggested to be the 'reproductive system', the 'urogenital system', and 'system to do with blood pressure', amongst others. So, there is likely a buildup of mass in such a manner that facilitates the creation of these systems. An assumption can be made that there is a 'red-mass' x-element and perhaps other y-elements that yield the different systems when built up. Equation 7.1.3, Reproductive System – Red Mass Possibility, suggests how this can be depicted:

$$Reproductive\ System - Red_{Mass_{Possibility}} = 'red - mass'.a + \overline{Yb_{0-n}}$$

$$where \begin{bmatrix} 'red - mass' \in [S_{System_{Pr}}] \\ Y \in [S_{System_{Pr}}, S_{System_P}, S_{System_K}, S_{System_N}] \\ a, b\ are\ integers;\ a > b \end{bmatrix}$$

Eq 7.1.3: Reproductive System - Red Mass Possibility

Note that (7.1.1 – 3) are proposed to be part of the material-fabric. As such they are of a pre-genetic type but also inform or even help form the cell-based genetic code. Further (7.1.1 – 3) are an illustration of the marriage between the electro-magnetic-wavearchetype-masspotential P-type being and P-type beings in the emergence of Life. Such a marriage is nothing other than an act of love in the 'vertical' as opposed to the 'horizontal' direction. While the horizontal direction implies integration of light's implicit fourfoldness, the vertical direction implies integration between progressively more complex P-type beings.

Orange

Orange is associated with the wavelength 590 to 635 nanometers.

In its association with life the energy-profile of 'orange' is suggested to engender qualities to do with 'invigorating' and 'adrenal stimulant' amongst possible others. Assuming that the primary or x-element has to do with 'invigorating', then various y-elements in combination can create the other engendered qualities. These relationships may be depicted in Equation 7.1.4, Orange Energy, in the following manner:

$$Orange_{Energy} = 'invigorating'.a + \overline{Yb_{0-n}}$$

$$where \begin{bmatrix} invigorating \in [S_{System_P}] \\ Y \in [S_{System_{Pr}}, S_{System_P}, S_{System_K}, S_{System_N}] \\ a, b\ are\ integers;\ a > b \end{bmatrix}$$

Eq 7.1.4: Orange Energy

The wave-archetype or intent has to do with a primary or x-element associated with the function 'joy'. Other y-elements in combination will engender intent such as 'antidepressant' and 'sociability'. Hence, these relationships may be depicted by Equation 7.1.5, Orange Intent:

$$Orange\ _{Intent} =' joy'.a + \overline{Yb_{0-n}}$$

$$where \begin{bmatrix} joy \in [S_{System_K}] \\ Y \in [S_{System_{Pr}}, S_{System_P}, S_{System_K}, S_{System_N}] \\ a, b\ are\ integers; a > b \end{bmatrix}$$

Eq 7.1.5: Orange Intent

The mass-potential or body-part most linked to orange is suggested to be the 'adrenal glands', 'vasomotricity', and 'musculoskeletal system', amongst possible others. So, there is likely a buildup of mass in such a manner that facilitates the creation of these systems. An assumption can be made that there is a 'orange-mass' x-element and perhaps other y-elements that yield the different systems when built up. Equation 7.1.6, Adrenal Gland - Orange Mass Possibility, suggests how this can be depicted:

$$Adrenal\ Glands - Orange_{Mass_{Possibility}} =' orange - mass'.a + \overline{Yb_{0-n}}$$

$$where \begin{bmatrix} 'orange - mass' \in [S_{System_{Pr}}] \\ Y \in [S_{System_{Pr}}, S_{System_P}, S_{System_K}, S_{System_N}] \\ a, b\ are\ integers; a > b \end{bmatrix}$$

Eq 7.1.6: Adrenal Glands - Orange Mass Possibility

Note that (7.1.4 – 6) are proposed to be part of the material-fabric. As such they are of a pre-genetic type but also inform or even help form the cell-based genetic code.

Yellow

Yellow is associated with the wavelength 570 to 590 nanometers.

In its association with life the energy-profile of 'yellow' is suggested to engender qualities to do with 'digestive stimulant', and 'lymphatic stimulant' amongst possible others. Assuming that the primary or x-element has to do with 'stimulant', then various y-elements in combination can create the other engendered qualities. These relationships may be depicted in Equation 7.1.7, Yellow Energy, in the following manner:

$$Yellow_{Energy} =' stimulant'.a + \overline{Yb_{0-n}}$$

$$where \begin{bmatrix} stimulant \in [S_{System_P}] \\ Y \in [S_{System_{Pr}}, S_{System_P}, S_{System_K}, S_{System_N}] \\ a, b \ are \ integers; a > b \end{bmatrix}$$

Eq 7.1.7: Yellow Energy

The wave-archetype or intent has to do with a primary or x-element associated with the function 'optimism'. then various y-elements in combination can create the other engendered qualities. Hence, these relationships may be depicted by Equation 7.1.8, Yellow Intent:

$$Yellow_{Intent} =' optimism'.a + \overline{Yb_{0-n}}$$

$$where \begin{bmatrix} optimism \in [S_{System_K}] \\ Y \in [S_{System_{Pr}}, S_{System_P}, S_{System_K}, S_{System_N}] \\ a, b \ are \ integers; a > b \end{bmatrix}$$

Eq 7.1.8: Yellow Intent

The mass-potential or body-part most linked to yellow is suggested to be the 'digestive system', the 'intestine', the 'spleen', the 'pancreas' amongst possible others. So, there is likely a buildup of mass in such a manner that facilitates the creation of these systems. An assumption can be made that there is a 'yellow-mass' x-element and perhaps other y-elements that yield the different systems when built up. Equation 7.1.9, Digestive System - Yellow Mass Possibility, suggests how this can be depicted:

$$Digestive \ System - Yellow_{Mass_{Possibility}} =' yellow - mass'.a + \overline{Yb_{0-n}}$$

$$where \begin{bmatrix} 'yellow - mass' \in [S_{System_{Pr}}] \\ Y \in [S_{System_{Pr}}, S_{System_P}, S_{System_K}, S_{System_N}] \\ a, b \ are \ integers; a > b \end{bmatrix}$$

Eq 7.1.9: Digestive System - Yellow Mass Possibility

Note that (7.1.7 – 9) are proposed to be part of the material-fabric. As such they are of a pre-genetic type but also inform or even help form the cell-based genetic code.

Lemon

Lemon is associated with the wavelength 550 to 570 nanometers.

In its association with life the energy-profile of 'lemon' is suggested to engender qualities to do with 'antiallergy', 'detoxifying', and 'addresses chronic conditions', amongst possible others. Assuming that the primary or x-element has to do with 'detoxifying', then various y-elements in combination can create the other engendered qualities Energy, in the following manner:

$$Red_{Energy} =' detoxifying'.a + \overline{Yb_{0-n}}$$

$$where \begin{bmatrix} detoxifying \in [S_{System_P}] \\ Y \in [S_{System_{Pr}}, S_{System_P}, S_{System_K}, S_{System_N}] \\ a, b \ are \ integers; a > b \end{bmatrix}$$

Eq 7.1.10: Lemon Energy

The wave-archetype or intent has to do with a primary or x-element associated with the function 'peace'. Other y-elements in combination will engender intent such as 'flexibility' and 'opening', amongst possible others. Hence, these relationships may be depicted by Equation 7.1.11, Lemon Intent:

$$Red_{Intent} =' peace'.a + \overline{Yb_{0-n}}$$

$$where \begin{bmatrix} peace \in [S_{System_K}] \\ Y \in [S_{System_{Pr}}, S_{System_P}, S_{System_K}, S_{System_N}] \\ a, b \ are \ integers; a > b \end{bmatrix}$$

Eq 7.1.11: Lemon Intent

The mass-potential or body-part most linked to lemon is suggested to be 'joints', 'gallbladder', 'diaphragm' and 'vision', amongst possible others. So, there is likely a buildup of mass in such a manner that facilitates the creation of these systems. An assumption can be made that there is a 'lemon-mass' x-element and perhaps other y-elements that yield the different systems when built up. Equation 7.1.12, Joints – Lemon Mass Possibility, suggests how this can be depicted:

$$Joints - Lemon_{Mass_{Possibility}} \quad =' lemon - mass'.a + \overline{Yb_{0-n}}$$

$$where \begin{bmatrix} 'lemon - mass' \in [S_{System_{Pr}}] \\ Y \in [S_{System_{Pr}}, S_{System_P}, S_{System_K}, S_{System_N}] \\ a, b \ are \ integers; a > b \end{bmatrix}$$

Eq 7.1.12: Joints - Lemon Mass Possibility

Note that (7.1.10 – 12) are proposed to be part of the material-fabric. As such they are of a pre-genetic type but also inform or even help form the cell-based genetic code.

Green

Green is associated with the wavelength 520 to 550 nanometers.

In its association with life the energy-profile of 'green' is suggested to engender qualities to do with 'general balancing', 'neutralizing', and 'anti-infection' amongst possible others. Assuming that the primary or x-element has to do with 'balancing', then various y-elements in combination can create the other engendered qualities. These relationships may be depicted in Equation 7.1.13, Green Energy, in the following manner:

$$Green_{Energy} \quad =' balancing'.a + \overline{Yb_{0-n}}$$

$$where \begin{bmatrix} balancing \in [S_{System_P}] \\ Y \in [S_{System_{Pr}}, S_{System_P}, S_{System_K}, S_{System_N}] \\ a, b \ are \ integers; a > b \end{bmatrix}$$

Eq 7.1.13: Green Energy

The wave-archetype or intent has to do with a primary or x-element associated with the function 'love'. Other y-elements in combination will engender intent such as 'abundance' and 'adaptability', amongst possible others. Hence, these relationships may be depicted by Equation 7.1.14, Green Intent:

$$Green_{Intent} =' love'.a + \overline{Yb_{0-n}}$$

$$where \begin{bmatrix} love \in [S_{System_K}] \\ Y \in [S_{System_{Pr}}, S_{System_P}, S_{System_K}, S_{System_N}] \\ a, b \ are \ integers; a > b \end{bmatrix}$$

Eq 7.1.14: Green Intent

The mass-potential or body-part most linked to green is suggested to be the 'liver', 'lungs', 'thymus', and 'immune system', amongst possible others. So, there is likely a buildup of mass in such a manner that facilitates the creation of these systems. An assumption can be made that there is a 'green-mass' x-element and perhaps other y-elements that yield the different systems when built up. Equation 7.1.15, Liver – Green Mass Possibility, suggests how this can be depicted:

$$Liver - Green_{Mass_{Possibility}} =' green - mass'.a + \overline{Yb_{0-n}}$$

$$where \begin{bmatrix} 'green - mass' \in [S_{System_{Pr}}] \\ Y \in [S_{System_{Pr}}, S_{System_P}, S_{System_K}, S_{System_N}] \\ a, b \ are \ integers; a > b \end{bmatrix}$$

Eq 7.1.15: Liver - Green Mass Possibility

Note that (7.1.13 – 15) are proposed to be part of the material-fabric. As such they are of a pre-genetic type but also inform or even help form the cell-based genetic code.

Cyan

Cyan is associated with the wavelength 490 to 520 nanometers.

In its association with life the energy-profile of 'cyan' is suggested to engender qualities to do with 'cleansing', 'refreshing', and 'addressing acute conditions' amongst possible others. Assuming that the primary or x-element has to do with 'cleansing', then various y-elements in combination can create the other engendered qualities. These relationships may be depicted in Equation 7.1.16, Cyan Energy, in the following manner:

$$Cyan_{Energy} =' cleansing'.a + \overline{Yb_{0-n}}$$

$$where \begin{bmatrix} cleansing \in [S_{System_P}] \\ Y \in [S_{System_{Pr}}, S_{System_P}, S_{System_K}, S_{System_N}] \\ a, b\ are\ integers; a > b \end{bmatrix}$$

Eq 7.1.16: Cyan Energy

The wave-archetype or intent has to do with a primary or x-element associated with the function 'wholeness'. Other y-elements in combination will engender intent such as 'autonomy', and 'fulcrum of emotion vs. thought', amongst possible others. Hence, these relationships may be depicted by Equation 7.1.17, Cyan Intent:

$$Cyan_{Intent} =' wholeness'.a + \overline{Yb_{0-n}}$$

$$where \begin{bmatrix} wholeness \in [S_{System_K}] \\ Y \in [S_{System_{Pr}}, S_{System_P}, S_{System_K}, S_{System_N}] \\ a, b\ are\ integers; a > b \end{bmatrix}$$

Eq 7.1.17: Cyan Intent

The mass-potential or body-part most linked to cyan is suggested to be the 'arms and hands', 'parathyroid glands', and 'skin'. amongst others. So, there is likely a buildup of mass in such a manner that facilitates the creation of these systems. An assumption can be made that there is a 'cyan-mass' x-element and perhaps other y-elements that yield the different systems when built up. Equation 7.1.18, Skin – Cyan Mass Possibility, suggests how this can be depicted:

$$Skin - Cyan_{Mass_{Possibility}} =' cyan - mass'.a + \overline{Yb_{0-n}}$$

$$where \begin{bmatrix} 'cyan - mass' \in [S_{System_{Pr}}] \\ Y \in [S_{System_{Pr}}, S_{System_P}, S_{System_K}, S_{System_N}] \\ a, b\ are\ integers; a > b \end{bmatrix}$$

Eq 7.1.18: Skin - Cyan Mass Possibility

Note that (7.1.16 – 18) are proposed to be part of the material-fabric. As such they are of a pre-genetic type but also inform or even help form the cell-based genetic code.

Blue

Blue is associated with the wavelength 460 to 490 nanometers.

In its association with life the energy-profile of 'blue' is suggested to engender qualities to do with 'calming', 'anti-inflammation', and 'antipyretic', amongst possible others. Assuming that the primary or x-element has to do with 'calming', then various y-elements in combination can create the other engendered qualities. These relationships may be depicted in Equation 7.1.19, Blue Energy, in the following manner:

$$Blue_{Energy} =' calming'.a + \overline{Yb_{0-n}}$$

$$where \begin{bmatrix} calming \in [S_{System_P}] \\ Y \in [S_{System_{Pr}}, S_{System_P}, S_{System_K}, S_{System_N}] \\ a, b \ are \ integers; a > b \end{bmatrix}$$

Eq 7.1.19: Blue Energy

The wave-archetype or intent has to do with a primary or x-element associated with the function 'antistress'. Other y-elements in combination will engender intent such as 'introversion' and 'endorphins stimulant', amongst possible others. Hence, these relationships may be depicted by Equation 7.1.20, Blue Intent:

$$Blue_{Intent} =' antistress'.a + \overline{Yb_{0-n}}$$

$$where \begin{bmatrix} antistress \in [S_{System_K}] \\ Y \in [S_{System_{Pr}}, S_{System_P}, S_{System_K}, S_{System_N}] \\ a, b \ are \ integers; a > b \end{bmatrix}$$

Eq 7.1.20: Blue Intent

The mass-potential or body-part most linked to blue is suggested to be the 'nose, throat, neck', 'thyroid gland', breathing', 'ears', amongst possible others. So, there is likely a buildup of mass in such a manner that facilitates the creation of these systems. An assumption can be made that there is a 'blue-mass' x-element and perhaps other y-elements that yield the different systems when built up. Equation 7.1.21, Ears – Blue Mass Possibility, suggests how this can be depicted:

$$Ears - Blue_{Mass_{Possibility}} \quad =' \ blue - mass'.\,a + \overline{Yb_{0-n}}$$

$$where \begin{bmatrix} 'blue - mass' \in [S_{System_{Pr}}] \\ Y \in [S_{System_{Pr}}, S_{System_P}, S_{System_K}, S_{System_N}] \\ a, b \ are \ integers; a > b \end{bmatrix}$$

Eq 7.1.21: Ears - Blue Mass Possibility

Note that (7.1.19 – 21) are proposed to be part of the material-fabric. As such they are of a pre-genetic type but also inform or even help form the cell-based genetic code.

Indigo

Indigo is associated with the wavelength 430 to 460 nanometers.

In its association with life the energy-profile of 'indigo' is suggested to engender qualities to do with 'sleep-inducing' and 'muscle relaxant', amongst possible others. Assuming that the primary or x-element has to do with 'relaxant', then various y-elements in combination can create the other engendered qualities. These relationships may be depicted in Equation 7.1.22, Indigo Energy, in the following manner:

$$Indigo_{Energy} \quad =' \ relaxant'.\,a + \overline{Yb_{0-n}}$$

$$where \begin{bmatrix} relaxant \in [S_{System_P}] \\ Y \in [S_{System_{Pr}}, S_{System_P}, S_{System_K}, S_{System_N}] \\ a, b \ are \ integers; a > b \end{bmatrix}$$

Eq 7.1.22: Indigo Energy

The wave-archetype or intent has to do with a primary or x-element associated with the function 'intuition'. Other y-elements in combination will engender intent such as 'imagination' and 'anxiolytic', amongst possible others. Hence, these relationships may be depicted by Equation 7.1.23, Indigo Intent:

$$Indigo_{Intent} =' intuition'. a + \overline{Yb_{0-n}}$$

$$where \begin{bmatrix} intuition \in [S_{System_K}] \\ Y \in [S_{System_{Pr}}, S_{System_P}, S_{System_K}, S_{System_N}] \\ a, b \ are \ integers; a > b \end{bmatrix}$$

Eq 7.1.23: Indigo Intent

The mass-potential or body-part most linked to indigo is suggested to be the 'sinuses', 'pituitary', 'forehead and occiput', and 'limbic system', amongst possible others. So, there is likely a buildup of mass in such a manner that facilitates the creation of these systems. An assumption can be made that there is a 'indigo-mass' x-element and perhaps other y-elements that yield the different systems when built up. Equation 7.1.24, Pituitary – Indigo Mass Possibility, suggests how this can be depicted:

$$Pituitary - Indigo_{Mass_{Possibility}} =' indigo - mass'. a + \overline{Yb_{0-n}}$$

$$where \begin{bmatrix} 'indigo - mass' \in [S_{System_{Pr}}] \\ Y \in [S_{System_{Pr}}, S_{System_P}, S_{System_K}, S_{System_N}] \\ a, b \ are \ integers; a > b \end{bmatrix}$$

Eq 7.1.24: Pituitary - Indigo Mass Possibility

Note that (7.1.22 – 24) are proposed to be part of the material-fabric. As such they are of a pre-genetic type but also inform or even help form the cell-based genetic code.

Violet

Violet is associated with the wavelength 380 to 430 nanometers.

In its association with life the energy-profile of 'violet' is suggested to engender qualities to do with 'immunostimulant', amongst possible others. Assuming that the primary or x-element has to do with 'stimulant', then various y-elements in combination can create the other engendered qualities. These relationships may be depicted in Equation 7.1.25, Violet Energy, in the following manner:

$$Violet_{Energy} \quad =' stimulant'.a + \overline{Yb_{0-n}}$$

$$where \begin{bmatrix} stimulant \in [S_{System_P}] \\ Y \in [S_{System_{Pr}}, S_{System_P}, S_{System_K}, S_{System_N}] \\ a, b \text{ are integers}; a > b \end{bmatrix}$$

Eq 7.1.25: Violet Energy

The wave-archetype or intent has to do with a primary or x-element associated with the function 'spirituality'. Other y-elements in combination will engender intent such as 'trust' and 'dignity', amongst possible others. Hence, these relationships may be depicted by Equation 7.1.26, Violet Intent:

$$Violet_{Intent} =' spirituality'.a + \overline{Yb_{0-n}}$$

$$where \begin{bmatrix} spirituality \in [S_{System_K}] \\ Y \in [S_{System_{Pr}}, S_{System_P}, S_{System_K}, S_{System_N}] \\ a, b \text{ are integers}; a > b \end{bmatrix}$$

Eq 7.1.26: Violet Intent

The mass-potential or body-part most linked to violet is suggested to be the 'cerebral cortex', 'pineal gland', and 'memory, amongst possible others. So, there is likely a buildup of mass in such a manner that facilitates the creation of these systems. An assumption can be made that there is a 'violet-mass' x-element and perhaps other y-elements that yield the different systems when built up. Equation 7.1.27, Cerebral Cortex–Violet Mass Possibility, suggests how this can be depicted:

$$Cerebral\ Cortex - Violet_{Mass_{Possibility}} \quad =' violet - mass'.a + \overline{Yb_{0-n}}$$

$$where \begin{bmatrix} 'violet - mass' \in [S_{System_{Pr}}] \\ Y \in [S_{System_{Pr}}, S_{System_P}, S_{System_K}, S_{System_N}] \\ a, b \text{ are integers}; a > b \end{bmatrix}$$

Eq 7.1.27: Cerebral Cortex - Violet Mass Possibility

Note that (7.1.25 – 27) are proposed to be part of the material-fabric. As such they are of a pre-genetic type but also inform or even help form the cell-based genetic code.

This chapter on the effect of the electro-magnetic-wavearchetype-masspotential P-type being on life is perhaps a clear example of how the sphere of influence of emerging P-type beings increases. The electro-magnetic-wavearchetype-masspotential P-type being exercises increased yet cohesive complexification of four-foldness in the pre-genetic material-fabric, not marginalizing the comprehensive logic of the space-time-energy-gravity quadrumvirate P-type beings put into place at the time of the Big Bang. Variation in wavelength in the electro-magnetic-wavearchetype-masspotential P-type being is the mechanism by which antecedent functional possibility encoded in Light moving at speeds greater than c materializes in an initial pre-material medium at light traveling at c. Such variation in wavelength creates diverse functional possibility in the intent, energetics, and mass-possibility of visible light. As life-related P-type beings emerge, such pre-existing logic or law is automatically operative so that there is deep consistency with it in the surfacing of more complex tissues and even organs, as suggested by the preceding discussion on mass-possibility. Cross-over of such fourfold consistency from one type of P-type being to another, as in the case from the electro-magnetic-wavearchetype-masspotential P-type being, to matter-based P-type beings, to life-based P-type beings is the hallmark of light-based singularities indicating that all that emerges in such a manner is of one light-based edifice seeking always to expand its essential basis of love.

Chapter 7.2: Generation of the Cell P-Type Being

The living cell has a universe of adaptability embedded in it. In 'The Machinery of Life', Goodsell, an Associate Professor of Molecular Biology at the Scripps Research Institute (Goodsell, 2010) suggests that every living thing on Earth uses a similar set of molecules to eat, to breathe, to move, and to reproduce. There are molecular machines that do the myriad things that distinguish living organisms that are identical in all living cells. This nanoscale machinery of cells uses four basic molecular plans with unique chemical personalities: nucleic acids, proteins, lipids, and polysaccharides.

Extrapolating from previous chapters there will likely be a vast number of iterations of the Light-Space-Time Emergence Equation to generate the cell P-type being and the four basic molecular plans integral to it. Equation 4.2.4, Generation of Cell P-Type Being and Genetic Code, created in Chapter 4.2 that explored the post Big-Bang generation of pre-genetic code, genetic code, and post-genetic code, is reproduced here for convenience. The starting point, x_U, is 'Molecules', and the ending point, x_T, is 'Cells':

$$Light - Space - Time\ Emergence_{Cell\ P-Type\ Being} =$$

$$
\left\|
\begin{matrix}
\begin{bmatrix}
c_\infty : [Pr, Po, K, H] \\
\left(\downarrow R_{C_K} = f(R_{C_\infty}) \right) \\
c_K : [S_{Pr}, S_{Po}, S_K, S_H] \\
\left(\downarrow R_{C_N} = f(R_{C_K}) \right) \\
c_N : f(S_{Pr}\ x\ S_{Po}\ x\ S_K\ x\ S_H) \\
\left(\downarrow R_{C_U} = f(R_{C_N}) \right) \\
c_U : [P, V, M, C] \\
\Uparrow \\
c_{0 : [D, W, I, C]}
\end{matrix}_{Light}
&
\begin{bmatrix}
M_3 \rightarrow System_X \\
(\uparrow F \rightarrow I) \\
M_2 \rightarrow S_{System_X} \\
(\uparrow Sig \rightarrow F) \\
M_1 \rightarrow Sig_x \\
(\uparrow > P_x) \\
U \rightarrow x_{Molecules}
\end{bmatrix}_{Space}
\\
\begin{bmatrix}
M_3 : -\infty \leq t \leq \infty \\
\downarrow \\
M_2 : 0 \geq t > \infty \\
\downarrow \\
M_1 : 0 > t > \infty \\
\downarrow \\
U \rightarrow \begin{matrix} t \leq E_{Cell};\ TC: M_3 \rightarrow U \\ t \sim E_{Human};\ TC: U \rightarrow M_3 \end{matrix}
\end{bmatrix}_{Time}
&
TC \rightarrow x_{Cells}
\end{matrix}
\right\| \langle x_U | x_T \rangle \Rightarrow
$$

$$\langle Space - Time - Energy - Gravity \ P - Type \ Beings \ and \ Pre - Genetic \ Code \rangle +$$

$$\left(\begin{array}{c} Electro - magnetic - wavearchetype - masspotential \ P - Type \ Being \\ and \ Pre - Genetic \ Code \end{array} \right) +$$

$$\langle Quantum - Particle \ P - Type \ Being \ and \ Pre - Genetic \ Code \rangle +$$

$$\langle Atom \ P - Type \ Being \ and \ Pre - Genetic \ Code \rangle +$$

$$\langle Molecule \ P - Type \ Being \ and \ Pre - Genetic \ Code \rangle + LSTE \ \langle ... \rangle +$$

$$Cell \ P - Type \ Being \ and \ Genetic \ Code$$

There are many likely iterations of this equation between molecules and cells and this in general is depicted by '*LSTE* $\langle ... \rangle$', where LSTE signifies iterations(s) of the Light-Space-Time Emergence equation. Also note that an assumption is made that all the previous material-fabric code-segments are intimately active at the level of cellular genetic code. This may be through the device of entanglement, or perhaps even by the material-fabric code segments being embedded in cellular genetic code, or perhaps by some protein-structures that exist in each cell that give the cell the capacity to read the pre-genetic code embedded in the material-fabric.

The emergence of a quadrumvirate cell-based structure specified by four molecular plans is significant in that the genetic code is now being housed in a post material-fabric structure. It is likely the collective and simultaneous action of nucleic acids, polysaccharides, lipids, and proteins that facilitates this post material-fabric housing ability. While in this chapter the code generated for each of these molecular plans appears to be separate, as an emergence of properties of Light they should be thought of as happening in such a manner so as to encapsulate the essential unity of the four properties.

Strictly speaking (4.2.4) should reflect the transition between the pre-genetic to genetic housing structure, and this is depicted as 'PGGT <...>', Pre-Genetic to Genetic Transition, in Equation 7.2.1, Generation of Cell P-Type Being:

$$Light - Space - Time \ Emergence_{Cell \ P-Type \ Being} =$$

$$
\left.\left[\begin{array}{c}
\left[\begin{array}{c}
c_\infty : [Pr, Po, K, H] \\
\left(\downarrow R_{C_K} = f(R_{C_\infty})\right) \\
c_K : [S_{Pr}, S_{Po}, S_K, S_H] \\
\left(\downarrow R_{C_N} = f(R_{C_K})\right) \\
c_N : f(S_{Pr} \times S_{Po} \times S_K \times S_H) \\
\left(\downarrow R_{C_U} = f(R_{C_N})\right) \\
c_U : [P, V, M, C] \\
\Uparrow \\
c_{0 : [D,W,I,C]}
\end{array}\right]_{Light}
\\
\left[\begin{array}{c}
M_3 : -\infty \leq t \leq \infty \\
\downarrow \\
M_2 : 0 \geq t > \infty \\
\downarrow \\
M_1 : 0 > t > \infty \\
\downarrow \\
U \rightarrow \begin{array}{l} t \leq E_{Cell}; TC: M_3 \rightarrow U \\ t \sim E_{Human}; TC: U \rightarrow M_3 \end{array}
\end{array}\right]_{Time}
\end{array}
\right.
\left[\begin{array}{c}
M_3 \rightarrow System_X \\
(\uparrow F \rightarrow I) \\
M_2 \rightarrow S_{System_X} \\
(\uparrow Sig \rightarrow F) \\
M_1 \rightarrow Sig_X \\
(\uparrow > P_x) \\
U \rightarrow x_{Molecules}
\end{array}\right]_{Space}
$$

$$ TC \rightarrow x_{Cells} $$

$$ \Rightarrow \quad \langle x_U | x_T \rangle $$

$\langle Space - Time - Energy - Gravity \ P - Type \ Beings \ and \ Pre - Genetic \ Code \rangle +$

$\left(\begin{array}{c} Electro - magnetic - wavearchetype - masspotential \ P - Type \ Being \\ and \ Pre - Genetic \ Code \end{array}\right) +$

$\langle Quantum - Particle \ P - Type \ Being \ and \ Pre - Genetic \ Code \rangle +$

$\langle Atom \ P - Type \ Being \ and \ Pre - Genetic \ Code \rangle +$

$\langle Molecule \ P - Type \ Being \ and \ Pre - Genetic \ Code \rangle + LSTE \ \langle ... \rangle +$

$PGGT < \cdots > + Cell \ P - Type \ Being \ and \ Genetic \ Code$

Eq 7.2.1: Generation of Cell P-Type Being

Light's Emergence as Living Cells

As discussed previously the Light-Space-Time Emergence equation (3.1.3) being iterative, can be used to model emergence as it proceeds from simpler four-fold to more complex four-fold manifestations. Hence (3.1.3) has already been applied to suggest the emergence of the space-time-energy-gravity quadrumvirate P-type beings, the electro-magnetic-wavearchetype-masspotential P-type being, the quantum-particle P-type being, the boson P-type being as a further instance of a particular kind of quantum particle, and the atom P-type being. Here it will be applied to suggest the generation of the cell P-type being as summarized by (7.2.1). But further, as implied by (3.6.12), the Potential Effect of Levels of Light on Genetic-Type Information equation, any process of deeper organization, such as is responsible for the architecture and cohesiveness of living cells, has the possibility of altering the material-fabric or a post material-fabric housing structure so long as the bases involved are driven primarily by a meta-level.

Starting with the Light-Matrix, the top left-hand matrix in (7.2.1), the first line from the top, C_∞: $[Pr, Po, K, H]$, specifies the fundamental architecture of the cell P-type being. As will be explored shortly, Nucleic Acids are an emergence of Light's property of Knowledge, Polysaccharides are an emergence of Light's property of Power, Proteins are an emergence of Light's property of Presence, and Lipids are an emergence of Light's property of Harmony. The fundamental architecture of these aspects, hence, is an emergence of the properties of Light at ∞.

Line 3 in the Light-Matrix, C_K: $[S_{Pr}, S_{Po}, S_K, S_H]$, elaborates the sets for Presence, Power, Knowledge, and Harmony, each containing multiple elements. For example, as will be explored in the section on Proteins, various elements derived from the four sets define the behavior of Proteins and could be functions such as 'exist for service', 'to bring about perfection at the level of the cell', 'extreme diligence and perseverance', amongst others, hence collectively describing Proteins' way of being. Specifically, Line 5, C_N: $f(S_{Pr} \times S_{Po} \times S_K \times S_H)$, suggests that unique seeds are created from a combination of such elements from all four sets, with a particular element leading or having more weight, that in effect creates the distinctness of proteins possible in the cell P-type being.

Line 6, ($\downarrow R_{C_U} = f(R_{C_N})$), specifies quantization between the layer where the seeds are formed, and the physical layer, and as explored in Chapter 3.5 and 3.6, will result in Line 7, C_U: $[P, V, M, C]$, hence changing the post material-fabric housing structure. Note that as in the process describing the generation of the code-segments for space-time-energy-gravity quantization at the time of the Big Bang, the FBLEE (four-base logic-encoding ecosystem) process is apparently transparent. In reality one may say that any P-type being is first a FBLEE-P-type being and then a MF-P-type being.

The possibilities represented by Lines 1 through 5 hence concretize through the quantization represented by Line 6 to become or enhance the cell P-type being with its subtle physical (related to Presence), vital (related to Power), mental (related to Knowledge), and connection (related to Harmony) aspects now existing in material reality typified by Light moving at c. Note that just as Line 6 represents a process of quantization relating the layer of reality created by Light traveling at c with the antecedent layers, so too Lines 2 and 4 as previously discussed, also represent quantization of a more rarefied kind that ultimately plays a critical part in allowing the material-fabric to express infinite diversity.

Typically, it is the process as captured by the Space-Matrix that will determine if Line 6 is activated. Specifically, patterns at the untransformed layer, U, will need to be overcome, as specified by the second line from the bottom of the Space-Matrix: $(\uparrow > P_x)$. But as specified by the bottom-line of the Time-Matrix, reproduced below, it is only with the advent of the human-system that the automaticity of the action of meta-levels is reversed:

$$U \rightarrow \begin{array}{l} \mathrm{t} \leq E_{Cell}; \mathrm{TC}: M_3 \rightarrow \mathrm{U} \\ \mathrm{t} \sim E_{Human}; \mathrm{TC}: U \rightarrow M_3 \end{array}$$

Hence in the case of the cell P-type being, which in this emergence is a pre-human system, the fact that patterns do not need to be overcome means that quantization more or less happens automatically, and that in reality the cell P-type being would therefore more accurately be termed the cell PMU-type being.

Even though the cell P-type being is a pre-human system and therefore the action of meta-levels are modeled as being automatic, it is nonetheless useful to review (3.6.12), **Potential Effect of Levels of Light on Genetic-Type Information,** to understand how the relationship with different forms of mutation has been specified:

$Potential\ Effect\ of\ Levels\ of\ Light\ on\ Genetic - Type\ Information =$
$$\begin{bmatrix} STATIC\ \langle |[L][S][T]TC \rightarrow x_T|_{\langle x_U | x_T \rangle} \rangle \\ \times \\ \left((Y > U: Z_Q) \vee (Y \leq U: Z_F) \vee (Y = U: Z_R) \right) \\ \ni \\ \left(\begin{array}{c} Z \in \mathbb{U}\ (Space, Time, Energy, Gravity) \\ Q: Quantization;\ F: Fragmentation;\ R: Random \end{array} \right) \end{bmatrix} \rightarrow h \ni$$

$$h \in \begin{pmatrix} [Q]: Constructive\ zone, \\ [Q]: Constructive\ zone \wedge Constructive\ mutation, \\ [F]: Destructive\ mutation, \\ [R]: Random\ mutation \end{pmatrix}$$

Line 1 from the top in the matrix is simply a static form of (3.6.1) the Simplified Light-Space-Time Emergence equation. The static form is designated by 'STATIC' and implies that fundamental operations true of (3.6.1) are being highlighted in (3.6.12). In other words (3.6.1) already has all the operations highlighted in (3.6.12) in it, but by 'freezing' it by making it static, the essential dynamics leading to possible mutations at the genetic level can more clearly be highlighted.

Line 1 is then subjected ($\times$) to a determination of the dominant levels of light that may be active, designated by Line 2, $\Big((Y > U: Z_Q) \vee (Y \leq U: Z_F) \vee (Y = U: Z_R) \Big)$. Unpacking this, 'Y > U' implies meta-levels are active and as a result it is possible that Z_Q is going to take place (the subscript 'Q' implies quantum-level action). This also implies activation and potential change of FBLEE. The call from below, as it were, may invoke some function that already exists in the subtle-libraries 'above', so that some already existing function may influence FBLEE through ∞-entanglement, K-entanglement, or N-entanglement. This may be thought of as a key-and-lock mechanism, where a deep enough visceral urge from below acting as the key, opens an entangled lock to alter FBLEE as per the visceral urge. 'Y $\leq$ U' implies that only the untransformed levels are active, and therefore also the sub-level where the speed of light is 0 is active and as a result Z_F is going to take place (the subscript 'F' implies 'fragmentation'). 'Y = U' implies that all levels are active and as a result Z_R is going to take place (the subscript 'R' implies 'random').

Line 3 elaborates the significance of Z_Q, Z_F, and Z_R. Hence Z is the union of potential quantum-operations of space, time, energy, and gravity, designated by 'Z $\in$ $\mathbb{U}$ ($Space, Time, Energy, Gravity$)'. But the nature of the operations, as suggested in the previous paragraph, is designated by 'Q: $Quantization$; F: $Fragmentation$; R: $Random$'. Z_Q, then, implies that the full quantization originating from updated four-base logic-encoding ecosystems (FBLEE) can take place. Z_F implies that the essential set will be fragmented and that only libraries at the level of local cellular-level DNA or precipitated material-fabric logic can potentially be altered. Z_R

251

also precludes full quantization, and that some partial local-library constructive or destructive mutation may take place.

The '→ h ∋ h ∈' segment resolves the outcome of the operations implied by Lines 1 – 3, suggesting that the outcome will be 'h' such that (∋) 'h' is an element (∈) of the set specified by the members '[Q]: *Constructive zone*', '[Q}: *Constructive zone AND Constructive mutation*', '[F}; *Destructive mutation*', and '{R}: *Random mutation*'. '[Q]: *Constructive zone*' implies that the in-built buffer has been crossed and that access to the deeper four-base logic-encoding ecosystem (FBLEE) has been granted. '[Q}' in this segment implies that there is the possibility that full-quantization as specified by Line 3 of the previous matrix can take place. Access to this zone is a prerequisite for constructive mutation to occur, as designated by the element '[Q}: *Constructive zone* ∩ *Constructive mutation*', which implies that full-quantization is going to take place and will result in material change. The '[F]' specifies the relationship between 'Fragmentation' in Line 3 of the previous matrix and destructive mutation. The '[R]' specifies the relationship between 'Random' in Line 3 of the previous matrix and random mutation.

But as just summarized in the Time-Matrix in (7.2.1) Y is by definition greater than U and hence quantization is automatic. In terms of the cell P-type being such quantization implies that wholeness becomes fully active through specific space, time, energy, and gravity quantization to create an holistic "ecosystem" with its own "cell P-type logic" as it were. The wholeness has now precipitated into the genetic code in the cellular structure. Note that such precipitation is captured by the action of 'PGGT <...>' that can be assumed to be triggered by Z_Q. This "logic" or more specifically genetic code is elaborated in the following sub-sections and suggests the increasing sphere of influence of the cell P-type being that now also contains light-engendered laws for all cells.

Generation of 'Nucleic Acids' Aspect of the Cell P-Type Being

Nucleic acids basically encode information. They store and transmit the genome, the hereditary information needed to keep the cell alive. They function as the cell's librarians and contain information on how to make proteins and when to make them.

They are, hence, the keepers of a cell's knowledge, its wisdom, its ability to make laws, the vehicle to spread knowledge within cells and to the next generation of cells. Being so, one can see that there is similarity with the set for system-knowledge highlighted earlier in Chapter 2.3. Reproducing Equation 2.3.3:

$$S_{System_K} \ni [Wisdom, Law\ Making, Spread\ of\ Knowledge\ ...]$$

Nucleic acids can therefore be thought of as a precipitation of system-knowledge at the cellular level.

Hence, a nucleic acid will have a generalized signature, as in Equation 7.2.2, derived from the system-knowledge family:

$$Sig_{nucleic\ acid} = Xa + \overline{Yb_{0-n}}\ \ where \left[\begin{array}{c} X \in [S_{System_K}] \\ Y \in [S_{System_{Pr}}, S_{System_P}, S_{System_K}, S_{System_N}] \\ a, b\ are\ integers; a > b \end{array} \right]$$

Eq 7.2.2: Generalized Nucleic Acid Signature of the Cell P-Type Being

The primary element X could be an attribute or function such as 'keeper of knowledge'. Secondary elements Y could be 'protein laws', 'generational knowledge', amongst others. The collectivity of elements as per the equation would specify the character of a nucleic acid.

Note that (7.2.2) already implies that Lines 1 – 5 in the Light Matrix (7.2.1) have been activated, and that the logic of the nucleic-acid-ecosystem will precipitate into the post material-fabric cell-based genetic structure through the action of Line 6-7 of (7.2.1). This action of precipitating into the cell-based structure as opposed to the material-fabric is designated by 'PGGT <...>' as specified in (7.2.1).

DNA and RNA, two types of nucleic acids, would hence have the equations as specified by Equation 7.2.3 and 7.2.4 respectively.

$$Sig_{DNA} = Xa + \overline{Yb_{0-n}}\ \ where \left[\begin{array}{c} X \in [S_{System_K}] \\ Y \in [S_{System_{Pr}}, S_{System_P}, S_{System_K}, S_{System_N}] \\ a, b\ are\ integers; a > b \end{array} \right]$$

Eq 7.2.3: DNA

$$Sig_{RNA} = Xa + \overline{Yb_{0-n}} \quad where \quad \begin{bmatrix} X \in [S_{System_K}] \\ Y \in [S_{System_{Pr}}, S_{System_P}, S_{System_K}, S_{System_N}] \\ a, b \ are \ integers; a > b \end{bmatrix}$$

Eq 7.2.4: RNA

The primary element X in (7.2.3) and (7.2.4) would be the same as that for nucleic acids (7.2.2). The secondary elements Y however will be a larger and more specific set with many elements in common with (7.2.2). Note also that the genetic code as specified by the following sub-sections on proteins, polysaccharides, and lipids, would be translated into DNA-format and stored in nucleic acids as specified by (7.2.3).

Generation of 'Protein' Aspect of the Cell P-Type Being

Proteins are the cells work-horses. Look anywhere in a cell and one will see proteins at work. Proteins are built in thousands of shapes and sizes, each performing a different function. As Goodsell describes, "some are built simply to adopt a defined shape, assembling into rods, nets, hollow spheres, and tubes. Some are molecular motors, using energy to rotate, or flex, or crawl. Many are chemical catalysts that perform chemical reactions atom-by-atom, transferring and transforming chemical groups exactly as needed." With their wide potential for diversity, proteins are constructed to perform most of the everyday tasks of the cells. In fact human cells build around 30,000 different kinds of proteins to execute on the diverse array of cellular level tasks.

Proteins, hence, exist for service, to bring about perfection at the level of the cell, are characterized by extreme diligence and perseverance, and so on. Being so, one can see that there is similarity with the set for system-presence highlighted earlier in Chapter 2.3. Reproducing Equation 2.3.1:

$$S_{System_{Pr}} \ni [Service, Perfection, Diligence, Perseverance, ...]$$

Proteins can therefore be thought of as a precipitation of system-presence at the cellular level.

Hence, a protein could have a generalized signature, as in Equation 7.2.5, derived from the system-presence family:

$$Sig_{protein} = Xa + \overline{Yb_{0-n}} \quad where \begin{bmatrix} X \in [S_{System_{Pr}}] \\ Y \in [S_{System_{Pr}}, S_{System_{P}}, S_{System_{K}}, S_{System_{N}}] \\ a, b \ are \ integers; a > b \end{bmatrix}$$

Eq 7.2.5: Generalized Protein Signature of the Cell P-Type Being

This could yield a vast number of functional proteins. In fact, it may be possible that the 30,000 or so known proteins created by the human cell could each be specified by a signature equation of this nature. It may be possible to map existing proteins to functionality as suggested by the four sets of molecular plans.

Note that (7.2.5) already implies that Lines 1 – 5 in the Light Matrix (7.2.1) have been activated, and that the logic of the protein-ecosystem will precipitate into the post material-fabric cell-based genetic structure through the action of Line 6-7 of (7.2.1). This action of precipitating into the cell-based structure as opposed to the material-fabric is designated by 'PGGT <...>' as specified in (7.2.1).

Consider Insulin, for example. Insulin regulates the metabolism of carbohydrates, fats and protein by promoting the absorption of, especially, glucose from the blood into fat, liver and skeletal muscle cells. Equation 7.2.6 for Insulin would hence be:

$$Sig_{insulin} = Xa + \overline{Yb_{0-n}} \quad where \begin{bmatrix} X \in [S_{System_{Pr}}] \\ Y \in [S_{System_{Pr}}, S_{System_{P}}, S_{System_{K}}, S_{System_{N}}] \\ a, b \ are \ integers; a > b \end{bmatrix}$$

Eq 7.2.6: Insulin

The primary element X could be an attribute or function such as 'workhorse'. Secondary elements Y could be 'metabolic regulation', 'glucose absorption', 'blood to fat channel', amongst others. The collectivity of elements as per the equation would specify the character of insulin.

Consider Histones as another example. They are the chief protein components of chromatin, acting as spools around which DNA winds, and playing a role in gene regulation. Without histones, the unwound DNA in chromosomes would be very long (a length to width ratio of more than 10 million to 1 in human DNA). Equation 7.2.7 for Histones would hence be:

$$Sig_{histones} = Xa + \overline{Yb_{0-n}} \quad where \begin{bmatrix} X \in [S_{System_{Pr}}] \\ Y \in [S_{System_{Pr}}, S_{System_P}, S_{System_K}, S_{System_N}] \\ a, b \ are \ integers; a > b \end{bmatrix}$$

Eq 7.2.7: Histones

The secondary element Y would have elements such as 'gene regulation', 'unwound DNA management', amongst others.

Generation of 'Lipids' Aspect of the Cell P-Type Being

Lipids by themselves are tiny molecules, but when grouped together form the largest structures of the cell. When placed in water, lipid molecules aggregate to form huge waterproof sheets. These sheets easily form boundaries at multiple levels and allow concentrated interactions and work to be performed within a cell. Hence, the nucleus and the mitochondria are contained within lipid-defined compartments. Similarly, each cell itself is contained within a lipid-defined boundary.

Lipids are therefore promoters of relationship, of harmony in the cell, of nurturing the cell-level division of labor, of allowing specialization and uniqueness to emerge, hence perhaps of earlier forms of compassion and love, and so on. The notion of such early forms of compassion is consistent with the biologist's perspective that at some point a gene for compassion was developed in pre-human species (Wright, 2009). Being so, one can see that there is similarity with the set for system-nurturing highlighted earlier in Chapter 2.3. Reproducing Equation 2.3.4:

$$S_{System_N} \ni [Love, Compassion, Harmony, Relationship \ ...]$$

This function of harmonization suggests that lipids can therefore be thought of as a precipitation of system-nurturing at the cellular level.

Lipids could have a generalized signature, as in Equation 7.2.8, derived from the system-nurturing family:

$$Sig_{lipid} = Xa + \overline{Yb_{0-n}} \quad where \begin{bmatrix} X \in [S_{System_N}] \\ Y \in [S_{System_{Pr}}, S_{System_P}, S_{System_K}, S_{System_N}] \\ a, b \ are \ integers; a > b \end{bmatrix}$$

Eq 7.2.8: Generalized Lipid Signature of the Cell P-Type Being

Note that (7.2.8) already implies that Lines 1 – 5 in the Light Matrix (7.2.1) have been activated, and that the logic of the lipid-ecosystem will precipitate into the post material-fabric cell-based genetic structure through the action of Line 6-7 of (7.2.1). This action of precipitating into the cell-based structure as opposed to the material-fabric is designated by 'PGGT <…>' as specified in (7.2.1).

Specific lipids such as monoglycerides and phospholipids could have the following equations:

$$Sig_{monoglyceride} = Xa +$$

$$\overline{Yb_{0-n}} \quad where \begin{bmatrix} X \in [S_{System_N}] \\ Y \in [S_{System_{Pr}}, S_{System_P}, S_{System_K}, S_{System_N}] \\ a, b \ are \ integers; a > b \end{bmatrix}$$

Eq 7.2.9: Monoglyceride

$$Sig_{phospholipids} = Xa +$$

$$\overline{Yb_{0-n}} \quad where \begin{bmatrix} X \in [S_{System_N}] \\ Y \in [S_{System_{Pr}}, S_{System_P}, S_{System_K}, S_{System_N}] \\ a, b \ are \ integers; a > b \end{bmatrix}$$

Eq 7.2.10: Phospholipids

The primary element X shared by each of the lipids could be an attribute or function such as 'compartmentalization'. Secondary elements Y could be of the nature of 'work breakdown', 'intra-cell love', amongst others, and would vary with each different kind of lipid.

Generation of 'Polysaccharide' Aspect of the Cell P-Type Being

Polysaccharides are long, often branched chains of sugar molecules. Sugars are covered with hydroxyl groups, which associate to form storage containers. As a result, polysaccharides function as the storehouse of cell's energy. In addition, polysaccharides are also used to build some of the most durable biological structures. The stiff shell of insects, for example are made of long polysaccharides.

Polysaccharides function to create energy, power, courage, strength thereby readying the cell for adventure, and so on. Being so, one can see that there is similarity with the set for system-power highlighted previously in Chapter 2.3. Reproducing Equation 2.3.2:

$$S_{System_P} \ni [Power, Courage, Adventure, Justice, ...]$$

Providing energy and strength, polysaccharides can be thought of as a precipitation of system-power at the cellular level.

Polysaccharides could have a generalized signature, as in Equation 7.2.11, derived from the system-power family:

$$Sig_{polysaccharide} = Xa +$$
$$\overline{Yb_{0-n}} \quad where \left[\begin{array}{c} X \in [S_{System_P}] \\ Y \in [S_{System_{Pr}}, S_{System_P}, S_{System_K}, S_{System_N}] \\ a, b \ are \ integers; a > b \end{array} \right]$$

Eq 7.2.11: Generalized Polysaccharide Signature of the Cell P-Type Being

Note that (7.2.11) already implies that Lines 1 – 5 in the Light Matrix (7.2.1) have been activated, and that the logic of the polysaccharide-ecosystem will precipitate into the post material-fabric cell-based genetic structure through the action of Line 6-7 of (7.2.1). This action of precipitating into the cell-based structure as opposed to the material-fabric is designated by 'PGGT <...>' as specified in (7.2.1).

Glycogen is an example of a polysaccharide. Glycogen forms an energy reserve that can be quickly mobilized to meet a sudden need for glucose, but one that is less compact and more immediately available as an energy reserve than say triglycerides. Equation 7.2.12 for Glycogen follows:

$$Sig_{glycogen} = Xa + \overline{Yb_{0-n}} \quad where \left[\begin{array}{c} X \in [S_{System_P}] \\ Y \in [S_{System_{Pr}}, S_{System_P}, S_{System_K}, S_{System_N}] \\ a, b \ are \ integers; a > b \end{array} \right]$$

Eq 7.1.12: Glycogen

Cellulose is another example of a polysaccharide. Cellulose is a polymer made with repeated glucose units bonded together by beta-linkages. Humans and many animals lack an enzyme to break the beta-linkages, so they do not digest cellulose. Equation 7.2.13 for Cellulose follows:

$$Sig_{cellulose} = Xa + \overline{Yb_{0-n}} \quad where \begin{bmatrix} X \in [S_{System_P}] \\ Y \in [S_{System_{Pr}}, S_{System_P}, S_{System_K}, S_{System_N}] \\ a, b \ are \ integers; a > b \end{bmatrix}$$

Eq 7.2.13: Cellulose

The primary element for the preceding polysaccharides X would be along the lines of 'energy storage'. The secondary elements may vary, with a Y element for Glycogen being 'rapid energy deployment' for example, and a Y element for Cellulose being 'bonded energy', for example.

Summary of the Cell P-Type Being and Genetic Code

Summarizing, after the cellular stage iterations of the Light-Space-Time Emergence equation are complete, the following code-segments will have been generated as specified by Equation 7.2.14, Active Cell P-Type Being and Genetic Code:

$$Light - Space - Time \ Emergence_{Cell \ P-Type \ Being} =$$

$$\begin{Vmatrix} \begin{bmatrix} c_\infty: [Pr, Po, K, H] \\ \left(\downarrow R_{C_K} = f\left(R_{C_\infty}\right) \right) \\ c_K: [S_{Pr}, S_{Po}, S_K, S_H] \\ \left(\downarrow R_{C_N} = f\left(R_{C_K}\right) \right) \\ c_N: f(S_{Pr} \times S_{Po} \times S_K \times S_H) \\ \left(\downarrow R_{C_U} = f\left(R_{C_N}\right) \right) \\ c_U: [P, V, M, C] \\ \Uparrow \\ c_{0:[D,W,I,C]} \end{bmatrix}_{Light} \begin{bmatrix} M_3 \rightarrow System_X \\ (\uparrow F \rightarrow I) \\ M_2 \rightarrow S_{System_X} \\ (\uparrow Sig \rightarrow F) \\ M_1 \rightarrow Sig_x \\ (\uparrow > P_x) \\ U \rightarrow x_{Molecules} \end{bmatrix}_{Space} \\ \begin{bmatrix} M_3 : -\infty \leq t \leq \infty \\ \downarrow \\ M_2 : 0 \geq t > \infty \\ \downarrow \\ M_1 : 0 > t > \infty \\ \downarrow \\ U \rightarrow \begin{matrix} t \leq E_{Cell}; TC: M_3 \rightarrow U \\ t \sim E_{Human}; TC: U \rightarrow M_3 \end{matrix} \end{bmatrix}_{Time} \quad TC \rightarrow x_{Cells} \end{Vmatrix} \Rightarrow$$

$$\langle x_U | x_T \rangle$$

$\langle Space - Time - Energy - Gravity\ P - Type\ Beings\ and\ Pre - Genetic\ Code \rangle +$

$\left(\begin{matrix} Electro - magnetic - wavearchetype - masspotential\ P - Type\ Being \\ and\ Pre - Genetic\ Code \end{matrix} \right) +$

$\langle Quantum - Particle\ P - Type\ Being\ and\ Pre - Genetic\ Code \rangle +$

$\langle Atom\ P - Type\ Being\ and\ Pre - Genetic\ Code \rangle +$

$\langle Molecule\ P - Type\ Being\ and\ Pre - Genetic\ Code \rangle + LSTE \langle ... \rangle +$

$$PGGT < \cdots > + \left(\begin{array}{c} \sum Sig_{lipid} = Xa + \overline{Yb_{0-n}} \\[6pt] where \left[\begin{array}{c} X \in [S_{System_N}] \\ Y \in [S_{System_{Pr}}, S_{System_P}, S_{System_K}, S_{System_N}] \\ a, b \ are \ integers; a > b \end{array} \right] \\[18pt] \sum Sig_{nucleic \ acid} = Xa + \overline{Yb_{0-n}} \\[6pt] where \left[\begin{array}{c} X \in [S_{System_K}] \\ Y \in [S_{System_{Pr}}, S_{System_P}, S_{System_K}, S_{System_N}] \\ a, b \ are \ integers; a > b \end{array} \right] \\[18pt] \sum Sig_{polysaccharide} = Xa + \overline{Yb_{0-n}} \\[6pt] where \left[\begin{array}{c} X \in [S_{System_P}] \\ Y \in [S_{System_{Pr}}, S_{System_P}, S_{System_K}, S_{System_N}] \\ a, b \ are \ integers; a > b \end{array} \right] \\[18pt] \sum Sig_{protein} = Xa + \overline{Yb_{0-n}} \\[6pt] where \left[\begin{array}{c} X \in [S_{System_{Pr}}] \\ Y \in [S_{System_{Pr}}, S_{System_P}, S_{System_K}, S_{System_N}] \\ a, b \ are \ integers; a > b \end{array} \right] \end{array} \right)$$

Eq. 7.2.14, *Active Cell P-Type Being and Genetic Code*

Note that the final code segment following 'PGGT <...>' contains the code for all possible ($\sum x$) lipids, nucleic acids, polysaccharides, and proteins, respectively. As mentioned previously in the sub-section on nucleic acids, this final code segment would also be contained in the nucleic acid segment in some form.

Each equation-segment in (7.2.14) will generate a vast body of "code". As a reminder the language of this code is essentially four-fold, deriving from properties of light, and function-based, and depicts the growing biography by which the singular light-based edifice expresses its materialization, and at this stage, through the cell P-type being.

Chapter 7.3: Generation of Fundamental-Capacities-of-Self P-Type Being

As human beings we experience sensations, urges and desires and wills, feelings and emotions, and thought. These are key aspects of our being and becoming and critical aspects of how choice at both the individual and collective levels may be determined. Further, there is known to be a tight relationship between these fundamental capacities of being and the effect on bodies. This implies that there is a deeply embedded relationship between these capacities and the actual functioning of bodies and cells.

This chapter suggests that this deeply embedded relationship is founded on genetic code that is created with the emergence of these capacities. This chapter, hence, will go over the quantum-level computation that suggests how light emerges as these capacities, how these precipitate into the post material-fabric genetic housing structure, and what the specific equation-segment function-based code manifests as.

Note also that in the view being elaborated in this book, the human being is a not static but fundamentally dynamic. In that dynamism different kinds of beings, commonly P-type beings but not exclusively so, can integrate with existing P-type beings already operating in the human being, to further its abilities and capacities. Such enhancement is the basis of transhumanism, and this chapter illustrates the process by which this happens.

Light's Emergence as Fundamental Capacities of Self

As discussed previously the Light-Space-Time Emergence equation (3.1.3) being iterative, can be used to model emergence as it proceeds from simpler four-fold to more complex four-fold manifestations. Hence (3.1.3) has already been applied to suggest the emergence of the space-time-energy-gravity quadrumvirate P-type beings, the electro-magnetic-wavearchetype-masspotential P-type being, the quantum-particle P-type being, the boson P-type being as a further instance of a particular kind of quantum-particle P-type being, the atom P-type being, and the cell P-type being.

Here, a revised form of it building off the PGGT form of the generation of cell P-type being (7.2.1) will be applied to suggest the generation of the fundamental-capacities-of-self P-type being, Equation 7.3.1, Generation of

Fundamental-Capacities-of-Self P-Type Being. The architecture and details of the fundamental-capacities-of-self P-type being can be seen to be the result of the application of the Light, Space, and Time matrices as will be elaborated. But further, as implied by (3.6.12), the Potential Effect of Levels of Light on Genetic-Type Information equation, any process of deeper organization, such as is responsible for the architecture and cohesiveness of the fundamental-capacities-of-self P-type being, has the possibility of altering the material-fabric or a post material-fabric housing structure so long as the bases involved are driven primarily by a meta-level.

Hence:

$Light - Space - Time\ Emergence_{Fundamental-Capacities-of-Self\ P-Type\ Being} =$

$$
\left|\begin{array}{ll}
\left[\begin{array}{l}
c_\infty: [Pr, Po, K, H] \\
\left(\downarrow R_{C_K} = f(R_{C_\infty})\right) \\
c_K: [S_{Pr}, S_{Po}, S_K, S_H] \\
\left(\downarrow R_{C_N} = f(R_{C_K})\right) \\
c_N: f(S_{Pr} \times S_{Po} \times S_K \times S_H) \\
\left(\downarrow R_{C_U} = f(R_{C_N})\right) \\
c_U: [P, V, M, C] \\
\qquad \Uparrow \\
\qquad c_{0:[D,W,I,C]}
\end{array}\right]_{Light}
&
\left[\begin{array}{l}
M_3 \rightarrow System_X \\
(\uparrow F \rightarrow I) \\
M_2 \rightarrow S_{System_X} \\
(\uparrow Sig \rightarrow F) \\
M_1 \rightarrow Sig_X \\
(\uparrow > P_x) \\
U \rightarrow x_{Cells}
\end{array}\right]_{Space} \\[3em]
\left[\begin{array}{l}
M_3: -\infty \le t \le \infty \\
\qquad \downarrow \\
M_2: 0 \ge t > \infty \\
\qquad \downarrow \\
M_1: 0 > t > \infty \\
\qquad \downarrow \\
U \rightarrow \begin{array}{l} t \le E_{Cell}; TC: M_3 \rightarrow U \\ t \sim E_{Human}; TC: U \rightarrow M_3 \end{array}
\end{array}\right]_{Time}
&
\begin{array}{l}
TC \rightarrow x_{Capacities\ of\ Self} \\[2em]
\langle x_U | x_T \rangle
\end{array}
\end{array}\right| \Rightarrow
$$

$\langle Space - Time - Energy - Gravity\ P - Type\ Beings\ and\ Pre - Genetic\ Code \rangle +$

$\left(\begin{array}{c} Electro - magnetic - wavearchetype - masspotential\ P - Type\ Being \\ and\ Pre - Genetic\ Code \end{array}\right) +$

$\langle Quantum - Particle\ P - Type\ Being\ and\ Pre - Genetic\ Code \rangle +$

$\langle Atom\ P - Type\ Being\ and\ Pre - Genetic\ Code \rangle +$

$\langle Molecule\ P-Type\ Being\ and\ Pre-Genetic\ Code \rangle + LSTE \langle ... \rangle +$

$PGGT < \cdots > + \langle Cell\ P-Type\ Being\ and\ Genetic\ Code \rangle$

$+LSTE < \cdots > +Fundamental-Capacities-of-Self\ P-Type\ Being\ and\ Genetic\ Code$

Eq 7.3.1:   Generation of Fundamental-Capacities-of-Self P-Type Being and Genetic Code

Starting with the Light-Matrix, the top left-hand matrix in (7.3.1), the first line from the top, C_∞: $[Pr, Po, K, H]$, specifies the architecture of the fundamental-capacities-of-self P-type being to be introduced in this chapter.   Hence, Thoughts are an emergence of Light's property of Knowledge, Urges, Desires, and Wills are an emergence of Light's property of Power, Sensations are an emergence of Light's property of Presence, and Feelings and Emotions are an emergence of Light's property of Harmony. The fundamental architecture of these aspects, hence, is an emergence of the properties of Light at ∞.

Line 3 in the Light-Matrix, C_K: $[S_{Pr}, S_{Po}, S_K, S_H]$, elaborates the sets for Presence, Power, Knowledge, and Harmony, each containing multiple elements. For example, as will be explored in the section on Sensations, various elements derived from the four sets define the behavior of Sensations and could be functions such as 'tangible', 'take notice of', amongst others, hence collectively describing Sensations' way of being. Specifically, Line 5, $C_{N:}$ $f(S_{Pr}\ x\ S_{Po}\ x\ S_K\ x\ S_H)$, suggests that unique seeds are created from a combination of such elements from all four sets, with a particular element leading or having more weight, that in effect creates the distinctness possible at the level of sensations in the fundamental-capacities-of-self P-type being.

Line 6, ($\downarrow R_{C_U} = f(R_{C_N})$), specifies quantization between the layer where the seeds are formed, and the physical layer, and as explored in Chapter 3.5 and 3.6, will result in Line 7, C_U: $[P, V, M, C]$, hence changing the post material-fabric housing structure. Note that as in the process describing the generation of the code-segments for space-time-energy-gravity quantization at the time of the Big Bang, the FBLEE (four-base logic-encoding ecosystem) process is apparently transparent. In reality one may say that any P-type being is first a FBLEE-P-type being and then a MF-P-type being.

264

The possibilities represented by Lines 1 through 5 hence concretize through the quantization represented by Line 6 to become or further enhance the capacities of self with its subtle physical (related to Presence), vital (related to Power), mental (related to Knowledge), and connection (related to Harmony) aspects now existing in material reality typified by Light moving at c.   Note that just as Line 6 represents a process of quantization relating the layer of reality created by Light traveling at c with the antecedent layers, so too Lines 2 and 4 as previously discussed, also represent quantization of a more subtle kind that ultimately plays a critical part in allowing the material-fabric to express infinite diversity.

Typically, it is the process as captured by the Space-Matrix that will determine if Line 6 is activated.   Specifically, patterns at the untransformed layer, U, will need to be overcome, as specified by the second line from the bottom of the Space-Matrix: $(\uparrow > P_x)$. But as specified by the bottom-line of the Time-Matrix, reproduced below, it is only with the advent of the human-system that the automaticity of the action of meta-levels is reversed:

$$U \rightarrow \begin{array}{l} t \leq E_{Cell}; TC: M_3 \rightarrow U \\ t \sim E_{Human}; TC: U \rightarrow M_3 \end{array}$$

Hence in the case of the fundamental-capacities-of-self P-type being, which in this emergence is a post-human system, the fact that patterns do need to be overcome means that quantization requires effort to happen. Given this, it is useful to review **Equation 3.6.5, Potential Effect of Levels of Light on Genetic-Type Information:**

$$Potential\ Effect\ of\ Levels\ of\ Light\ on\ Genetic - Type\ Information =$$

$$\begin{bmatrix} STATIC\ \langle|[L][S][T]TC \rightarrow x_T|_{\langle x_U|x_T\rangle}\rangle \\ \times \\ \left((Y > U: Z_Q) \vee (Y \leq U: Z_F) \vee (Y = U: Z_R)\right) \\ \ni \\ \left(\begin{array}{c} Z \in \mathbb{U}\ (Space, Time, Energy, Gravity) \\ Q: Quantization; F: Fragmentation; R: Random \end{array} \right) \end{bmatrix} \rightarrow h \ni$$

$$h \in \left(\begin{array}{c} [Q]: Constructive\ zone, \\ [Q]: Constructive\ zone \wedge Constructive\ mutation, \\ [F]: Destructive\ mutation, \\ [R]: Random\ mutation \end{array} \right)$$

Line 1 from the top in the matrix is simply a static form of (3.6.1) the Simplified Light-Space-Time Emergence equation. The static form is designated by 'STATIC' and implies that fundamental operations true of (3.6.1) are being highlighted in (3.6.12). In other words (3.6.1) already has all the operations highlighted in (3.6.12) in it, but by 'freezing' it by making it static, the essential dynamics leading to possible mutations at the genetic level can more clearly be highlighted.

Line 1 is then subjected ($\times$) to a determination of the dominant levels of light that may be active, designated by Line 2, '$\big((Y > U: Z_Q) \vee (Y \leq U: Z_F) \vee (Y = U: Z_R)\big)$'. Unpacking this, '$Y > U$' implies meta-levels are active and as a result it is possible that Z_Q is going to take place (the subscript 'Q' implies quantum-level action). This also implies activation and potential change of FBLEE. The call from below, as it were, may invoke some function that already exists in the subtle-libraries 'above', so that some already existing function may influence FBLEE through ∞-entanglement, K-entanglement, or N-entanglement. This may be thought of as a key-and-lock mechanism, where a deep enough visceral urge from below acting as the key, opens an entangled lock to alter FBLEE as per the visceral urge. '$Y \leq U$' implies that only the untransformed levels are active, and therefore also the sub-level where the speed of light is 0 is active and as a result Z_F is going to take place (the subscript 'F' implies 'fragmentation'). '$Y = U$' implies that all levels are active and as a result Z_R is going to take place (the subscript 'R' implies 'random').

Line 3 elaborates the significance of Z_Q, Z_F, and Z_R. Hence Z is the union of potential quantum-operations of space, time, energy, and gravity, designated by '$Z \in \mathbb{U}\,(Space, Time, Energy, Gravity)$'. But the nature of the operations, as suggested in the previous paragraph, is designated by 'Q: Quantization; F: Fragmentation; R: Random'. Z_Q, then, implies that the full quantization originating from updated four-base logic-encoding ecosystems (FBLEE) can take place, and will result in lasting material change at the genetic level. Z_F implies that the essential set will be fragmented and that only libraries at the level of local cellular-level DNA or can potentially be altered. Z_R also precludes full quantization, and that some partial cellular-level DNA constructive or destructive mutation may take place.

The '→ h ∋ h ∈' segment resolves the outcome of the operations implied by Lines 1 – 3, suggesting that the outcome will be 'h' such that (∋) 'h' is an element (∈) of the set specified by the members '[Q]: *Constructive zone*', '[Q]: *Constructive zone AND Constructive mutation*', '[F]; *Destructive mutation*', and '{R}: *Random mutation*'. '[Q]: *Constructive zone*' implies that the in-built buffer has been crossed and that access to the deeper four-base logic-encoding ecosystem (FBLEE) has been granted. '[Q]' in this segment implies that there is the possibility that full-quantization as specified by Line 3 of the previous matrix can take place. Access to this zone is a prerequisite for constructive mutation to occur, as designated by the element '[Q]: *Constructive zone* ∩ *Constructive mutation*', which implies that full-quantization is going to take place and will result in material change. The '[F]' specifies the relationship between 'Fragmentation' in Line 3 of the previous matrix and destructive mutation. The '[R]' specifies the relationship between 'Random' in Line 3 of the previous matrix and random mutation.

But as just summarized in the Time-Matrix in (7.3.1) Y is by definition not greater than U and hence quantization is not automatic. In terms of the fundamental-capacities-of-self P-type being, such quantization implies that increasing wholeness can become fully active through specific space, time, energy, and gravity quantization to create an holistic "ecosystem" with its own "fundamental-capacities-of-self P-type logic" as it were. Possible precipitation into the post material-fabric housing structure is captured by the action of 'PGGT <…>' that can be assumed to be triggered by Z_Q. This "logic" or more specifically genetic code is elaborated in the following sub-sections and suggests the increasing sphere of influence of the fundamental-capacities-of-self P-type being that now also contains light-engendered laws for all selves.

Generation of 'Sensation' Aspect of the Fundamental-Capacities-of-Self P-Type Being

Sensations are those things we experience with our senses. We see things, hear things, and smell things, taste things, can touch things. This ability to enter into relationship with objects through sensation is nothing other than a result of the emergence of Light's property of Presence. We become present to Presence through the device of sensation. Sensation can be thought of as the means by which this property of Light – Presence - molds or ingrains itself in us as human beings. Its potentiality, all which is

267

contained in this aspect of Light, becomes available to us through the power of sensation. Hence an equation, Equation 7.3.2, will generally represent the family of sensations. Some elements that it would comprise of may be 'tangible', 'take notice of', amongst others.

$$Sig_{sensation} = Xa + \overline{Yb_{0-n}} \quad where \quad \begin{bmatrix} X \in [S_{System_{Pr}}] \\ Y \in [S_{System_{Pr}}, S_{System_P}, S_{System_K}, S_{System_N}] \\ a, b \ are \ integers; a > b \end{bmatrix}$$

Eq 7.3.2: Generalized Sensation Signature of the Fundamental-Capacities-of-Self P-Type Being

There could also be equations for hearing, seeing, tasting, touching, and smelling.

But there is also a deeper experience of sensation that is possible. When we see things, for instance, what are we seeing? Is it just the surface rendering of the play of matter, or do we see that the fullness of Light is still there, with all its potentiality and possibility, in the smallest thing we look at? Do we see that the whole universe and more is present in all its fullness in the least thing that we easily ignore, or belittle, or loathe? When we touch things is it the seeming concreteness of the play of the particles or atoms or chains of molecules that we touch? Or is it the Love and Light and the vastness of all that IS that allows itself to be as a small corner that we touch so as to make infinity be felt by something so finite?

Such a deeper contact offered through sensation suggests a subset of (7.3.2) with secondary elements perhaps described as 'fullness of Light', 'contacting infinity', amongst others, thus also yielding an equation form (7.3.3):

$$Sig_{deeper-sensation} = Xa +$$

$$\overline{Yb_{0-n}} \quad where \quad \begin{bmatrix} X \in [S_{System_{Pr}}] \\ Y \in [S_{System_{Pr}}, S_{System_P}, S_{System_K}, S_{System_N}] \\ a, b \ are \ integers; a > b \end{bmatrix}$$

Eq 7.3.3: Generalized Deeper Sensation Signature of the Fundamental-Capacities-of-Self P-Type Being

Note that (7.3.2 and 7.3.3) already implies that Lines 1 – 5 in the Light Matrix (7.3.1) have been activated, and that the logic of the sensations-ecosystem will precipitate into the post material-fabric cell-based genetic structure through the action of Line 6-7 of (7.3.1). This action of precipitating into the cell-based structure as opposed to the material-fabric is designated by 'PGGT <...>' as specified in (7.3.1).

Generation of 'Urges, Desires & Wills' Aspect of the Fundamental-Capacities-of-Self P-Type Being

Urges and desires and wills are similarly a play of the emergence of Light's property of Power. In the mystery of focus, the vastness of Light has projected itself in us into an apparent smallness that is in reality everything that is. And this smallness is trying through urge and desire and will to connect viscerally or even intentionally to other smallnesses that similarly are nothing other than the fullness of Light projected into a small smorgasbord of selected function. So the urge or desire for food, or companionship, or of possession, or of climbing a peak, is nothing other than Light's compressed property of Power, trying to reach more of the fullness that it is through a fulfillment of the urge or desire or will that it masquerades as. Hence urges can be represented as Equation 7.3.4:

$$Sig_{urges} = Xa + \overline{Yb_{0-n}} \quad where \quad \begin{bmatrix} X \in [S_{System_P}] \\ Y \in [S_{System_{Pr}}, S_{System_P}, S_{System_K}, S_{System_N}] \\ a, b \ are \ integers; a > b \end{bmatrix}$$

Eq 7.3.4: Generalized Urges Signature of the Fundamental-Capacities-of-Self P-Type Being

Elements may be of the type of 'grasp', 'possess', 'deeply connect', amongst others.

Note that (7.3.4) already implies that Lines 1 – 5 in the Light Matrix (7.3.1) have been activated, and that the logic of the urges-ecosystem will precipitate into the post material-fabric cell-based genetic structure through the action of Line 6-7 of (7.3.1). This action of precipitating into the cell-based structure as opposed to the material-fabric is designated by 'PGGT <...>' as specified in (7.3.1).

Generation of 'Feelings & Emotions' Aspect of the Fundamental-Capacities-of-Self P-Type Being

Feelings and emotions are a play of the emergence of Light's property of Harmony or Nurturing. Its instrument is the Heart and it generates an array of emotions that are an indication or active radar of whether we are moving toward or away from a reality of harmony, whether based on our small self or some larger Self of Light. Gradually, by navigating with these emotions and feelings we can get to a state where we always feel positive emotions which basically means we have more truly entered into relationship with some larger continent of Light. An equation for feelings is as represented by Equation 7.3.5:

$$Sig_{feelings} = Xa + \overline{Yb_{0-n}} \quad where \begin{bmatrix} X \in [S_{System_N}] \\ Y \in [S_{System_{Pr}}, S_{System_P}, S_{System_K}, S_{System_N}] \\ a, b \ are \ integers; a > b \end{bmatrix}$$

Eq 7.3.5: Generalized Feeling Signature of the Fundamental-Capacities-of-Self P-Type Being

Note that (7.3.5) already implies that Lines 1 – 5 in the Light Matrix (7.3.1) have been activated, and that the logic of the feelings-ecosystem will precipitate into the post material-fabric cell-based genetic structure through the action of Line 6-7 of (7.3.1). This action of precipitating into the cell-based structure as opposed to the material-fabric is designated by 'PGGT <...>' as specified in (7.3.1).

Generation of 'Thought' Aspect of the Fundamental-Capacities-of-Self P-Type Being

Thoughts are a play of the emergence of Light's property of Knowledge. Through the thought we can become greater or conceptualize things greater or begun to enter into relationship with some things other than our small self. Thought allows us to connect to more "othernesses" or even the oneness of the reality of Light. An equation for thoughts is the following:

$$Sig_{thoughts} = Xa + \overline{Yb_{0-n}} \quad where \begin{bmatrix} X \in [S_{System_K}] \\ Y \in [S_{System_{Pr}}, S_{System_P}, S_{System_K}, S_{System_N}] \\ a, b \ are \ integers; a > b \end{bmatrix}$$

Eq 7.3.6: Generalized Thought Signature of the Fundamental-Capacities-of-Self P-Type Being

Note that (7.3.5) already implies that Lines 1 – 5 in the Light Matrix (7.3.1) have been activated, and that the logic of the thoughts-ecosystem will precipitate into the post material-fabric cell-based genetic structure through the action of Line 6-7 of (7.3.1). This action of precipitating into the cell-based structure as opposed to the material-fabric is designated by 'PGGT <...>' as specified in (7.3.1).

Summary of the Fundamental-Capacities-of-Self P-Type Being and Genetic Code

Summarizing, after the fundamental capacities of self iterations of the Light-Space-Time Emergence equation are complete, the following code-segments will have been generated as specified by Equation 7.3.7, Active Fundamental-Capacities-of-Self P-Type Being Pre-Genetic and Genetic Code:

$$Light - Space - Time\ Emergence_{Fundamental-Capacities-of-Self\ P-Type\ Being} =$$

$$\left\| \begin{bmatrix} \begin{matrix} c_\infty: [Pr, Po, K, H] \\ \left(\downarrow R_{C_K} = f(R_{C_\infty})\right) \\ c_K: [S_{Pr}, S_{Po}, S_K, S_H] \\ \left(\downarrow R_{C_N} = f(R_{C_K})\right) \\ c_N: f(S_{Pr} \times S_{Po} \times S_K \times S_H) \\ \left(\downarrow R_{C_U} = f(R_{C_N})\right) \\ c_U: [P, V, M, C] \\ \Uparrow \\ c_{0:[D,W,I,C]} \end{matrix} \end{bmatrix}_{Light} \begin{bmatrix} M_3 \rightarrow System_X \\ (\uparrow F \rightarrow I) \\ M_2 \rightarrow S_{System_X} \\ (\uparrow Sig \rightarrow F) \\ M_1 \rightarrow Sig_x \\ (\uparrow > P_x) \\ U \rightarrow x_{Cells} \end{bmatrix}_{Space} \right.$$
$$\left. \begin{bmatrix} M_3 : -\infty \le t \le \infty \\ \downarrow \\ M_2 : 0 \ge t > \infty \\ \downarrow \\ M_1 : 0 > t > \infty \\ \downarrow \\ U \rightarrow \begin{matrix} t \le E_{Cell}; TC: M_3 \rightarrow U \\ t \sim E_{Human}; TC: U \rightarrow M_3 \end{matrix} \end{bmatrix}_{Time} \quad TC \rightarrow x_{Capacities\ of\ Self} \right\|_{\langle x_U | x_T \rangle} \Rightarrow$$

$$\langle Space - Time - Energy - Gravity\ P - Type\ Beings\ and\ Pre - Genetic\ Code \rangle +$$

$$\left(\begin{matrix} Electro - magnetic - wavearchetype - masspotential\ P - Type\ Being \\ and\ Pre - Genetic\ Code \end{matrix} \right) +$$

$$\langle Quantum - Particle\ P - Type\ Being\ and\ Pre - Genetic\ Code \rangle +$$

$\langle Atom\ P-Type\ Being\ and\ Pre-Genetic\ Code \rangle +$

$\langle Molecule\ P-Type\ Being\ and\ Pre-Genetic\ Code \rangle + LSTE\ \langle ... \rangle +$

$PGGT < \cdots > + \langle Cell\ P-Type\ Being\ and\ Genetic\ Code \rangle$

$+LSTE < \cdots > +$

$$
\left(
\begin{array}{l}
where \left[\begin{array}{c} \sum Sig_{feelings} = Xa + \overline{Yb_{0-n}} \\ X \in [S_{System_N}] \\ Y \in [S_{System_{Pr}}, S_{System_P}, S_{System_K}, S_{System_N}] \\ a, b\ are\ integers; a > b \end{array} \right] \\
where \left[\begin{array}{c} \sum Sig_{thoughts} = Xa + \overline{Yb_{0-n}} \\ X \in [S_{System_K}] \\ Y \in [S_{System_{Pr}}, S_{System_P}, S_{System_K}, S_{System_N}] \\ a, b\ are\ integers; a > b \end{array} \right] \\
where \left[\begin{array}{c} \sum Sig_{urges} = Xa + \overline{Yb_{0-n}} \\ X \in [S_{System_P}] \\ Y \in [S_{System_{Pr}}, S_{System_P}, S_{System_K}, S_{System_N}] \\ a, b\ are\ integers; a > b \end{array} \right] \\
where \left[\begin{array}{c} \sum Sig_{sensation} = Xa + \overline{Yb_{0-n}} \\ X \in [S_{System_{Pr}}] \\ Y \in [S_{System_{Pr}}, S_{System_P}, S_{System_K}, S_{System_N}] \\ a, b\ are\ integers; a > b \end{array} \right]
\end{array}
\right)
$$

Eq. 7.3.7, Active Fundamental-Capacities- of-Self P-Type Being Pre-Genetic and Genetic Code

Note that the final code segment following 'LSTE <…>' contains the code for all possible ($\sum x$) sensations, urges, feelings, and thought, respectively, and depicts the growing biography by which the singular light-based edifice expresses its materialization, and at this stage, through the fundamental-capacities-of self P-type being.

At the core individuals can be thought of as projections of seeds formed in a vast continent of Light. So that core must always be there, but it is often covered by surface dynamics of the physical, the vital, and the mental nature. These too are formed from properties of Light, but since the light has been separated from its source these movements and dynamics are incomplete in their nature.

It is easy for these dynamics and movements to completely occupy all individual processing power. This can happen to such an extent that the individual acting on the surface can become subject to these surface movements. But so long as there is the remembrance that even in separation and smallness - whether selfishness, myopia, fundamentalism, or any other ism - these movements are still light, then that can become the means by which the apparent chains around individuals can be loosened.

For in its essence these small movements are trying to extend into something other than what they are. By connecting the light in them with the larger Light this extension can yield something different. And in the process the hold that these smaller movements have is diminished and one can begin to see or experience larger vistas of light.

Such a passage where one is always moving to larger vistas of light, more easily allows interaction, or the influence, or even entry into a field of original seeds and it is so that each can begin to enter into conscious communion with truer individuality.

Light's Emergence as Truer Individuality

As discussed previously the Light-Space-Time Emergence equation (3.1.3) being iterative, can be used to model emergence as it proceeds from simpler four-fold to more complex four-fold manifestations. Hence (3.1.3) has already been applied to suggest the emergence of the space-time-energy-gravity quadrumvirate P-type beings, the electro-magnetic-wavearchetype-masspotential P-type being, the quantum-particle P-type being, the boson P-type being as a further instance of a particular kind of quantum-particle P-type being, the atom P-type being, the cell P-type being, and the fundamental-capacities-of-self P-type being.

Here, a revised form of the truer individuality equation (4.2.6) will be applied to suggest the emergence of truer individuality, Equation 7.4.1, Generation of Truer-Individuality P-Type Being. The architecture and details of the truer-individuality P-type being can be seen to be the result of the application of the Light, Space, and Time matrices as will be elaborated. But further, as implied by (3.6.12), the Potential Effect of Levels of Light on Genetic-Type Information equation, any process of deeper organization, such as is responsible for the architecture and cohesiveness of the truer-individuality P-type being, has the possibility of altering FBLEE and post material-fabric housing structures so long as the bases involved are driven primarily by a meta-level.

$$Light - Space - Time\ Emergence_{Truer\ Individuality\ P-Type\ Being} =$$

$$\left| \left[\begin{array}{c} \left[\begin{array}{c} c_\infty: [Pr, Po, K, H] \\ \left(\downarrow R_{C_K} = f(R_{C_\infty}) \right) \\ c_K: [S_{Pr}, S_{Po}, S_K, S_H] \\ \left(\downarrow R_{C_N} = f(R_{C_K}) \right) \\ c_N: f(S_{Pr} \times S_{Po} \times S_K \times S_H) \\ \left(\downarrow R_{C_U} = f(R_{C_N}) \right) \\ c_U: [P, V, M, C] \\ \Uparrow \\ c_{0:[D,W,I,C]} \end{array} \right]_{Light} \quad \left[\begin{array}{c} M_3 \rightarrow System_x \\ (\uparrow F \rightarrow I) \\ M_2 \rightarrow S_{System_x} \\ (\uparrow Sig \rightarrow F) \\ M_1 \rightarrow Sig_x \\ (\uparrow > P_x) \\ U \rightarrow x_{Humans} \end{array} \right]_{Space} \\ \left[U \rightarrow \begin{array}{c} M_3: -\infty \le t \le \infty \\ \downarrow \\ M_2: 0 \ge t > \infty \\ \downarrow \\ M_1: 0 > t > \infty \\ \downarrow \\ t \le E_{Cell}; TC: M_3 \rightarrow U \\ t \sim E_{Human}; TC: U \rightarrow M_3 \end{array} \right]_{Time} \quad TC \rightarrow x_{Truer\ Individuality} \end{array} \right] \langle x_U | x_T \rangle \Rightarrow$$

$$\langle Space - Time - Energy - Gravity\ P - Type\ Beings\ and\ Pre - Genetic\ Code \rangle +$$

$$\left(\begin{array}{c} Electro - magnetic - wavearchetype - masspotential\ P - Type\ Being \\ and\ Pre - Genetic\ Code \end{array} \right) +$$

$$\langle Quantum - Particle\ P - Type\ Being\ and\ Pre - Genetic\ Code \rangle +$$

$$\langle Atom\ P - Type\ Being\ and\ Pre - Genetic\ Code \rangle +$$

$$\langle Molecule\ P - Type\ Being\ and\ Pre - Genetic\ Code \rangle + LSTE \langle ... \rangle +$$

$\langle Cell\ P - Type\ Being\ and\ Genetic\ Code \rangle + LSTE\ \langle ... \rangle + FBLEEE\ \langle ... \rangle +$

$\langle Human\ P - Type\ Being\ and\ Genetic\ Code \rangle + LSTE\ \langle ... \rangle + FBLEEE\ \langle ... \rangle +$

$\langle Fundamental - Capacities - of - Self\ P - Type\ Being\ and\ Genetic\ Code \rangle + LSTE\ \langle ... \rangle +$

$FBLEEE\ \langle ... \rangle + Truer - Individuality\ P - Type\ Being\ and\ Genetic\ Code$

Eq 7.4.1: Generation of Truer-Individuality P-Type Being

Starting with the Light-Matrix, the top left-hand matrix in (7.4.1), the first line from the top, C_{∞}: $[Pr, Po, K, H]$, specifies the architecture of the truer-individuality P-type being as will be further discussed in this chapter. Hence, Knowledge-type individuals are an emergence of Light's property of Knowledge, Power-type individuals are an emergence of Light's property of Power, Service-type individuals are an emergence of Light's property of Presence, and Harmony-type individuals are an emergence of Light's property of Harmony. The fundamental architecture of these aspects, hence, is an emergence of the properties of Light at ∞.

Line 3 in the Light-Matrix, C_K: $[S_{Pr}, S_{Po}, S_K, S_H]$, elaborates the sets for Presence, Power, Knowledge, and Harmony, each containing multiple elements. For example, as will be explored in the subsequent subsection, various elements derived from the four sets would define the architecture of Harmony-type individuals and could be functions such as 'driven to connect people together', amongst others, hence collectively describing Harmony-type individuals' way of being. Specifically, Line 5, C_N: $f(S_{Pr}\ x\ S_{Po}\ x\ S_K\ x\ S_H)$, suggests that unique seeds are created from a combination of such elements from all four sets, with a particular element leading or having more weight, that in effect creates the distinctness possible in the truer-individuality P-type being.

Line 6, ($\downarrow R_{CU} = f(R_{C_N})$), specifies quantization between the layer where the seeds are formed, and the physical layer, and as explored in Chapter 3.5 and 3.6, will result in Line 7, C_U: $[P, V, M, C]$, hence changing the post material-fabric housing structure. Note that as in the process describing the generation of the code-segments for space-time-energy-gravity quantization at the time of the Big Bang, the FBLEE (four-base logic-encoding ecosystem) process is apparently transparent. In reality one may say that any P-type being is first a FBLEE-P-type being and then a MF-P-type being.

The possibilities represented by Lines 1 through 5 hence concretize through the quantization represented by Line 6 to become or further enhance the truer-individuality P-type being with its subtle physical (related to Presence), vital (related to Power), mental (related to Knowledge), and connection (related to Harmony) aspects now existing in material reality typified by Light moving at c.　Note that just as Line 6 represents a process of quantization relating the layer of reality created by Light traveling at c with the antecedent layers, so too Lines 2 and 4 as previously discussed, also represent quantization of a more subtle kind that ultimately plays a critical part in allowing FBLEE and post material-fabric housing structure to express infinite diversity.

Typically, it is the process as captured by the Space-Matrix that will determine if Line 6 is activated.　Specifically, patterns at the untransformed layer, U, will need to be overcome, as specified by the second line from the bottom of the Space-Matrix: $(\uparrow > P_x)$.　But as specified by the bottom-line of the Time-Matrix, reproduced below, it is only with the advent of the human-system that the automaticity of the action of meta-levels is reversed:

$$U \rightarrow \begin{array}{l} t \leq E_{Cell};\, \text{TC: } M_3 \rightarrow U \\ t \sim E_{Human};\, \text{TC: } U \rightarrow M_3 \end{array}$$

Hence in the case of the truer-individuality P-type being, which in this emergence is a post-human system, the fact that patterns do need to be overcome means that quantization requires effort to happen.　Given this, it is useful to review **Equation 3.6.5, Potential Effect of Levels of Light on Genetic-Type Information:**

$$Potential\ Effect\ of\ Levels\ of\ Light\ on\ Genetic - Type\ Information =$$

$$\left[\begin{array}{c} STATIC\ \langle |[L][S][T]TC \rightarrow x_T|_{\langle x_U | x_T \rangle}\rangle \\ \times \\ \left((Y > U: Z_Q) \vee (Y \leq U: Z_F) \vee (Y = U: Z_R) \right) \\ \ni \\ Z \in \mathbb{U}\ (Space, Time, Energy, Gravity) \\ (Q:\ Quantization; F:\ Fragmentation; R:\ Random) \end{array} \right] \rightarrow h \ni$$

$$h \in \begin{pmatrix} [Q]: \textit{Constructive zone,} \\ [Q]: \textit{Constructive zone} \wedge \textit{Constructive mutation,} \\ [F]: \textit{Destructive mutation,} \\ [R]: \textit{Random mutation} \end{pmatrix}$$

Line 1 from the top in the matrix is simply a static form of (3.6.1) the Simplified Light-Space-Time Emergence equation. The static form is designated by 'STATIC' and implies that fundamental operations true of (3.6.1) are being highlighted in (3.6.12). In other words (3.6.1) already has all the operations highlighted in (3.6.12) in it, but by 'freezing' it by making it static, the essential dynamics leading to possible mutations at the genetic level can more clearly be highlighted.

Line 1 is then subjected ($\times$) to a determination of the dominant levels of light that may be active, designated by Line 2, $'\big((Y > U : Z_Q) \vee (Y \leq U : Z_F) \vee (Y = U : Z_R) \big)'$. Unpacking this, '$Y > U$' implies meta-levels are active and as a result it is possible that Z_Q is going to take place (the subscript 'Q' implies quantum-level action). This also implies activation and potential change of FBLEE. The call from below, as it were, may invoke some function that already exists in the subtle-libraries 'above', so that some already existing function may influence FBLEE through ∞-entanglement, K-entanglement, or N-entanglement. This may be thought of as a key-and-lock mechanism, where a deep enough visceral urge from below acting as the key, opens an entangled lock to alter FBLEE as per the visceral urge. '$Y \leq U$' implies that only the untransformed levels are active, and therefore also the sub-level where the speed of light is 0 is active and as a result Z_F is going to take place (the subscript 'F' implies 'fragmentation'). '$Y = U$' implies that all levels are active and as a result Z_R is going to take place (the subscript 'R' implies 'random').

Line 3 elaborates the significance of Z_Q, Z_F, and Z_R. Hence Z is the union of potential quantum-operations of space, time, energy, and gravity, designated by '$Z \in \mathbb{U} \, (Space, Time, Energy, Gravity)$'. But the nature of the operations, as suggested in the previous paragraph, is designated by 'Q: $Quantization$; F: $Fragmentation$; R: $Random$'. Z_Q, then, implies that the full quantization originating from updated four-base logic-encoding ecosystems (FBLEE) can take place. Z_F implies that the essential set will be fragmented and that only libraries at the level of local cellular-level DNA or precipitated material-fabric logic can potentially be altered. Z_R

also precludes full quantization, and that some partial local-library constructive or destructive mutation may take place.

The '$\rightarrow h \ni h \in$' segment resolves the outcome of the operations implied by Lines 1 – 3, suggesting that the outcome will be 'h' such that ($\ni$) 'h' is an element ($\in$) of the set specified by the members '[Q]: *Constructive zone*', '[Q}: *Constructive zone AND Constructive mutation*', '[F]; *Destructive mutation*', and '{R}: *Random mutation*'. '[Q]: *Constructive zone*' implies that the in-built buffer has been crossed and that access to the deeper four-base logic-encoding ecosystem (FBLEE) has been granted. '[Q]' in this segment implies that there is the possibility that full-quantization as specified by Line 3 of the previous matrix can take place. Access to this zone is a prerequisite for constructive mutation to occur, as designated by the element '[Q}: *Constructive zone $\cap$ Constructive mutation*', which implies that full-quantization is going to take place and will result in material change. The '[F]' specifies the relationship between 'Fragmentation' in Line 3 of the previous matrix and destructive mutation. The '[R]' specifies the relationship between 'Random' in Line 3 of the previous matrix and random mutation.

But as just summarized in the Time-Matrix in (7.4.1) Y is by definition not greater than U and hence quantization is not automatic. In terms of truer individuality such quantization would imply that increasing wholeness can become fully active through specific space, time, energy, and gravity quantization to create an holistic "ecosystem" with its own "truer-individuality P-type being logic" as it were. Possible precipitation into the post material-fabric housing structure is captured by the action of 'PGGT $<...>$' that can be assumed to be triggered by Z_Q. This "logic" or more specifically genetic code is elaborated in the following sub-section and suggests the increasing sphere of influence of the truer-individuality P-type being that now also contains light-engendered laws for deeper individuality.

Generation of the Truer-Individuality P-Type Being

The truer-individuality P-type being itself is some mathematical function of the four properties of Light - Harmony, Power, Knowledge, and Service, which in turn, are practically infinite sets of qualities related to one of the main properties of Light.

278

So, it may be that one individual is primarily driven to connect people together, itself a variation of the property of Harmony. The individual may further want to do this by deeply understanding what makes these people tick, itself a variation of the property of Knowledge. So the seed or the truer individuality of this person can be thought of as a mathematical function consisting of some element or property from the set of Harmony and some element or quality from the set of Knowledge, combined together with possibly different elements from the same or different sets, all with possibly different weights, but with the first weight of the need to connect people, being the strongest.

Such an individual may have an essential form as expressed in Equation 7.4.2, Harmony-Type Individual:

$$Sig_{Harmony-type} = Xa +$$
$$\overline{Yb_{0-n}} \quad where \begin{bmatrix} X \in [S_{System_N}] \\ Y \in [S_{System_{Pr}}, S_{System_P}, S_{System_K}, S_{System_N}] \\ a, b \ are \ integers; a > b \end{bmatrix}$$

Eq 7.4.2: Generalized Harmony-Type Signature of the Truer-Individuality P-Type Being

But equally, and consistent with the emergent properties of Light there could be essential Service-Type, Power-Type, and Knowledge-Type individuals as well, with infinite variation in the precise nature of the core seed, as captured in Equations 7.4.3-5:

$$Sig_{Service-type} = Xa + \overline{Yb_{0-n}}$$

$$where \begin{bmatrix} X \in [S_{System_{Pr}}] \\ Y \in [S_{System_{Pr}}, S_{System_P}, S_{System_K}, S_{System_N}] \\ a, b \ are \ integers; a > b \end{bmatrix}$$

Eq 7.4.3: Generalized Service-Type Signature of the Truer-Individuality P-Type Being

$$Sig_{Power-type} = Xa + \overline{Yb_{0-n}}$$

$$where \begin{bmatrix} X \in [S_{System_P}] \\ Y \in [S_{System_{Pr}}, S_{System_P}, S_{System_K}, S_{System_N}] \\ a, b \ are \ integers; a > b \end{bmatrix}$$

Eq 7.4.4: Generalized Power-Type Signature of the Truer-Individuality P-Type Being

$$Sig_{Knowledge-type} = Xa + \overline{Yb_{0-n}}$$

$$where \begin{bmatrix} X \in [S_{System_K}] \\ Y \in [S_{System_{Pr}}, S_{System_P}, S_{System_K}, S_{System_N}] \\ a, b \ are \ integers; a > b \end{bmatrix}$$

Eq 7.4.5: Generalized Knowledge-Type Signature of the Truer-Individuality P-Type Being

Note that (7.4.2 - 5) already implies that Lines 1 – 5 in the Light Matrix (7.4.1) have been activated, and that the logic of the truer-individuality-ecosystem will precipitate into the post material-fabric cell-based genetic structure through the action of Line 6-7 of (7.4.1). This action of precipitating into the cell-based structure as opposed to the material-fabric is designated by 'PGGT <...>' as specified in (7.4.1).

Summary of Truer-Individuality P-Type Being and Genetic Code

Summarizing, after the truer individuality iteration of the Light-Space-Time Emergence equation is complete, the following code-segments will have been generated as specified by Equation 7.4.6, Active Truer-Individuality P-Type Being Pre-Genetic and Genetic Code:

$$Light - Space - Time \ Emergence_{Truer-Individuality \ P-Type \ Being} =$$

$$\left|\begin{bmatrix} \begin{bmatrix} c_\infty: [Pr, Po, K, H] \\ \left(\downarrow R_{C_K} = f(R_{C_\infty})\right) \\ c_K: [S_{Pr}, S_{Po}, S_K, S_H] \\ \left(\downarrow R_{C_N} = f(R_{C_K})\right) \\ c_N: f(S_{Pr} \times S_{Po} \times S_K \times S_H) \\ \left(\downarrow R_{C_U} = f(R_{C_N})\right) \\ c_U: [P, V, M, C] \\ \Uparrow \\ c_{0:[D,W,I,C]} \end{bmatrix}_{Light} \begin{bmatrix} M_3 \to System_X \\ (\uparrow F \to I) \\ M_2 \to S_{System_X} \\ (\uparrow Sig \to F) \\ M_1 \to Sig_x \\ (\uparrow > P_x) \\ U \to x_{Human} \end{bmatrix}_{Space} \\ \begin{bmatrix} M_3 : -\infty \le t \le \infty \\ \downarrow \\ M_2 : 0 \ge t > \infty \\ \downarrow \\ M_1 : 0 > t > \infty \\ \downarrow \\ U \to \begin{array}{l} t \le E_{Cell}; TC: M_3 \to U \\ t \sim E_{Human}; TC: U \to M_3 \end{array} \end{bmatrix}_{Time} \quad TC \to x_{Truer\ Individuality} \end{bmatrix} \right| \Rightarrow \langle x_U | x_T \rangle$$

$\langle Space - Time - Energy - Gravity\ P - Type\ Beings\ and\ Pre - Genetic\ Code \rangle +$

$\left(\begin{array}{c} Electro - magnetic - wavearchetype - masspotential\ P - Type\ Being \\ and\ Pre - Genetic\ Code \end{array}\right) +$

$\langle Quantum - Particle\ P - Type\ Being\ and\ Pre - Genetic\ Code \rangle +$

$\langle Atom\ P - Type\ Being\ and\ Pre - Genetic\ Code \rangle +$

$\langle Molecule\ P - Type\ Being\ and\ Pre - Genetic\ Code \rangle + LSTE\ \langle ... \rangle +$

$\langle Cell\ P - Type\ Being\ and\ Genetic\ Code \rangle + LSTE\ \langle ... \rangle + FBLEEE\ \langle ... \rangle +$

$\langle Human\ P - Type\ Being\ and\ Genetic\ Code \rangle + LSTE\ \langle ... \rangle + FBLEEE\ \langle ... \rangle +$

$\langle Fundamental - Capacities - of - Self\ P - Type\ Being\ and\ Genetic\ Code \rangle + LSTE\ \langle ... \rangle +$

$$FBLEEE \langle ... \rangle + \left(\begin{array}{l} \sum Sig_{Harmony-type} = Xa + \overline{Yb_{0-n}} \\ \left[\begin{array}{l} X \in [S_{System_N}] \\ where \;\; Y \in [S_{System_{Pr}}, S_{System_P}, S_{System_K}, S_{System_N}] \\ a, b \; are \; integers; a > b \end{array} \right] \\ \sum Sig_{Knowledge-type} = Xa + \overline{Yb_{0-n}} \\ \left[\begin{array}{l} X \in [S_{System_K}] \\ where \;\; Y \in [S_{System_{Pr}}, S_{System_P}, S_{System_K}, S_{System_N}] \\ a, b \; are \; integers; a > b \end{array} \right] \\ \sum Sig_{Power-type} = Xa + \overline{Yb_{0-n}} \\ \left[\begin{array}{l} X \in [S_{System_P}] \\ where \;\; Y \in [S_{System_{Pr}}, S_{System_P}, S_{System_K}, S_{System_N}] \\ a, b \; are \; integers; a > b \end{array} \right] \\ \sum Sig_{Service-type} = Xa + \overline{Yb_{0-n}} \\ \left[\begin{array}{l} X \in [S_{System_{Pr}}] \\ where \;\; Y \in [S_{System_{Pr}}, S_{System_P}, S_{System_K}, S_{System_N}] \\ a, b \; are \; integers; a > b \end{array} \right] \end{array} \right)$$

Eq. 7.4.6. *Active Truer-Individuality P-Type Being and Genetic Code*

Note that the final code segment following 'FBLEEE <...>' (recall that FBLEEE refers to FBLEE-entanglement) contains the code for all possible ($\sum x$) human-types, and depicts the growing biography by which the singular light-based edifice expresses its materialization, and at this stage, through the truer-individuality P-type being.

SECTION 8: GENERATION OF P-TYPE BEINGS IN THE SURFACING OF COMPLEX ORGANIZATION

Having traced the computations resulting in the generation the of space-time-energy-gravity P-type beings, the electro-magnetic-wavearchetype-masspotential P-type being, and the P-type beings associated with the surfacing of matter and life, we now turn our attention to study the computation resulting in the generation of P-type beings related to complex organizations.

In the previous sections we have seen that there is a gradual addition to P-type being biographies as pre-genetic and then genetic code capturing more complex fourfold-functionality is generated. In these biographies the condition for constructive mutation is that some meta-level is primarily active. When this happens, there is a possibility that the existing four-base logic-encoding ecosystems (FBLEE) are enhanced or that new ones are created. Note that it is such FBLEE action that is envisioned to alter the quantum-level interface between the material and antecedent layers of light in effect changing the basis by which matter materializes. In other words, FBLEE changes what matter can be.

But at its heart it is good to get clear that it is the complexification of four-foldness that is the driver of changes to FBLEE, and therefore of possible changes to the material-fabric, the genetic substance of living cells, and of any post-genetic substance that were to arise. This is so because such complexification essentially necessitates that existing habits, of any kind, are continually broken and in the process can admit of greater and more sweeping actions of light from the meta-levels. Such a dual process of complexification in which change in existing patterns at one level allow greater light from another level to be admitted into that level, is perhaps similar to Teilhard de Chardin's conceptualization of a law of complexification in which there is a simultaneous dual-evolution of physical reality and consciousness itself, toward levels of greater complexity (Chardin, 2008). All of light is ever-present but requires the material receptacle to alter in order that it may precipitate more of its infinite possibility. It is such action that makes clear how the emergence of complex organization actually links to changes such as to matter and to the make-up of genetic content.

Such a line of development also makes clear how being, becoming, life itself, and therefore possible transhumanism is intimately related to larger

283

forms that we create. The path to transhumanism is clear and apparent from this point of view. There are no additional technologies that need to be developed. There are no minds that need to be uploaded into a silicon-based computing infrastructure. Instead we have to become more present to what is already before us. We need to change our habitual reactions to things and master ourselves and our relationship with things, so that in this process more of the infinite possibility in light can effectively materialize to even change the very bases of matter through the action of FBLEE. The path to any transhumanism is therefore already staring at us in the face. We have only to recognize this, grasp it, and let Light play more fully, materially, so that the infinity in it can firmly express itself through us.

Chapter 8.1 examines the generation of the stable-mega-organization FBLEE-based P-type being. Similarly, Chapter 8.2 examines the generation of the FBLEE-based sustainable-global-civilization P-type being. In both these cases the nature of the P-type being is FBLEE-based, as opposed to being both FBLEE and MF-based. This is so because both the stable mega-organization and the sustainable global civilization are in process of being worked out. Their becoming realities means that there is a clear material and therefore genetic-type foundation within the being, so that in effect these forms of collectivity are nothing other than extensions of being. Hence there is a post-human structure and possibly a post-cell genetic-type structure to house their respective ecosystem logic materially that will need to be worked out. FBLEE is the realm where human aspiration meets with a response from functions and libraries of possibility in meta-levels of light, and gets housed in logic, that then becomes available to the human being. The precipitation of such logic alters matter, and it becomes clearer through the generation of such P-type beings that house such enhanced genetic-type information that in its reality matter is nothing other than a crystallization of possibilities of light, and will one day house light so much more completely that it will be entirely different from what we perceive it as being at present time.

Such post-cell genetic structure, as will be further explored in Section 9 on super-matter-based P-type beings, requires a much higher threshold of functional richness to come into material reality.

Also, to be noted is that the generation of P-type beings is a variable process, meaning it is not inevitable that particular kinds of P-type beings be developed. This variability implies that in a Cosmology of Light in which light-based-singularities are the foundation for materialization of infinite possibility in matter, there is a variable path to transhumanism. The important thing is the increase in functional richness dues to a deep urge or aspiration or willing as the case may be.

Chapter 8.1: Generation of FBLEE-Based Stable-Mega-Organization P-Type Being

We have traced the emergence of the four properties of Light through the primary fourfold emergence of the space-time-energy-gravity P-type beings, the electro-magnetic-wavearchetype-masspotential P-type being, and through P-type beings in the surfacing of matter and life. At each of these levels it seems it is a balanced combination of all four of the properties that creates stability and makes that structure or organism or being, sustainable. And this has to be since the four properties occur simultaneously in light and must be impelled to hold to that relationship even when projected from it. Hence it should be possible to deductively arrive at an equation for organizational sustainability given the four-fold emergence of properties of light through subsequent layers of manifestation.

To triangulate, though, we briefly trace the four-fold emergence, reinforce the significance of this and inductively arrive at an equation for sustainability that we would have arrived at deductively.

Starting with the primary emergence of the space-time-energy-gravity P-type beings we see that space is defined as that in which seeds of knowledge are planted – hence, existing as a means to express Light's property of knowledge. Time is defined as the inevitability of the seeds of knowledge maturing into fullness, in spite of any opposition. Hence time is seen as an emergence of Light's property of power. Energy is seen as that which accumulates to create matter. Hence it is an emergence of Light's property of presence. Gravity is that by which relationship between objects comes into being – hence an emergence of Light's property of harmony.

With the electro-magnetic-wavearchetype-masspotential P-type being we see that the being as a whole is defined or architected by the speed of light c, as an emergence of nurturing, the wavelength λ, as an emergence of knowledge, the frequency f, as an emergence of power, and masspotential as an emergence of presence. These occur simultaneously and constitute the wholeness of the electro-magnetic-wavearchetype-masspotential P-type being.

Similarly, at the level of the quantum-particle P-type being it can be seen that the integrity and functioning of the atom also depends on the integration of all four properties of light. Hence quarks, an emergence of knowledge, leptons, an emergence of power, bosons, an emergence of harmony or nurturing, and the Higgs-boson, an emergence of presence, act together to create the structure and functionality of every single atom.

What this reinforces is that every single thing must have been created through some integration and balance of the four properties of light.

Going up the organizational scale to the next level of complexity, to the level of the atom P-type being, the same pattern is found again: every single atom belongs to one of four groups, and in unison these four groups orchestrate the set of combinations of known compounds. Hence, Alkali Metals and Alkali Earth Metals configured by the s-Group, emanate from the family of system-power. Metals, Metalloid, Non-Metals, Halogens, and Noble Gases configured by the p-Group, emanate from the family of system-knowledge. Transition Metals configured by the d-Group, emanate from the family of system-presence. Lanthanides and Actinides configured by the f-Group, emanate from the family of system-nurturing. These groups act together to create the complex array of compounds that form the entire material from which the physical and subsequently even the organic world around us is constructed.

Going up the organizational scale to another level of complexity, to the level of the cell P-type being, the same pattern is found again: basically every single cell of every single living creature that has ever been studied by humankind also has a similar balance and integration of these four sets of forces acting together. Hence, proteins, from the family of presence, nucleic acids, from the family of knowledge, polysaccharides, from the family of power, and lipids, from the family of nurturing, work together to create the balanced functioning of every living cell.

The hypothesis, hence, is that even at larger scale, in considering the most innovative organization at the level of teams, corporations, or markets, a similar integration and balance of the four sets of properties or forces may yield the best results. Recent developments in sustainability investment models ranging from Socially Responsible Investing (Logue, 2008), the Global Reporting Initiative (Willis, 2003), the Dow Jones Sustainability Index (Hope & Fowler, 2007), the Principles of Responsible Investing

(Harvard-Edu, 2014), all reinforce the concept of investment criteria becoming broader, to be based not only on economic, but on environmental, social, and governance factors as well. Further evidence suggests that financial returns on such broad-based investment models continually beat financial returns on regular funds such as the S&P 500, for instance (Openshaw, 2015). In their study of major transitions in evolution Smith and Szathmary (Smith & Szathmary, 1995) chronicle the development of life toward increasing complexity as an application of successful collaboration and even co-evolution by which species evolve by changing together as system pressures increase.

This hypothesis hence can be summarized by the following graph:

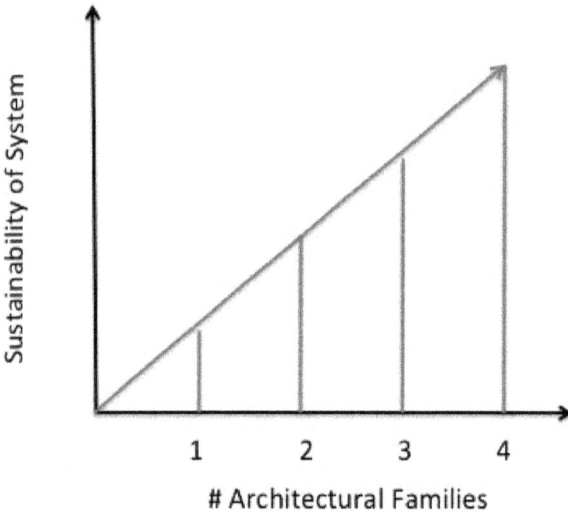

Figure 8.1.1 Sustainability of Systems

An equation for the sustainability of systems, $Sustainability_{Systems}$, where the interaction between the four families of forces is instrumental can be created. Hence, as in Equation 8.1.1:

$$Sustainability_{Systems} \propto Interaction \ (S_{System_{Pr}}, S_{System_P}, S_{System_K}, S_{System_N})$$

Eq 8.1.1: Sustainability of Systems

This notion of the balance of the fourfold properties of Light as fundamental in creating stable mega-organization will be borne out in the next subsections of this chapter.

Light's Emergence as Stable Mega-Organization

As discussed previously the Light-Space-Time Emergence equation (3.1.3) being iterative, can be used to model emergence as it proceeds from simpler four-fold to more complex four-fold manifestations. Hence (3.1.3) has already been applied to suggest the emergence of the space-time-energy-gravity quadrumvirate P-type beings, the electro-magnetic-wavearchetype-masspotential P-type being, the quantum-particle P-type being in general, the boson P-type being as a further instance of a particular kind of quantum particle, the atom P-type being, the cell P-type being, the fundamental-capacities-of-self P-type being, and the truer-individuality P-type being.

Here, a revised form of it leveraging the sustainable global civilization equation (4.2.7) introduced in Chapter 4.2 on post-Big Bang generation of P-type beings, will be applied to suggest the emergence of the stable-mega-organization P-type being, Equation 8.1.2, Generation of FBLEE-Based Stable-Mega-Organization P-Type Being. The architecture and details of the FBLEE-based stable-mega-organization P-type being can be seen to be the result of the application of the Light, Space, and Time matrices as will be elaborated. But further, as implied by (3.6.12), the Potential Effect of Levels of Light on Genetic-Type Information equation, any process of deeper organization, such as is responsible for the architecture and cohesiveness of the FBLEE-based stable-mega-organization P-type being, has the possibility of altering FBLEE, and in this case through FBLEEE influence, post-material fabric housing structures, so long as the bases involved are driven primarily by a meta-level.

$$L - S - T \; Emergence_{FBLEE-Based \; Stable-Mega-Orgnization \; P-Type \; Being} =$$

$$\begin{vmatrix} \begin{bmatrix} c_\infty : [Pr, Po, K, H] \\ \left(\downarrow R_{C_K} = f(R_{C_\infty}) \right) \\ c_K : [S_{Pr}, S_{Po}, S_K, S_H] \\ \left(\downarrow R_{C_N} = f(R_{C_K}) \right) \\ c_N : f(S_{Pr} \times S_{Po} \times S_K \times S_H) \\ \left(\downarrow R_{C_U} = f(R_{C_N}) \right) \\ c_U : [P, V, M, C] \\ \Uparrow \\ c_{0 : [D, W, I, C]} \end{bmatrix}_{Light} \begin{bmatrix} M_3 \rightarrow System_X \\ (\uparrow F \rightarrow I) \\ M_2 \rightarrow S_{System_X} \\ (\uparrow Sig \rightarrow F) \\ M_1 \rightarrow Sig_x \\ (\uparrow > P_x) \\ U \rightarrow x_{Humans} \end{bmatrix}_{Space} \\ \begin{bmatrix} M_3 : -\infty \leq t \leq \infty \\ \downarrow \\ M_2 : 0 \geq t > \infty \\ \downarrow \\ M_1 : 0 > t > \infty \\ \downarrow \\ U \rightarrow \begin{array}{l} t \leq E_{Cell}; TC: M_3 \rightarrow U \\ t \sim E_{Human}; TC: U \rightarrow M_3 \end{array} \end{bmatrix}_{Time} \quad TC \rightarrow x_{Stable\ Mega\ Organization} \end{vmatrix} \Rightarrow \langle x_U | x_T \rangle$$

$\langle Space - Time - Energy - Gravity\ P - Type\ Beings\ and\ Pre - Genetic\ Code \rangle +$

$\left(\begin{array}{c} Electro - magnetic - wavearchetype - masspotential\ P - Type\ Being \\ and\ Pre - Genetic\ Code \end{array} \right) +$

$\langle Quantum - Particle\ P - Type\ Being\ and\ Pre - Genetic\ Code \rangle +$

$\langle Atom\ P - Type\ Being\ and\ Pre - Genetic\ Code \rangle +$

$\langle Molecule\ P - Type\ Being\ and\ Pre - Genetic\ Code \rangle + LSTE \langle \dots \rangle +$

$\langle Cell\ P - Type\ Being\ and\ Genetic\ Code \rangle + LSTE \langle \dots \rangle + FBLEEE \langle \dots \rangle +$

$\langle Human\ P - Type\ Being\ and\ Genetic\ Code \rangle + LSTE \langle \dots \rangle + FBLEEE \langle \dots \rangle +$

$\langle FBLEE - Based\ Stable\ Mega - Organization\ P - Type\ Being\ and\ Post - Genetic\ Code \rangle$

$+ PGCI \langle \dots \rangle$

Eq 8.1.2: Generation of FBLEE-Based Stable-Mega FBLEE-Based P-Type Being

Note that as indicated by the '$\langle FBLEE - Based\ Stable\ Mega - Organization\ P - Type\ Being\ and\ Post - Genetic\ Code \rangle$' code-segment, the code generated is at the level of FBLEE only. Further, as indicated by the '$PGCI \langle \dots \rangle$' code-segment, a process referred to as Post Genetic Code Initiation is activated, by which some possibly hybrid genetic-type information housing-structure that integrates the subtle-libraries with matter more intimately, and allows an easy going back and forth between the two may materialize sometime in the future.

Starting with the Light-Matrix, the top left-hand matrix in (8.1.2), the first line from the top, C_∞: $[Pr, Po, K, H]$, specifies the architecture of the stable-mega-organization P-type being as will be discussed shortly. Hence, Academic organizations are an emergence of Light's property of Knowledge, Political organizations are an emergence of Light's property of Power, Social organizations are an emergence of Light's property of Presence, and Commercial organizations can be thought of as an emergence of Light's property of Harmony. The fundamental architecture of these aspects, hence, is an emergence of the properties of Light at ∞.

Line 3 in the Light-Matrix, C_K: $[S_{Pr}, S_{Po}, S_K, S_H]$, elaborates the sets for Presence, Power, Knowledge, and Harmony, each containing multiple elements. For example, as will be explored in the following subsection, various elements derived from the four sets would define the architecture of a Silicon Valley type mega-commercial organization and could be functions such as 'commercial immunity', 'power of aesthetics', 'ease of informal meetings', amongst others, hence collectively describing a stable mega-commercial organization's way of being. Specifically, Line 5, C_N: $f(S_{Pr} \times S_{Po} \times S_K \times S_H)$, suggests that unique seeds are created from a combination of such elements from all four sets, with a particular element leading or having more weight, that in effect creates the distinctness possible at the level of the stable-mega-organization P-type being.

Line 6, ($\downarrow R_{C_U} = f(R_{C_N})$), normally specifies quantization between the layer where the seeds are formed, and the physical layer, and as explored in Chapter 3.5 and 3.6, will normally result in Line 7, C_U: $[P, V, M, C]$, hence potentially changing the post material-fabric housing structure. In a post-human organization such as the stable-mega-organization P-type being as just suggested, however, the beginnings of a post-genetic hybrid structure will likely only begin to be initiated and would require far more functional-richness to materialize before it were to become a reality. Because such a post-genetic housing structure has yet to come into existence (8.1.2) can strictly only be a FBLEE-based P-type being.

291

The possibilities represented by Lines 1 through 5 hence concretize through the quantization represented by Line 6 to become or further enhance the stable-mega-organization P-type being with its subtle physical (related to Presence), vital (related to Power), mental (related to Knowledge), and connection (related to Harmony) aspects now at least existing in FBLEE in the quantum-layer behind U   As a reminder FBLEE exists at that barrier where the speed of light is c, but at the quantum-level only. Note that just as Line 6 represents a process of quantization relating the layer of reality created by Light traveling at c with the antecedent layers, so too Lines 2 and 4 as previously discussed, also represent quantization of a more subtle kind that ultimately plays a critical part in allowing FBLEE to express infinite diversity.

Typically, it is the process as captured by the Space-Matrix that will determine if Line 6 is activated.   Specifically, patterns at the untransformed layer, U, will need to be overcome, as specified by the second line from the bottom of the Space-Matrix: $(\uparrow > P_x)$. But as specified by the bottom-line of the Time-Matrix, reproduced below, it is only with the advent of the human-system that the automaticity of the action of meta-levels is reversed:

$$U \rightarrow \begin{array}{l} t \leq E_{Cell};\, TC: M_3 \rightarrow U \\ t \sim E_{Human};\, TC: U \rightarrow M_3 \end{array}$$

Hence in the case of the stable-mega-organization P-type being, which in this emergence is a post-human system, the fact that patterns do need to be overcome means that quantization requires effort to happen. Given this, it is useful to review **Equation 3.6.5, Potential Effect of Levels of Light on Genetic-Type Information:**

$Potential\ Effect\ of\ Levels\ of\ Light\ on\ Genetic - Type\ Information =$

$$\begin{bmatrix} STATIC\ \langle |[L][S][T]TC \rightarrow x_T|_{\langle x_U|x_T \rangle} \rangle \\ \times \\ \left((Y > U: Z_Q) \vee (Y \leq U: Z_F) \vee (Y = U: Z_R) \right) \\ \ni \\ \left(\begin{array}{c} Z \in \mathbb{U}\ (Space, Time, Energy, Gravity) \\ Q:\ Quantization;\ F:\ Fragmentation;\ R:\ Random \end{array} \right) \end{bmatrix} \rightarrow h \ni$$

$$h \in \left(\begin{array}{c} [Q]:\ Constructive\ zone, \\ [Q]:\ Constructive\ zone \wedge Constructive\ mutation, \\ [F]:\ Destructive\ mutation, \\ [R]:\ Random\ mutation \end{array} \right)$$

Line 1 from the top in the matrix is simply a static form of (3.6.1) the Simplified Light-Space-Time Emergence equation. The static form is designated by 'STATIC' and implies that fundamental operations true of (3.6.1) are being highlighted in (3.6.12). In other words (3.6.1) already has all the operations highlighted in (3.6.12) in it, but by 'freezing' it by making it static, the essential dynamics leading to possible mutations at the genetic level can more clearly be highlighted.

Line 1 is then subjected ($\times$) to a determination of the dominant levels of light that may be active, designated by Line 2, $'\big((Y > U : Z_Q) \vee (Y \leq U : Z_F) \vee (Y = U : Z_R) \big)'$. Unpacking this, '$Y > U$' implies meta-levels are active and as a result it is possible that Z_Q is going to take place (the subscript 'Q' implies quantum-level action). This also implies activation and potential change of FBLEE. The call from below, as it were, may invoke some function that already exists in the subtle-libraries 'above', so that some already existing function may influence FBLEE through ∞-entanglement, K-entanglement, or N-entanglement. This may be thought of as a key-and-lock mechanism, where a deep enough visceral urge from below acting as the key, opens an entangled lock to alter FBLEE as per the visceral urge. '$Y \leq U$' implies that only the untransformed levels are active, and therefore also the sub-level where the speed of light is 0 is active and as a result Z_F is going to take place (the subscript 'F' implies 'fragmentation'). '$Y = U$' implies that all levels are active and as a result Z_R is going to take place (the subscript 'R' implies 'random').

Line 3 elaborates the significance of Z_Q, Z_F, and Z_R. Hence Z is the union of potential quantum-operations of space, time, energy, and gravity, designated by '$Z \in \mathbb{U} \, (Space, Time, Energy, Gravity)$'. But the nature of the operations, as suggested in the previous paragraph, is designated by 'Q: Quantization; F: Fragmentation; R: Random'. Z_Q, then, implies that the full quantization originating from updated four-base logic-encoding ecosystems (FBLEE) can take place. Z_F implies that only a local library, as opposed to FBLEE, can potentially be altered. Z_R also precludes full quantization, and that some partial local-library constructive or destructive mutation may take place.

The '$\rightarrow h \ni h \in$' segment resolves the outcome of the operations implied by Lines 1 – 3, suggesting that the outcome will be 'h' such that ($\ni$) 'h' is

293

an element (∈) of the set specified by the members '[Q]: *Constructive zone*', '[Q}: *Constructive zone AND Constructive mutation*', '[F}; *Destructive mutation*', and '{R}: *Random mutation*'. '[Q]: *Constructive zone*' implies that the in-built buffer has been crossed and that access to the deeper four-base logic-encoding ecosystem (FBLEE) has been granted. '[Q}' in this segment implies that there is the possibility that full-quantization as specified by Line 3 of the previous matrix can take place. Access to this zone is a prerequisite for constructive mutation to occur, as designated by the element ' [Q}: *Constructive zone ∩ Constructive mutation* ', which generally implies that full-quantization is going to take place. In the case of an early post-human organization though, as already suggested in the discussion on (8.1.2), since there is not yet a post-genetic structure securely in place to house such full-quantization at the material level, this likely initiates a call to antecedent layers for such a structure to begin to materialize through the activation of PGCI. In the meanwhile, quantization will affect FBLEE, and through FBLEEE may affect human-level genetic-code. The '[F]' specifies the relationship between 'Fragmentation' in Line 3 of the previous matrix and destructive mutation. The '[R]' specifies the relationship between 'Random' in Line 3 of the previous matrix and random mutation.

Generation of FBLEE-Based Stable-Mega-Organization P-Type Being

As suggested by inductively derived (8.1.1), a sustainable organization is the result of two dynamics: maturity along each of the architectural force dimensions, and a robust interaction between all four sets of forces. These dynamics are implicit in (8.1.2) as borne out in the previous subsection.

Consider the example of Silicon Valley.

While Silicon Valley may be primarily a commercial organization, yet we begin to glimpse something of what may become possible when the four emergent forces of light can begin to interact with one another in free fashion. That is, without the motive of doing everything for the sake of generating money only. In the case of Silicon Valley, we see that the climate and beauty of the Bay Area, thereby likely representing the impetus of nurturing and harmony, began to attract many talented people into the vicinity. Over time the talent pool became progressively diversified. Educational institutes, such as Stanford University and University of Berkeley cropped up and became centers of cutting-edge research, representing the impetus of knowledge. The Armed Forces were attracted to the area for the same reason and came with their huge requirements for research for research's sake, and their funds in support of the area, representing the impetus of power. Graduates from the universities started companies that began to in turn support the universities with funds. Talent moved around from company to company like people from department to department. Thus, we see that even when individual companies failed, Silicon Valley functioned as a larger organization and was able to retain the talent in the area, was able further, to buffer the shocks to some extent, hence preparing the ground for future waves of innovation. This reality of becoming a container for dynamic experimentation perhaps represents the impetus of service in that the container is serving the dynamic components existing within it.

Pragmatically speaking this dynamic of functioning as a larger mega-organization meant that some level of commercial immunity began to develop, so that people losing their jobs was not necessarily considered as that stressful an event. There was a freeing, thus, from the purely commercial element. If the various powers of beauty, aesthetics, knowledge, power in the form of money, plus the myriad streams of talent – from engineers, scientists, managers, lawyers etc., did not exist in such close proximity, and further, if the various professionals could not thus support each other through their continued informal meetings, through the urge to create the next wave of innovation, through the urge to pursue progress for the sake of progress, then they would have left for other areas and the phenomenon of Silicon Valley would never have been.

Usually modern-day organizations cannot provide these various buffers and opportunities for interaction that Silicon Valley provides, and hence when hit with adversity, more often than not, simply crumble. This is perhaps due to the fact that it is always only one-dimension that drives them, and when the health of that is threatened or falls, the organization resorts to tactics that will ensure that that dimension looks good at any cost, in the bargain sacrificing the development of the other dimensions, and therefore its longer term health.

The four-forces led organization will be something like a community. Having attained freedom from an exaggerated commercial impetus, people will 'live' their 'jobs' because it is what fulfills them. Such a freedom is what will allow the four primary powers to manifest to greater

295

and greater degree within them and their environment. Seeing thus, their ability to become centers of knowledge and wisdom, or mutuality and harmony, or power and leadership and energy, or perfection and service, or some unique combination of these primary forces, so increase, a sense of satisfaction with life will more easily accompany all that they continue to do. Under the freer flow of these powers their uniqueness will be refined and flourish, and correspondingly, so too will the uniqueness of their respective organizations.

An organization may be political and hence be primarily driven by the power of leadership and courage, or it may be social and hence be primarily driven by the power of service and perfection, or it may be commercial and hence be primarily driven by the power of mutuality and harmony, or it may be research-oriented and academic and hence be primarily driven by the powers of knowledge and wisdom, but always each of the other powers will also be behind it, fulfilling and completing its primary urge. Thus, the dynamic of unique primary powers being supported by an increasing plethora of secondary powers is represented by the following four organizational equations, (8.1.3 – 6) and illustrates how the four powers of light emerge even in complex human-based collectivities:

$$Sig_{Political} = Xa + \overline{Yb_{0-n}}$$

$$where \begin{bmatrix} X \in [S_{System_P}] \\ Y \in [S_{System_{Pr}}, S_{System_P}, S_{System_K}, S_{System_N}] \\ a, b \ are \ integers; a > b \end{bmatrix}$$

Eq 8.1.3: *Generalized Political Organization Signature of the FBLEE-Based Stable-Mega-Organization P-Type Being*

$$Sig_{Social} = Xa + \overline{Yb_{0-n}}$$

$$where \begin{bmatrix} X \in [S_{System_{Pr}}] \\ Y \in [S_{System_{Pr}}, S_{System_P}, S_{System_K}, S_{System_N}] \\ a, b \ are \ integers; a > b \end{bmatrix}$$

Eq 8.1.4: *Generalized Social Organization Signature of the FBLEE-Based Stable-Mega-Organization P-Type Being*

$$Sig_{Commercial} = Xa + \overline{Yb_{0-n}}$$

$$where \begin{bmatrix} X \in [S_{System_N}] \\ Y \in [S_{System_{Pr}}, S_{System_P}, S_{System_K}, S_{System_N}] \\ a, b \ are \ integers; a > b \end{bmatrix}$$

Eq 8.1.5: Generalized Commercial Organization Signature of the FBLEE-Based Stable-Mega-Organization P-Type Being

$$Sig_{Academic} = Xa + \overline{Yb_{0-n}}$$

$$where \begin{bmatrix} X \in [S_{System_K}] \\ Y \in [S_{System_{Pr}}, S_{System_P}, S_{System_K}, S_{System_N}] \\ a, b \ are \ integers; a > b \end{bmatrix}$$

Eq 8.1.6: Generalized Academic Organization Signature of the FBLEE-Based Stable-Mega-Organization P-Type Being

Note that (8.1.3 - 6) already implies that Lines 1 – 5 in the Light Matrix (3.1.3) have been activated, and that the logic of the stable mega-organization- ecosystem will precipitate into FBLEE and may also impact post material-fabric cell-based genetic structure through FBLEEE.

Summary of FBLEE-Based Stable-Mega-Organization P-Type Being and Genetic-Type Code

Summarizing, after the stable mega-organization iteration of the Light-Space-Time Emergence equation is complete, the following code-segments will have been generated as specified by Equation 8.1.7, Active FBLEE-Based Stable-Mega-Organization P-Type Being and Genetic-Type Code:

$$L - S - T \ Emergence_{FBLEE-Based \ Stable-Mega-Orgnization \ P-Type \ Being} =$$

$$\left|\begin{matrix}\begin{bmatrix}c_\infty:[Pr,Po,K,H]\\\left(\downarrow R_{C_K}=f(R_{C_\infty})\right)\\c_K:[S_{Pr},S_{Po},S_K,S_H]\\\left(\downarrow R_{C_N}=f(R_{C_K})\right)\\c_N:f(S_{Pr}\ x\ S_{Po}\ x\ S_K\ x\ S_H)\\\left(\downarrow R_{C_U}=f(R_{C_N})\right)\\c_U:[P,V,M,C]\\\Uparrow\\c_{0:[D,W,I,C]}\end{bmatrix}_{Light}\quad\begin{bmatrix}M_3\ \rightarrow\ System_X\\(\uparrow F\ \rightarrow I)\\M_2\ \rightarrow\ S_{System_X}\\(\uparrow Sig\ \rightarrow F)\\M_1\ \rightarrow\ Sig_x\\(\uparrow>P_x)\\U\ \rightarrow\ x_{Humans}\end{bmatrix}_{Space}\\\begin{bmatrix}M_3:\ -\infty\ \leq t\ \leq\ \infty\\\downarrow\\M_2:0\geq t>\infty\\\downarrow\\M_1:\ 0>t>\infty\\\downarrow\\U\ \rightarrow\ \begin{matrix}t\leq E_{Cell};\text{TC}:M_3\ \rightarrow U\\t\sim E_{Human};\text{TC}:U\ \rightarrow M_3\end{matrix}\end{bmatrix}_{Time}\quad TC\ \rightarrow\ x_{Stable\ Mega-Organization}\end{matrix}\right|_{\langle x_U|x_T\rangle}\ \Rightarrow$$

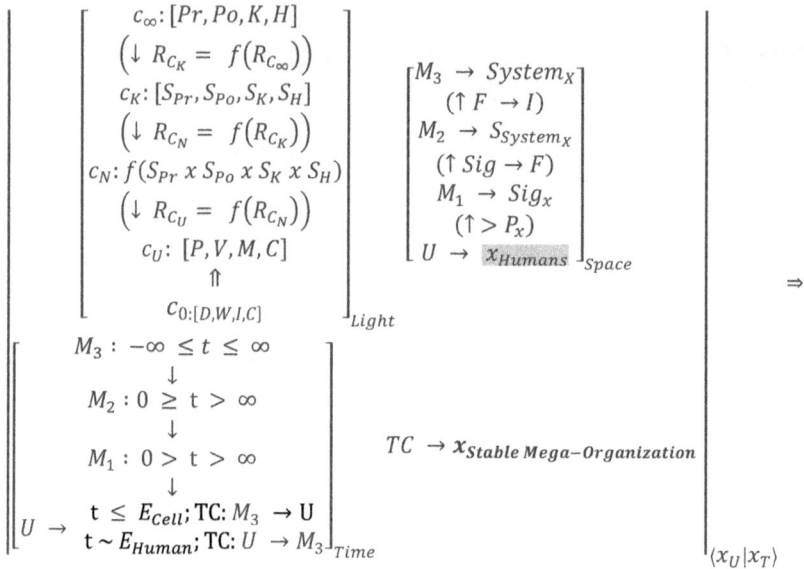

$\langle Space-Time-Energy-Gravity\ P-Type\ Beings\ and\ Pre-Genetic\ Code\rangle+$

$\left(\begin{matrix}Electro-magnetic-wavearchetype-masspotential\ P-Type\ Being\\and\ Pre-Genetic\ Code\end{matrix}\right)+$

$\langle Quantum-Particle\ P-Type\ Being\ and\ Pre-Genetic\ Code\rangle+$

$\langle Atom\ P-Type\ Being\ and\ Pre-Genetic\ Code\rangle+$

$\langle Molecule\ P-Type\ Being\ and\ Pre-Genetic\ Code\rangle+LSTE\ \langle...\rangle+$

$\langle Cell\ P-Type\ Being\ and\ Genetic\ Code\rangle+\ LSTE\ \langle...\rangle+FBLEEE\ \langle...\rangle+$

$\langle Human\ P-Type\ Being\ and\ Genetic\ Code\rangle+LSTE\ \langle...\rangle+FBLEEE\ \langle...\rangle+$

$$\left(\begin{array}{c} \sum Sig_{Commercial} = Xa + \overline{Yb_{0-n}} \\ \left[\begin{array}{c} X \in [S_{System_N}] \\ Y \in [S_{System_{Pr}}, S_{System_P}, S_{System_K}, S_{System_N}] \\ a, b \ are \ integers; a > b \end{array}\right] \\ \sum Sig_{Academic} = Xa + \overline{Yb_{0-n}} \\ \left[\begin{array}{c} X \in [S_{System_K}] \\ Y \in [S_{System_{Pr}}, S_{System_P}, S_{System_K}, S_{System_N}] \\ a, b \ are \ integers; a > b \end{array}\right] \\ \sum Sig_{Political} = Xa + \overline{Yb_{0-n}} \\ \left[\begin{array}{c} X \in [S_{System_P}] \\ Y \in [S_{System_{Pr}}, S_{System_P}, S_{System_K}, S_{System_N}] \\ a, b \ are \ integers; a > b \end{array}\right] \\ \sum Sig_{Social} = Xa + \overline{Yb_{0-n}} \\ \left[\begin{array}{c} X \in [S_{System_{Pr}}] \\ Y \in [S_{System_{Pr}}, S_{System_P}, S_{System_K}, S_{System_N}] \\ a, b \ are \ integers; a > b \end{array}\right] \end{array}\right) + PGCI \langle ... \rangle$$

where (applies to each segment above)

Eq. 8.1.7, Active FBLEE-Based Stable-Mega-Organization P-Type Being and Genetic-Type Code

Note that the final code segment preceding 'PGCI <...>' contains the code for all possible $(\sum x)$ mega-organization types, and depicts the growing biography by which the singular light-based edifice expresses its materialization at FBLEE, and at this stage, through the FBLEE-based stable-mega-organization P-type being.

Chapter 8.2: Generation of FBLEE-Based Sustainable-Global-Civilization P-Type Being

This chapter will consider the quantum-level computation necessary in the generation of the FBLEE-based sustainable-global-civilization P-type being and the genetic-type code segments that accompany that.

Light's Emergence as Sustainable Global Civilization

As discussed in previous chapters the Light-Space-Time Emergence equation (3.1.3) being iterative, can be used to model emergence as it proceeds from simpler four-fold to more complex four-fold manifestations. Hence (3.1.3) has already been applied to suggest the emergence of the space-time-energy-gravity quadrumvirate P-type beings, the electro-magnetic-wavearchetype-masspotential P-type being, the quantum-particle P-type being, the boson P-type being as a further instance of a particular kind of quantum particle, the atom P-type being, the cell P-type being, the fundamental-capacities-of-self P-type being, the truer-individuality P-type being, and the stable-mega-organization P-type being.

The FBLEE-based sustainable-global-civilization P-type being equation (4.2.7) introduced in Chapter 4.2 on post-Big Bang generation of P-type beings, will be leveraged as a starting point, and is reproduced below for convenience. The architecture and details of the FBLEE-based sustainable-global-civilization p-type being can be seen to be the result of the application of the Light, Space, and Time matrices as will be elaborated. But further, as implied by (3.6.12), the Potential Effect of Levels of Light on Genetic-Type Information equation, any process of deeper organization, such as is responsible for the architecture and cohesiveness of the FBLEE-based sustainable-global-civilization P-type being, has the possibility of altering FBLEE, and in this case through FBLEEE influence of altering post material fabric housing structures, so long as the bases involved are driven primarily by a meta-level.

$$L - S - T \ Emergence_{FBLEE-Based \ Sustainable-Global-Civilization \ P-Type \ Being} =$$

$$
\left| \begin{bmatrix} \begin{bmatrix} c_\infty : [Pr, Po, K, H] \\ \left(\downarrow R_{C_K} = f(R_{C_\infty}) \right) \\ c_K : [S_{Pr}, S_{Po}, S_K, S_H] \\ \left(\downarrow R_{C_N} = f(R_{C_K}) \right) \\ c_N : f(S_{Pr} \times S_{Po} \times S_K \times S_H) \\ \left(\downarrow R_{C_U} = f(R_{C_N}) \right) \\ c_U : [P, V, M, C] \\ \Uparrow \\ c_{0:[D,W,I,C]} \end{bmatrix}_{Light} \begin{bmatrix} M_3 \rightarrow System_X \\ (\uparrow F \rightarrow I) \\ M_2 \rightarrow S_{System_X} \\ (\uparrow Sig \rightarrow F) \\ M_1 \rightarrow Sig_x \\ (\uparrow > P_x) \\ U \rightarrow x_{Humans} \end{bmatrix}_{Space} \\ \begin{bmatrix} M_3 : -\infty \le t \le \infty \\ \downarrow \\ M_2 : 0 \ge t > \infty \\ \downarrow \\ M_1 : 0 > t > \infty \\ \downarrow \\ U \rightarrow \begin{matrix} t \le E_{Cell}; TC: M_3 \rightarrow U \\ t \sim E_{Human}; TC: U \rightarrow M_3 \end{matrix} \end{bmatrix}_{Time} \quad TC \rightarrow x_{Sustainbale\ Global\ Civilization} \end{bmatrix} \right\rangle_{\langle x_U | x_T \rangle} \Rightarrow
$$

$\langle Space - Time - Energy - Gravity\ P - Type\ Beings\ and\ Pre - Genetic\ Code \rangle +$

$\left(\begin{matrix} Electro - magnetic - wavearchetype - masspotential\ P - Type\ Being \\ and\ Pre - Genetic\ Code \end{matrix} \right) +$

$\langle Quantum - Particle\ P - Type\ Being\ and\ Pre - Genetic\ Code \rangle +$

$\langle Atom\ P - Type\ Being\ and\ Pre - Genetic\ Code \rangle +$

$\langle Molecule\ P - Type\ Being\ and\ Pre - Genetic\ Code \rangle + LSTE \langle \dots \rangle +$

$\langle Cell\ P - Type\ Being\ and\ Genetic\ Code \rangle + LSTE \langle \dots \rangle + FBLEEE \langle \dots \rangle +$

$\langle Human\ P - Type\ Being\ and\ Genetic\ Code \rangle + LSTE \langle \dots \rangle + FBLEEE \langle \dots \rangle +$

$\langle FBLEE - Based\ Stable\ Mega - Organization\ P - Type\ Being\ and\ Post - Genetic\ Code \rangle +$

$PGCI \langle \dots \rangle +$

$FBLEE - Based\ Sustainable\ Global\ Civilization\ P - Type\ Being\ and\ Post - Genetic\ Code$

301

Note that as indicated by the 'FBLEE-Based *Sustainable – Global – Civilization P –* *Type Being and Post – Genetic Code'* code-segment, the code generated is at the level of FBLEE only.

Starting with the Light-Matrix, the top left-hand matrix in (4.2.7), the first line from the top, $C_\infty: [Pr, Po, K, H]$, specifies the architecture of the FBLEE-Based sustainable-global-civilization P-type being as will be further discussed in this chapter. A key to such a civilization is also pointed to by the previously derived equation (8.1.1), reproduced here for convenience:

$$Sustainability_{Systems} \propto Interaction\ (S_{System_{Pr}}, S_{System_P}, S_{System_K}, S_{System_N})$$

This requires maturity by large organized parts of the world, whether nations or regional blocs. The maturity is such that the fourfold properties of Light are adequately expressed. Hence in the example of nations to be illustrated shortly, a country like India is in its deeper essence perhaps an emergence of Light's property of Knowledge, a country like Japan is in its deeper essence perhaps an emergence of Light's property of Power, a country like Thailand is in its deeper essence perhaps an emergence of Light's property of Presence, and a country like UK is in in its deeper essence perhaps an emergence of Light's property of Harmony. The fundamental architecture of the combination of such emergences, which as per (8.1.1) is required for sustainability, is hence an emergence of the properties of Light at ∞.

Line 3 in the Light-Matrix, $C_K: [S_{Pr}, S_{Po}, S_K, S_H]$, elaborates the sets for Presence, Power, Knowledge, and Harmony, each containing multiple elements. For example, as will be explored, various elements derived from the four sets define the architecture of a knowledge-essence country like India and could be functions such as 'exceptional capacity for penetrating behind the surface', 'meaningfully synthesizing many streams of development', amongst others, hence collectively describing different aspects of a knowledge-essence country's way of being. Specifically, Line 5, $C_N: f(S_{Pr}\ x\ S_{Po}\ x\ S_K\ x\ S_H)$, suggests that unique seeds are created from a combination of such elements from all four sets, with a particular element leading or having more weight, that in effect creates the distinctness possible at the level of FBLEE-based sustainable-global-civilization P-type being.

Line 6, ($\downarrow R_{C_U} = f(R_{C_N})$), normally specifies quantization between the layer where the seeds are formed, and the physical layer, and as explored in Chapter 3.5 and 3.6, will normally result in Line 7, C_U: $[P, V, M, C]$, hence potentially changing the post material-fabric housing structure. In a post-human organization such as a FBLEE-based sustainable-global-civilization P-type being, however, the beginnings of a post-genetic hybrid structure will likely be incrementally reinforced and would require far more functional-richness to materialize before it were to become a reality. Because such a post-genetic housing structure has yet to come into existence (4.2.7) can strictly only be a FBLEE-based partial-singularity.

The possibilities represented by Lines 1 through 5 hence concretize through the quantization represented by Line 6 to enhance the FBLEE-based sustainable-global-civilization P-type being with subtle physical (related to Presence), vital (related to Power), mental (related to Knowledge), and connection (related to Harmony) aspects now at least existing in FBLEE in the quantum-layer behind U. As a reminder FBLEE exists at that barrier where the speed of light is c, but at the quantum-level only. Note that just as Line 6 represents a process of quantization relating the layer of reality created by Light traveling at c with the antecedent layers, so too Lines 2 and 4 as previously discussed, also represent quantization of a more subtle kind that ultimately plays a critical part in allowing FBLEE to express infinite diversity.

Typically, it is the process as captured by the Space-Matrix that will determine if Line 6 is activated. Specifically, patterns at the untransformed layer, U, will need to be overcome, as specified by the second line from the bottom of the Space-Matrix: $(\uparrow > P_x)$. But as specified by the bottom-line of the Time-Matrix, reproduced below, it is only with the advent of the human-system that the automaticity of the action of meta-levels is reversed:

$$U \rightarrow \begin{matrix} t \leq E_{Cell}; \text{TC}: M_3 \rightarrow U \\ t \sim E_{Human}; \text{TC}: U \rightarrow M_3 \end{matrix}$$

Hence in the case of the FBLEE-based sustainable-global-civilization P-type being, which in this emergence is a post-human system, the fact that patterns do need to be overcome means that quantization requires effort to happen. Given this, it is useful to review **Equation 3.6.5, Potential Effect of Levels of Light on Genetic-Type Information:**

$$Potential\ Effect\ of\ Levels\ of\ Light\ on\ Genetic-Type\ Information=$$

$$\begin{bmatrix} STATIC\ \langle|[L][S][T]TC \rightarrow x_T|_{\langle x_U|x_T\rangle}\rangle \\ \times \\ \left((Y > U: Z_Q) \vee (Y \leq U: Z_F) \vee (Y = U: Z_R)\right) \\ \ni \\ \left(\begin{array}{c} Z \in \mathbb{U}\ (Space, Time, Energy, Gravity) \\ Q: Quantization;\ F:\ Fragmentation;\ R:\ Random \end{array}\right) \end{bmatrix} \rightarrow h \ni$$

$$h \in \left(\begin{array}{c} [Q]: Constructive\ zone, \\ [Q]: Constructive\ zone \wedge Constructive\ mutation, \\ [F]: Destructive\ mutation, \\ [R]: Random\ mutation \end{array}\right)$$

Line 1 from the top in the matrix is simply a static form of (3.6.1) the Simplified Light-Space-Time Emergence equation. The static form is designated by 'STATIC' and implies that fundamental operations true of (3.6.1) are being highlighted in (3.6.12). In other words (3.6.1) already has all the operations highlighted in (3.6.12) in it, but by 'freezing' it by making it static, the essential dynamics leading to possible mutations at the genetic level can more clearly be highlighted.

Line 1 is then subjected ($\times$) to a determination of the dominant levels of light that may be active, designated by Line 2, $'\left((Y > U: Z_Q) \vee (Y \leq U: Z_F) \vee (Y = U: Z_R)\right)'$. Unpacking this, '$Y > U$' implies meta-levels are active and as a result it is possible that Z_Q is going to take place (the subscript 'Q' implies quantum-level action). This also implies activation and potential change of FBLEE. The call from below, as it were, may invoke some function that already exists in the subtle-libraries 'above', so that some already existing function may influence FBLEE through ∞-entanglement, K-entanglement, or N-entanglement. This may be thought of as a key-and-lock mechanism, where a deep enough visceral urge from below acting as the key, opens an entangled lock to alter FBLEE as per the visceral urge. '$Y \leq U$' implies that only the untransformed levels are active, and therefore also the sub-level where the speed of light is 0 is active and as a result Z_F is going to take place (the subscript 'F' implies 'fragmentation'). '$Y = U$' implies that all levels are active and as a result Z_R is going to take place (the subscript 'R' implies 'random').

Line 3 elaborates the significance of Z_Q, Z_F, and Z_R. Hence Z is the union of potential quantum-operations of space, time, energy, and gravity, designated by '$Z \in \mathbb{U}\ (Space, Time, Energy, Gravity)$'. But the nature of the operations, as suggested in the previous paragraph, is designated by 'Q: *Quantization*; F: *Fragmentation*; R: *Random*'. Z_Q, then, implies that the full quantization originating from updated four-base logic-encoding ecosystems (FBLEE) can take place. Z_F implies that only a local library, as opposed to FBLEE, can potentially be altered. Z_R also precludes full quantization, and that some partial local-library constructive or destructive mutation may take place.

The '$\rightarrow h \ni h \in$' segment resolves the outcome of the operations implied by Lines 1 – 3, suggesting that the outcome will be 'h' such that ($\ni$) 'h' is an element ($\in$) of the set specified by the members '[Q]: *Constructive zone*', '[Q]: *Constructive zone AND Constructive mutation*', '[F]; *Destructive mutation*', and '{R}: *Random mutation*'. '[Q]: *Constructive zone*' implies that the in-built buffer has been crossed and that access to the deeper four-base logic-encoding ecosystem (FBLEE) has been granted. '[Q]' in this segment implies that there is the possibility that full-quantization as specified by Line 3 of the previous matrix can take place. Access to this zone is a prerequisite for constructive mutation to occur, as designated by the element ' [Q]: *Constructive zone $\cap$ Constructive mutation* ', which generally implies that full-quantization is going to take place. In the case of an early post-human organization though, as previously suggested, since there is not yet a post-genetic structure securely in place to house such full-quantization, this likely reinforces the call to antecedent layers for such a structure to begin to materialize through the activation of PGCI. In the meanwhile, quantization will affect FBLEE, and through FBLEEE may affect human genetic-code. The '[F]' specifies the relationship between 'Fragmentation' in Line 3 of the previous matrix and destructive mutation. The '[R]' specifies the relationship between 'Random' in Line 3 of the previous matrix and random mutation.

Generation of FBLEE-Based Sustainable-Global-Civilization P-Type Being and Genetic-Type Code

A brief look at history will reinforce the idea that it is typically maturity along multiple and distinct dimensions, represented by the four properties of light, and their rich interaction that allows civilizations to endure.

Thus, those civilizations that have endured typically have a balance of all four families (Sri Aurobindo, 1971). Civilizations that have become extinct

305

typically have had a focus on few drivers of innovation. Jared Diamond proposes five interconnected causes of collapse that may reinforce each other: non-sustainable exploitation of resources, climate changes, diminishing support from friendly societies, hostile neighbors, and inappropriate attitudes for change (Diamond, 2005). But these five sources may also be thought of as symptoms that arise due to the failure to adopt the catholicity of the sources of innovation emanating from each of the four sets or families. Further, the historian Toynbee suggested that societies decay because of their over-reliance on structures that helped them solve old problems (Toynbee, 1961). It can be interpreted that being thus biased they are unable to adopt the catholicity of the sources of innovation emanating from each of the four sets of families.

Approaching Civilization from a big-picture, global basis though, the sustainability of humankind will be ensured by a balance of development amongst the four sets of forces. This means that countries and global regions must be unique and in such a way that their primary emergence is distributed amongst all four sets of forces. Further, and based on this uniqueness, there must be an open and healthy interaction amongst these centers of uniqueness.

Hence, India, Japan, Thailand, and UK will be considered as representative examples of the four distinct and emergent properties that must be balanced in the whole.

Historically at least, India, for example, in its essence, may be thought of as having an exceptional capacity for penetrating behind the surface, and further of meaningfully synthesizing many streams of development. Its primary power may thus be from the family of knowledge, with a strong secondary driver being its ability to create living harmonies. The equation for India will then likely be represented by Equation 8.2.1:

$$Sig_{India} = Xa + \overline{Yb_{0-n}} \quad where \begin{bmatrix} X \in [S_{System_K}] \\ Y \in [S_{System_{Pr}}, S_{System_P}, S_{System_K}, S_{System_N}] \\ a, b \ are \ integers; a > b \end{bmatrix}$$

Eq 8.2.1: India (Knowledge Family)

Japan, in its essence, may be thought of as having a strong and noble warrior nature, along with a refined sense of aesthetics, amongst other qualities. Its equation, Equation 8.2.2 would be of the form:

306

$$Sig_{Japon} = Xa + \overline{Yb_{0-n}} \quad where \quad \begin{bmatrix} X \in [S_{System_P}] \\ Y \in [S_{System_{Pr}}, S_{System_P}, S_{System_K}, S_{System_N}] \\ a, b \ are \ integers; a > b \end{bmatrix}$$

Eq 8.2.2: Japan (Power Family)

UK, in its essence, may be thought of as having a strong ability to create practical, materialistic harmonies, resulting in such things as working parliaments and advanced democracy, for example. Its equation, Equation 8.2.3, would be of the form:

$$Sig_{UK} = Xa + \overline{Yb_{0-n}} \quad where \quad \begin{bmatrix} X \in [S_{System_N}] \\ Y \in [S_{System_{Pr}}, S_{System_P}, S_{System_K}, S_{System_N}] \\ a, b \ are \ integers; a > b \end{bmatrix}$$

Eq 8.2.3: UK (Nurturing Family)

Thailand, in its essence, may be characterized by an exceptional sense of hospitality and sweet service, with an attention to detail in the practical arrangement of things. Its equation, Equation 8.2.4, would be of the form:

$$Sig_{Thailand} = Xa + \overline{Yb_{0-n}} \quad where \quad \begin{bmatrix} X \in [S_{System_N}] \\ Y \in [S_{System_{Pr}}, S_{System_P}, S_{System_K}, S_{System_N}] \\ a, b \ are \ integers; a > b \end{bmatrix}$$

Eq 8.2.4: Thailand (Service Family)

Note that (8.2.1 - 4) already implies that Lines 1 – 5 in the Light Matrix (3.1.3) have been activated, and that the logic of the FBLEE-based sustainable-global-civilization P-type being ecosystem will precipitate into FBLEE and may also impact post material-fabric cell-based genetic structure through FBLEEE.

Similarly, every nation on earth will have a uniqueness that can be represented by equations belonging to one of the four families.

As discussed in Chapter 2.5, A Process of Becoming Involving Mutations for the P-Type Being, for uniqueness to emerge and mature, is a process. The process is represented by Equation 2.5.3 reproduced here for convenience:

$$Sig_E = X \begin{vmatrix} C: Sig * mod \left(\int = 1 \right) \\ F: Sig \; mod \; (c) \\ I: Sig \; mod \left(\int \overline{G, e, \pi} \right) \\ M: Sig * mod \; (G) \\ V: Sig * mod \; (e) \\ P: Sig * mod \; (\pi) \end{vmatrix}$$

Equations 8.2.1 – 4 represent example of the essence of uniqueness. For these to become living practicalities requires work at many different levels within a nation. If a nation has not really done the work to animate itself with its uniqueness then it may exist in the physical or P-state, or vital or V-state, or mental or M-state, which by definition lack sufficient maturity, and interaction with other nations is then going to be compromised. Moving to the integral or I-state, will allow a nation to at least not be locked into a point of view. A consistent I-state practiced by each nation then is the minimum requirement for a sustainable global civilization. But a consistent I-state also means that restricting patterns are always being broken and that the quantization effect of organization on the post material-fabric is actively in play. In other words, separations of light are continually uniting with the larger continent of Light and changing the very nature of possibility on larger and larger scale.

Summary of FBLEE-Based Sustainable-Global-Civilization P-Type Being and Genetic-Type Code

Summarizing, after the sustainable global civilization stage iteration the Light-Space-Time Emergence equation is complete, the following code-segments will have been generated as specified by Equation 8.2.5, Active FBLEE-Based Sustainable-Global-Civilization P-Type Being and Genetic-Type Code:

$Light - Space - Time \; Emergence_{FBLEE \; Sustainable-Global-Civilization \; P-Type \; Being} =$

$$\left[\begin{array}{l} \left[\begin{array}{c} c_{\infty}: [Pr, Po, K, H] \\ \left(\downarrow R_{C_K} = f(R_{C_\infty})\right) \\ c_K: [S_{Pr}, S_{Po}, S_K, S_H] \\ \left(\downarrow R_{C_N} = f(R_{C_K})\right) \\ c_N: f(S_{Pr} \times S_{Po} \times S_K \times S_H) \\ \left(\downarrow R_{C_U} = f(R_{C_N})\right) \\ c_U: [P, V, M, C] \\ \Uparrow \\ c_{0:[D,W,I,C]} \end{array}\right]_{Light} \\ \left[\begin{array}{l} M_3: -\infty \le t \le \infty \\ \qquad \downarrow \\ M_2: 0 \ge t > \infty \\ \qquad \downarrow \\ M_1: 0 > t > \infty \\ \qquad \downarrow \\ U \to \begin{array}{l} t \le E_{Cell}; \text{TC}: M_3 \to U \\ t \sim E_{Human}; \text{TC}: U \to M_3 \end{array} \end{array}\right]_{Time} \end{array}\right.$$

$$\left.\begin{array}{c} \left[\begin{array}{c} M_3 \to System_X \\ (\uparrow F \to I) \\ M_2 \to S_{System_X} \\ (\uparrow Sig \to F) \\ M_1 \to Sig_x \\ (\uparrow > P_x) \end{array}\right] \\ U \to x_{Stable\ Mega-Organization} \end{array}\right]_{Space}$$

$$TC \to x_{Sust.Global\ Civilization}$$

$$\Rightarrow \quad \langle x_U | x_T \rangle$$

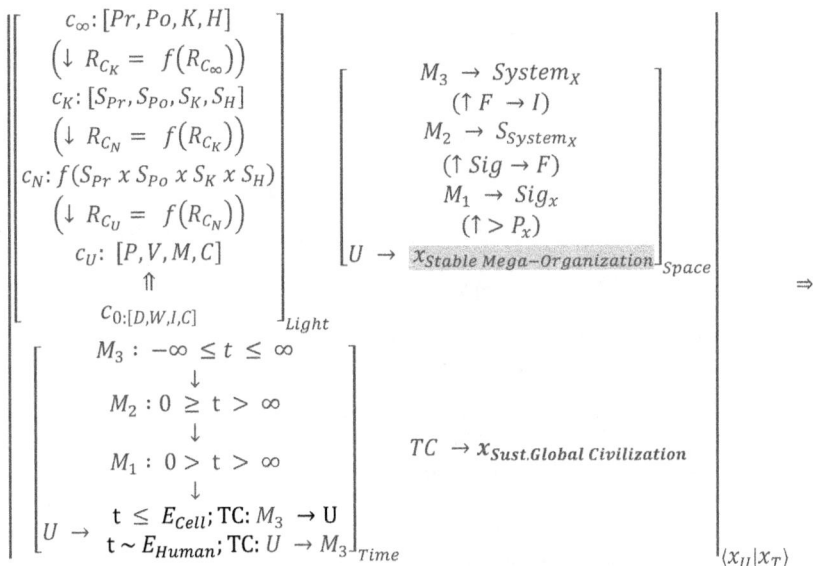

$\langle Space - Time - Energy - Gravity\ P - Type\ Beings\ and\ Pre - Genetic\ Code\rangle +$

$\left(\begin{array}{c} Electro - magnetic - wave archetype - masspotential\ P - Type\ Being \\ and\ Pre - Genetic\ Code \end{array}\right) +$

$\langle Quantum - Particle\ P - Type\ Being\ and\ Pre - Genetic\ Code\rangle +$

$\langle Atom\ P - Type\ Being\ and\ Pre - Genetic\ Code\rangle +$

$\langle Molecule\ P - Type\ Being\ and\ Pre - Genetic\ Code\rangle + LSTE\ \langle ... \rangle +$

$\langle Cell\ P - Type\ Being\ and\ Genetic\ Code\rangle + LSTE\ \langle ... \rangle + FBLEEE\ \langle ... \rangle +$

$\langle Human\ P - Type\ Being\ and\ Genetic\ Code\rangle + LSTE\ \langle ... \rangle + FBLEEE\ \langle ... \rangle +$

$\langle FBLEE - Based\ Stable\ Mega - Organization\ P - Type\ Being\ and\ Post - Genetic\ Code\rangle +$

PGCI $\langle ... \rangle + LSTE < \cdots > +$

$$
\left(
\begin{array}{l}
\sum Sig_{Nurturing-Nation} = Xa + \overline{Yb_{0-n}} \\[4pt]
where \left[
\begin{array}{l}
X \in [S_{System_N}] \\
Y \in [S_{System_{Pr}}, S_{System_P}, S_{System_K}, S_{System_N}] \\
a, b \ are \ integers; a > b
\end{array}
\right] \\[10pt]
\sum Sig_{Knowlegde-Nation} = Xa + \overline{Yb_{0-n}} \\[4pt]
where \left[
\begin{array}{l}
X \in [S_{System_K}] \\
Y \in [S_{System_{Pr}}, S_{System_P}, S_{System_K}, S_{System_N}] \\
a, b \ are \ integers; a > b
\end{array}
\right] \\[10pt]
\sum Sig_{Power-Nation} = Xa + \overline{Yb_{0-n}} \\[4pt]
where \left[
\begin{array}{l}
X \in [S_{System_P}] \\
Y \in [S_{System_{Pr}}, S_{System_P}, S_{System_K}, S_{System_N}] \\
a, b \ are \ integers; a > b
\end{array}
\right] \\[10pt]
\sum Sig_{Service-Nation} = Xa + \overline{Yb_{0-n}} \\[4pt]
where \left[
\begin{array}{l}
X \in [S_{System_{Pr}}] \\
Y \in [S_{System_{Pr}}, S_{System_P}, S_{System_K}, S_{System_N}] \\
a, b \ are \ integers; a > b
\end{array}
\right]
\end{array}
\right)
$$

Eq. 8.2.5, Active FBLEE-Based Sustainable-Global-Civilization **P-Type Being and** *Genetic-Type Code*

Note that the final code segment contains the code for all possible ($\sum x$) nation types and depicts the growing biography by which the singular light-based edifice expresses its materialization, and at this stage, through the FBLEE-Based sustainable-global-civilization P-type being.

SECTION 9: EXTRAPOLATION TO SUPER-MATTER-BASED P-TYPE BEINGS

Sections 2 and 3 proposed a mathematical framework for being and becoming. Sections 4 through 8 leveraged the mathematical framework to explore the biographies of seed-singularity based and P-type beings. Such biographies are entirely generated by key aspects of the light-space-time emergence equation. Components of such biographies include the essential architecture of the P-type being generated from the four principal characteristics of light - which is an action of the UPI-type being, the four sets of properties related to the principal characteristics of light - which is an action of the K-type being, unique seeds created from combinations of elements derived from the four sets - which is an action of the N-type being, and the essential dynamics of maturation, or becoming, culminating in possible quantization. Hence, in the biography of any P-type being there is active participation by all seed-singularity based beings – the UPI-type being, the K-type being, and the N-type being - as summarized by different levels of light in the Light-Matrix, and by the Space-Matrix and Time-Matrix that detail the conditions of becoming.

Previous sections elaborated the broad lines of emergent biographies of the P-type being. This section will go a step further and examine micro-level dynamics due to the action of the space-time-energy-gravity micro-being (Chapter 9.1 will look at insect adaptation leveraging a discussion on cells in Chapter 7.2, and Chapter 9.2 will look at sustainable global civilization adaptation leveraging a discussion in Chapter 8.2) leading to meta-level induced change in specific biographies, and macro-implications that will result in possible creation of super-matter (Chapter 9.3).

Super-matter is suggested to be that type of matter created through the intervention of conscious will or cohesive want. While complexification of matter as summarized in previous sections is often an "automatic" process of Nature, super-matter is a foundation based on will or cohesive want and sets the stage for a potentially unending willed development in which functional-richness existing in Light in its native state traveling at infinite speed, can manifest in this material universe. From a macro-level, meta-level bases or more conscious integration with seed-singularity beings in an increasing triumph of love, is what will ensure the development of super-matter.

Creation of such super-matter is tied to the meaning of space, and to cosmic observations of the expansion of the universe (Chapter 9.4), hence involving the space-time-energy-gravity macro-being. Chapter 9.4 hence focuses on the nature of matter by drawing on observations from the field of astrophysics.

The gestalt of such an analysis may engender a visceral sense of the intimacy and oneness of the light-based adventure that results in the vastness of the cosmos tied to human-precipitated determinable dynamics at the quantum-level. As such the persistent quantum-level computation resulting in a constant stream of genetic-type information alters the code and the very light-based computational machinery emergent as more and more sophisticated pre-matter, matter, and post or super-matter structures. Genetics then has to be viewed more broadly as also integrating instructions or insights from our larger and larger collective endeavors to hook or connect us more intimately with the world around us. It is foreseeable that such genetic-type information expands the action of nucleic acids, proteins, lipids, and polysaccharides to even allow the cell and therefore a body to act or morph in unforeseen ways. The boundaries between the world outside and the world inside may thin and dissolve, and all life progressively become one unbroken material though intensely diverse totality precipitating the oneness and infinity of light. Matter then, or super-matter, becomes the medium in which this grand adventure of light is experienced. Any transhumanism traces the steps by which the adventure is pursued.

Leveraging the discussion on the broad lines of the emerging cell p-type being from Chapter 7.2, Generation of Cell P-Type Being, this chapter will specifically consider how aspects of this biography are determined in the case of the adaptation of an insect. Such adaptation signals the successful change in biography of the insect P-type being, and in the case elaborated in this chapter, due to the interaction of sufficient urge that precipitates a process of quantization. Such an urge-to-quantization process sheds insight into the dynamics of becoming of the P-type being.

Recall that the 'quantum' is a window into layers of reality behind the surface layer U. This insight was captured by Schrodinger's Equation (discussed in Chapter 3.3) and by Heisenberg's Uncertainty Principle (discussed in Chapter 3.4), the former suggesting that matter, or what is going to appear materially, is a function of some incredible number of superposed states containing infinite possibility, and the latter suggesting that quantum fluctuation due to an always buzzing pregnant-infinity, lies behind everything even when seemingly still.

Recall that the basis of cell adaptability was discussed in Chapter 7.2, and (7.2.1) modeling generation of the cell P-type being is reproduced here:

$$Light - Space - Time\ Emergence_{Cell\ P-Type\ Being} =$$

$$
\begin{bmatrix}
\begin{bmatrix}
c_\infty: [Pr, Po, K, H] \\
\left(\downarrow R_{C_K} = f\left(R_{C_\infty}\right) \right) \\
c_K: [S_{Pr}, S_{Po}, S_K, S_H] \\
\left(\downarrow R_{C_N} = f\left(R_{C_K}\right) \right) \\
c_N: f(S_{Pr} \times S_{Po} \times S_K \times S_H) \\
\left(\downarrow R_{C_U} = f\left(R_{C_N}\right) \right) \\
c_U: [P, V, M, C] \\
\Uparrow \\
c_{0:[D,W,I,C]}
\end{bmatrix}_{Light}
&
\begin{bmatrix}
M_3 \rightarrow System_X \\
(\uparrow F \rightarrow I) \\
M_2 \rightarrow S_{System_X} \\
(\uparrow Sig \rightarrow F) \\
M_1 \rightarrow Sig_x \\
(\uparrow > P_x) \\
U \rightarrow x_{Molecules}
\end{bmatrix}_{Space}
& \Rightarrow \\[2em]
\begin{bmatrix}
M_3 : -\infty \leq t \leq \infty \\
\downarrow \\
M_2 : 0 \geq t > \infty \\
\downarrow \\
M_1 : 0 > t > \infty \\
\downarrow \\
U \rightarrow \begin{matrix} t \leq E_{Cell}; \text{TC}: M_3 \rightarrow U \\ t \sim E_{Human}; \text{TC}: U \rightarrow M_3 \end{matrix}
\end{bmatrix}_{Time}
&
\begin{matrix} TC \rightarrow x_{Cells} \end{matrix}
&
\langle x_U | x_T \rangle
\end{bmatrix}
$$

$\langle Space - Time - Energy - Gravity \ P - Type \ Beings \ and \ Pre - Genetic \ Code \rangle +$

$\left(\dfrac{Electro - magnetic - wavearchetype - masspotential \ P - Type \ Being}{and \ Pre - Genetic \ Code} \right) +$

$\langle Quantum - Particle \ P - Type \ Being \ and \ Pre - Genetic \ Code \rangle +$

$\langle Atom \ P - Type \ Being \ and \ Pre - Genetic \ Code \rangle +$

$\langle Molecule \ P - Type \ Being \ and \ Pre - Genetic \ Code \rangle + LSTE \ \langle ... \rangle +$

$PGGT < \cdots > + Cell \ P - Type \ Being \ and \ Genetic \ Code$

The Light-Matrix, the top-left matrix in (7.2.1), suggests the possibilities in the pregnant-infinity. These possibilities, recall, have been set up by realities so created by light traveling at different speeds. While prevalent dynamics at the surface layer, U, are always influenced by dynamics from the deeper layers, there is a state that can be created that will allow a more focused and intentional activation of the meta-functions resident in the deeper layers. Such activation will potentially trigger a series of processes culminating in a state of quantum-certainty (Malik, 2017e), introduced in Chapter 3.5 when taking a deeper look at STEG quantization, and as will be explored through the rest of this chapter.

The Space-Matrix, the top-right matrix in (7.2.1), suggests the dynamics of becoming involved in creating such an activation-state. Essentially this involves the overcoming of patterns at U.

Insect Adaptation Case

Consider the case of a hypothetical insect, for example, that is always prey to other predators. At some point a visceral urge to overcome some of these predators may arise in this species of insect. This visceral urge is a breaking of habitual patterns common to that species of insect. If it is deep enough and pervasive enough, it can be thought of as an activation-state and will allow the opening of a quantum-window so that there is now a connection with the quantum worlds, Q (as introduced in Chapter 3.2), behind.

It is proposed that for successful adaptability to come about it is only such a pervasive state that will open a quantum-window to effectively stimulate the interaction between layers that will ultimately allow adaptability to occur. This activation-state can be further specified by leveraging (3.7.4) – The Generalized Equation for Direction of P-Type Becoming, that elaborates some Space-Matrix dynamics and was derived in Chapter 3.7, on the application of qualified determinism in the becoming of the P-type being.

Reproducing Equation 3.7.4:

$$P - Type\_Becoming\_Dir = DI \left(\begin{bmatrix} M_3 \rightarrow System_X \\ (\uparrow F \rightarrow I) \\ M_2 \rightarrow S_{System_X} \\ (\uparrow Sig \rightarrow F) \\ M_1 \rightarrow Sig_X \\ (\uparrow > P_P) \\ U \rightarrow x_U \end{bmatrix}^{x=p,v,m,i} \right) \rightarrow$$

$x\_matrix_{strongest} @ level_{strongest}$

The activation-state can be thought of as being invoked when $level_{strongest}$ is at least M_1. Hence, as in Equation 9.1.1:

$$Activation - State_{condition} \geq M_1$$

Eq 9.1.1: Condition for Activation-State

If the activation-state is invoked, that implies access to the four-base logic-encoding ecosystem (FBLEE).

Note that the Time-Matrix, in the lower-left side of (7.2.1), specifies that approximately, adaptability may tend to be more automatic up to the emergence of cellular organisms. This is specified by the direction $M_3 \rightarrow$ U, which indicates that it is the meta-level that organizes activity at U. At approximately the human level the direction is flipped as specified by $U \rightarrow M_3$, and suggests that will, intention, feeling and the like are more important in stimulating the process of adaptability.

The visceral urge, then, may stimulate interaction with a collective-intelligence or 'specific-species-intelligence meta-function' at FBLEE, which allows the generation of a new and specific 'predator-overcoming meta-function' so that the hypothetical insect in question can now go through an adaptation to survive at least some kinds of predator attacks. This may take the form of this species of insect creating a hard-shell around it. Mathematically this new meta-function may take the following form as suggested by Equation 9.1.2:

$$Sig_{hard-protective-shell} = Xa + \overline{Yb_{0-n}}$$

$$where \begin{bmatrix} X \in [S_{System_P}] \\ Y \in [S_{System_{Pr}}, S_{System_P}, S_{System_K}, S_{System_N}] \\ a, b \ are \ integers; a > b \end{bmatrix}$$

Eq 9.1.2: Hard Protective Shell Meta-Function

In this example, S_{System_P} relates to a set of Polysaccharides, as introduced in Chapter 7.2, and suggests that the chains of sugar molecules will adapt to become a shell to protect the insect. The primary element X is therefore an element of the set or Polysaccharides. $S_{System_{Pr}}$ refers to the set of Proteins. S_{System_K} refers to the set of Nucleic Acids. S_{System_N} refers to the set of Lipids. Y as the secondary element will invoke the action of some proteins, some existing polysaccharides, some nucleic acids, and some lipids in bringing about the adaptation as specified by $Sig_{hard-protective-shell}$. Perhaps the nucleic acids will code or coordinate how the proteins will work with the existing polysaccharides and lipids to create a new arrangement of polysaccharides that becomes the protective layer for the insect. This new arrangement of polysaccharides then becomes the purpose of a particular new type of specialized cell that exists to protect the insect against certain types of predators.

So, while the breaking of patterns as specified by the Space-Matrix was a first step in the process, the subsequent creation of a new meta-function can be thought of as a second step. This new meta-function can be thought of as happening due to some combined functioning of the Light and Space Matrices. Specifically, in the Space-Matrix, the breaking of patterns, ($\uparrow > P_x$), allows the dynamics of M_1 to become active: $M_1 \rightarrow Sig_x$. This allows the ever-present layers of light, and specifically $C_N: f(S_{Pr} \times S_{Po} \times S_K \times S_H)$ in the Light-Matrix to become more consciously active so that a new meta-function that may add to the insect-specific FBLEE is generated.

The third step is suggested by ($\downarrow R_{C_U} = f(R_{C_N})$) that allows specific quantization to occur. It is proposed that the hard-protective-shell adaptability becomes real through a series of tightly coordinated space, time, energy, and gravity quantization as specified by (3.5.2-7) derived in Chapter 3.5, in other words, through the action of the space-time-energy-gravity micro-being. A relevant subset of equations is reproduced here for convenience and suggests the mechanics for the adaptability along this specific line of development.

Equation 3.5.2 models space-quantization:

$$Space_{quantization} = h_{UK}(Xa + \overline{Yb_{0-n}})$$

$$where: \begin{bmatrix} X \in [S_{System_K}] \\ Y \in [S_{System_{Pr}}, S_{System_P}, S_{System_K}, S_{System_N}] \\ a, b\ are\ integers; a > b \end{bmatrix}$$

Recall that space-quantization further structures space so that seeds of knowledge or knowledge-potential to do with the new meta-function are now resident in the material ecosystem associated with the meta-function. In this case, the material ecosystem is the cell-based genetic structure, as indicated by the action of PGGT in (7.2.1). In such a manner the species that set this meta-function into action will more and more tap into the possibilities to bring about the new protective mechanism it seeks.

Equation 3.5.4 models time-quantization:

$$Time_{quantization} = h_{UP}(Xa + \overline{Yb_{0-n}})$$

$$where: \begin{bmatrix} X \in [S_{System_P}] \\ Y \in [S_{System_{Pr}}, S_{System_P}, S_{System_K}, S_{System_N}] \\ a, b \text{ } are \text{ } integers; a > b \end{bmatrix}$$

Recall that time-quantization alters the structure of time so that phases of maturity associated with seeds in space are now encoded in the material ecosystem. Slow and faster periods of change are in this way connected to the species seeking to protect itself against predators.

Equation 3.5.6 models energy-quantization:

$$Energy_{quantization} = h_{UPr}(Xa + \overline{Yb_{0-n}})$$

$$where: \begin{bmatrix} X \in [S_{System_{Pr}}] \\ Y \in [S_{System_{Pr}}, S_{System_P}, S_{System_K}, S_{System_N}] \\ a, b \text{ } are \text{ } integers; a > b \end{bmatrix}$$

Recall that energy-quantization allows matter to be formed, and in this case, will have to do with the incremental material changes that the species will go through in forming a hard-protective shell.

Equation 3.5.8 models gravity-quantization:

$$Gravity_{quantization} = h_{UN}(Xa + \overline{Yb_{0-n}})$$

$$where: \begin{bmatrix} X \in [S_{System_N}] \\ Y \in [S_{System_{Pr}}, S_{System_P}, S_{System_K}, S_{System_N}] \\ a, b \text{ } are \text{ } integers; a > b \end{bmatrix}$$

Recall that gravity-quantization has to do with inter-relation between the species and surrounding objects and will change the very nature of gravity to allow a subtle new balance in the species interaction with its surrounding so that the deep urge of the species can more easily be fulfilled.

As a result of such adaptation FBLEE is changed and this changes the way in which matter can materialize. It is such changes initiated by deep urge or will made at the margin that accumulate to enhance the materialization of fourfold richness and therefore of the functionality of matter. An accumulated or threshold volume of such changes will gradually change the very way matter materializes, approaching the reality of materialization of functionally rich super-matter.

Chapter 9.2: Space-Time-Energy-Gravity Micro-Being Based Development Dynamics in a FBLEE-Based Sustainable-Global-Civilization P-Type Being

Leveraging the discussion on the broad lines of the emerging sustainable global civilization P-type being from Chapter 8.2, Generation of a FBLEE-Based Sustainable-Global-Civilization P-Type Being, this chapter will specifically consider how aspects of this biography are determined in the case of the development of a sustainable global civilization. Such development signals the successful change in biography of the sustainable global civilization P-type being, and in the case elaborated in this chapter, due to the interaction of sufficient urge that precipitates a process of quantization to FBLEE. Such an urge-to-quantization process sheds insight into an increasingly important aspect of the dynamics of becoming of P-type beings, especially when humans begin to play a part.

The Light-Space-Time Emergence equation (4.2.7) leveraged in Chapter 8.2 is reproduced here for convenience:

$$Light - Space - Time\ Emergence_{FBLEE\ Sustainable\ Global\ Civilization\ P-Type\ Being} =$$

$$\left| \begin{array}{l} \begin{bmatrix} \begin{array}{c} c_\infty : [Pr, Po, K, H] \\ \left(\downarrow R_{C_K} = f\left(R_{C_\infty}\right) \right) \\ c_K : [S_{Pr}, S_{Po}, S_K, S_H] \\ \left(\downarrow R_{C_N} = f\left(R_{C_K}\right) \right) \\ c_N : f(S_{Pr} \times S_{Po} \times S_K \times S_H) \\ \left(\downarrow R_{C_U} = f\left(R_{C_N}\right) \right) \\ c_U : [P, V, M, C] \\ \Uparrow \\ c_{0:[D,W,I,C]} \end{array} \end{bmatrix}_{Light} \begin{bmatrix} M_3 \rightarrow System_X \\ (\uparrow F \rightarrow I) \\ M_2 \rightarrow S_{System_X} \\ (\uparrow Sig \rightarrow F) \\ M_1 \rightarrow Sig_X \\ (\uparrow > P_x) \\ U \rightarrow x_{Human} \end{bmatrix}_{Space} \\ \\ \left[U \rightarrow \begin{array}{c} M_3 : -\infty \leq t \leq \infty \\ \downarrow \\ M_2 : 0 \geq t > \infty \\ \downarrow \\ M_1 : 0 > t > \infty \\ \downarrow \\ t \leq E_{Cell}; TC: M_3 \rightarrow U \\ t \sim E_{Human}; TC: U \rightarrow M_3 \end{array} \right]_{Time} \qquad TC \rightarrow x_{FBLEE\ Sust.Global\ Civilization} \end{array} \right| \Rightarrow$$

$$|x_U|x_T\rangle$$

$$\langle Space - Time - Energy - Gravity\ P - Type\ Beings\ and\ Pre - Genetic\ Code \rangle +$$

$$\left(\begin{array}{c} Electro-magnetic-wavearchetype-masspotential\ P-Type\ Being \\ and\ Pre-Genetic\ Code \end{array} \right) +$$

$$\langle Quantum-Particle\ P-Type\ Being\ and\ Pre-Genetic\ Code \rangle +$$

$$\langle Atom\ P-Type\ Being\ and\ Pre-Genetic\ Code \rangle +$$

$$\langle Molecule\ P-Type\ Being\ and\ Pre-Genetic\ Code \rangle + LSTE\ \langle ... \rangle +$$

$$\langle Cell\ P-Type\ Being\ and\ Genetic\ Code \rangle + LSTE\ \langle ... \rangle + FBLEEE\ \langle ... \rangle +$$

$$\langle Human\ P-Type\ Being\ and\ Genetic\ Code \rangle + LSTE\ \langle ... \rangle + FBLEEE\ \langle ... \rangle +$$

$$\langle FBLEE-Based\ Stable\ Mega-Organization\ P-Type\ Being\ and\ Post-Genetic\ Code \rangle +$$

$$PGCI\ \langle ... \rangle +$$

$$FBLEE-Based\ Sustainable\ Global\ Civilization\ P-Type\ Being\ and\ Post-\text{Genetic Code}$$

The Light-Matrix, the top-left matrix in (4.2.7), recall, has been set up by realities so created by light traveling at different speeds. While prevalent dynamics at the surface layer, U, are always influenced by dynamics from the deeper layers, there is an activation-state that can be created that will allow a more focused and intentional activation or creation of the meta-functions in the deeper layers.

The Space-Matrix, the top-right matrix in (4.2.7), suggests the dynamics involved in creating such an activation-state. Essentially this involves the overcoming of patterns at U. Note that by definition such an activation-state necessitates moving in the direction where the speed of light increases. Note that speed of light increasing implies that the sense of separation is diminishing, and entities can have a more complete access to the fullness that they are. This can only happen when patterns get elevated. If patterns are driven primarily by the U realm, being therefore untransformed, that will by definition not allow a quantum-window to open.

This activation-state can be further specified by leveraging (3.7.4) – The Generalized Equation for Direction of P-Type Becoming, that elaborates some Space-Matrix dynamics and was derived in Chapter 3.7, on the application of qualified determinism in the becoming of the P-type being.

Reproducing Equation 3.7.4:

$$P - Type\_Becoming\_Dir = DI \left(\begin{bmatrix} M_3 \rightarrow System_X \\ (\uparrow F \rightarrow I) \\ M_2 \rightarrow S_{System_X} \\ (\uparrow Sig \rightarrow F) \\ M_1 \rightarrow Sig_X \\ (\uparrow > P_P) \\ U \rightarrow x_U \end{bmatrix}_{x=p,v,m,i} \right) \rightarrow$$

$x\_matrix_{strongest} @ level_{strongest}$

Activation-state can be thought of as being invoked when $level_{strongest}$ is at least M_1. Hence, as suggested by the previously derived (9.1.1), reproduced here for convenience:

$Activation - state_{condition} \geq M_1$

If the activation-state is invoked, that implies access to the four-base logic-encoding ecosystem (FBLEE).

Deeper Nation Essence Foundation Case

Similar to the discussion in the previous chapter, assume that there is a deep wanting felt by a threshold number of people. Perhaps this wanting, a step towards a global sustainable civilization, is for each nation to operate more from their true essence, or I-state, as opposed to fundamentalist form. In other words a meta-function, $Sig_{Nation-essence-foundation}$, must be created that will then organize its own reality, and as indicated by Equation 9.2.1.

$Sig_{Nation-essence-foundatiuon} = Xa + \overline{Yb_{0-n}}$

$$where \begin{bmatrix} X \in [S_{System_{Pr}}] \\ Y \in [S_{System_{Pr}}, S_{System_P}, S_{System_K}, S_{System_N}] \\ a, b \text{ are integers}; a > b \end{bmatrix}$$

Eq 9.2.1: Nation-Essence-Foundation Meta-Function

While the key aspect, or primary X-element in the nation-essence-foundation meta-function may be an element from the set of Presence, such as 'nation-essence appreciation', there will be many parts or sub-functions, that could be represented by Y-elements, such as the structure, processes, institutions, culture, cohesion that will also need to be created to ensure such nation essence foundation. Cohesion, as a sub-function, represents an important denominator in breaking patterns at the individual and collective levels and as in Equation 9.2.2, Nation-Essence-Cohesion, is represented with the primary element belonging to the set of Harmony or Nurturing:

$$Sig_{Nation-essence-cohesion} = Xa + \overline{Yb_{0-n}}$$

$$where \begin{bmatrix} X \in [S_{System_N}] \\ Y \in [S_{System_{Pr}}, S_{System_P}, S_{System_K}, S_{System_N}] \\ a, b \ are \ integers; a > b \end{bmatrix}$$

Eq 9.2.2: Nation-Essence-Cohesion Meta-Function

Hence, S_{System_K} relates to the set of Thoughts required to contextualize and frame nation-essence. $S_{System_{Pr}}$ refers to the set of Sensations required to assess and interpret the constantly arising signs of the new development – the sensory cues as it were that will allow any individual or collectivity to sense that they were on the right path. S_{System_P} refers to the set of Urges and Wills constantly required to ensure that the goal of cohesion was attained. S_{System_N} refers to the set of Feelings and Emotions that will need to be generated to attain cohesion.

As such cohesion becomes a reality, older patterns will break down and there will be easier and longer activation-state periods, potentially allowing the additional related sub-functions to also become increasingly active.

As suggested by the previous chapter, while the breaking of patterns as specified by the Space-Matrix was a first step in the process to making 'cohesion' a reality, the subsequent creation of a new meta-function of cohesion can be thought of as a second step. This new meta-function can be thought of as happening due to some combined functioning of the Light and Space Matrices. Specifically, in the Space-Matrix, the breaking of patterns, $(\uparrow > P_x)$, allows the dynamics of M_1 to become active: $M_1 \to Sig_x$. This allows the ever-present layers of light, and specifically $C_{N:} f(S_{Pr} \ x \ S_{Po} \ x \ S_K \ x \ S_H)$ in the Light-Matrix to become more consciously active so that new meta-function that may add to the FBLEE-based sustainable-global-civilization P-type being.

The third step is suggested by ($\downarrow R_{C_U} = f(R_{C_N})$) that allows specific quantization to occur. Quantization is of fundamental importance because in this model it is what allows the fabric of experienced reality to change. The new thought to do with cohesion is quantized as seeds of knowledge in space, fundamentally changing the structure of space. Hence, reproducing Equation 3.5.2, space-quantization is modeled as:

$$Space_{quantization} = h_{UK}(Xa + \overline{Yb_{0-n}})$$

$$where: \begin{bmatrix} X \in [S_{System_K}] \\ Y \in [S_{System_{Pr}}, S_{System_P}, S_{System_K}, S_{System_N}] \\ a, b \ are \ integers; a > b \end{bmatrix}$$

The phases of maturity that the seeds of knowledge will go through give time a new reality or structure. In other words, time too is re-oriented to promote the outcome of the new cohesion meta-function. This quantization of time, essentially giving time in this new fabric a different meaning is modeled by Equation 3.5.4, reproduced here for convenience:

$$Time_{quantization} = h_{UP}(Xa + \overline{Yb_{0-n}})$$

$$where: \begin{bmatrix} X \in [S_{System_P}] \\ Y \in [S_{System_{Pr}}, S_{System_P}, S_{System_K}, S_{System_N}] \\ a, b \ are \ integers; a > b \end{bmatrix}$$

The new meta-function changes the very nature of matter, which is quantized to allow the material fabric of existence to promote the new meta-function of cohesion in unimaginable ways. This quantization of energy, resulting in a subtly different matter is modeled by Equation 3.5.6, reproduced here for convenience:

$$Energy_{quantization} = h_{UPr}(Xa + \overline{Yb_{0-n}})$$

$$where: \begin{bmatrix} X \in [S_{System_{Pr}}] \\ Y \in [S_{System_{Pr}}, S_{System_P}, S_{System_K}, S_{System_N}] \\ a, b \ are \ integers; a > b \end{bmatrix}$$

As a result, the very force of gravity is altered locally as well, so that its quantization promotes subtle interaction in the related ecosystem that will be predisposed to making the meta-function a reality. This quantization of gravity is modeled by Equation 3.5.8, reproduced here for convenience:

$$Gravity_{quantization} = h_{UN}\left(Xa + \overline{Yb_{0-n}}\right)$$

$$where: \begin{bmatrix} X \in [S_{System_N}] \\ Y \in [S_{System_{Pr}}, S_{System_P}, S_{System_K}, S_{System_N}] \\ a, b \text{ are integers}; a > b \end{bmatrix}$$

As a result of such space-time-energy-gravity micro-being adaptation FBLEE is changed and this changes the way in which matter can materialize. It was proposed in Section 8 that in the case of post-human organizations, such as a sustainable global civilization, the genetic structure of cells might change primarily through FBLEEE. This will have to be the case until such time as some future hybridized structure allows more sophisticated multi-layer genetic-type information to materialize.

It is such changes initiated by deep urge or will made at the margin that accumulate to enhance the materialization of fourfold richness and therefore of the functionality of matter. An accumulated or threshold volume of such changes will gradually change the very way matter materializes, approaching the reality of materialization of functionally rich super-matter.

Chapter 9.3: Macro-Conditions for Transitioning to Super-Matter-Based P-Type Beings

Super-Matter is positioned as being fundamentally different from matter. While matter can be thought of as the result of Nature's automatic working, super-matter can be thought of as the result of a conscious will and cohesive wanting that causes a deliberate process of quantization by which the very fabric of matter is changed. Quantization occurs in any case even when Nature's automatic working drives change. But with conscious will this process can be accelerated, broadened, and heightened, so that a wider and higher possibility of function may consciously materialize.

As suggested in the previous sections such quantization is essentially related to enhancing the functional-richness of matter. Imagine through a cohesive-wanting matter being able to become more flexible, more durable, more stretchable, light at will, heavy at will, change color at will, expand in size, contract in size, amongst infinite other functions and possibilities contained in Light in its native state at infinite speed.

The previous sections explored numerous examples of the process by which matter becomes functionally rich. Such increase in functional richness of matter is synonymous with an increase in a P-type being's sphere of influence as recorded by its genetic-type information based biography. In cases getting closer to, and even beyond the human-system, such increase in functional richness involves a visceral urge that can open the entity to the influence of meta-levels of Light. Such opening also gives access to FBLEE, and as further elaborated in the previous two chapters, an enhanced function can then materialize through a process of quantization. This increasing possibility encoded into FBLEE means that matter can transition to richer versions of itself.

Further, in considering the vast information embedded in the ubiquitous-point-instant pre-genetic ∞-entanglement library, the architectural forces pre-genetic K-entanglement library, and the organizational-uniqueness pre-genetic N-entanglement library, discussed in Chapters 2.2, 2.3, and 2.4 respectively, it is clear that many possible functions already exist in layers of light. The call from below, whether a pattern-breaking will, urge, or desire, likely causes some meta-function or precipitation into FBLEE, from where materialization becomes possible through the process of quantization.

Another way to look at this is summarized in the previously derived (4.2.8) which models the generation of the super-matter-based P-type being:

$$Light - Space - Time \; Emergence_{Suoer-Matter \; Based \; P-Type \; Being} =$$

$$\left\| \begin{bmatrix} \begin{bmatrix} c_\infty: [Pr, Po, K, H] \\ \left(\downarrow R_{C_K} = f(R_{C_\infty}) \right) \\ c_K: [S_{Pr}, S_{Po}, S_K, S_H] \\ \left(\downarrow R_{C_N} = f(R_{C_K}) \right) \\ c_N: f(S_{Pr} \; x \; S_{Po} \; x \; S_K \; x \; S_H) \\ \left(\downarrow R_{C_U} = f(R_{C_N}) \right) \\ c_U: [P, V, M, C] \\ \Uparrow \\ c_{0:[D,W,I,C]} \end{bmatrix}_{Light} \begin{bmatrix} M_3 \rightarrow System_X \\ (\uparrow F \rightarrow I) \\ M_2 \rightarrow S_{System_X} \\ (\uparrow Sig \rightarrow F) \\ M_1 \rightarrow Sig_x \\ (\uparrow > P_x) \\ U \rightarrow x_{Molecule} \end{bmatrix}_{Space} \\ \\ \begin{bmatrix} U \rightarrow \begin{matrix} M_3: -\infty \leq t \leq \infty \\ \downarrow \\ M_2: 0 \geq t > \infty \\ \downarrow \\ M_1: 0 > t > \infty \\ \downarrow \\ t \leq E_{Cell}; TC: M_3 \rightarrow U \\ t \sim E_{Human}; TC: U \rightarrow M_3 \end{matrix} \end{bmatrix}_{Time} \quad TC \rightarrow x_{Super-Matter} \end{bmatrix} \langle x_U | x_T \rangle \right\| \Rightarrow$$

$\langle Space - Time - Energy - Gravity \; P - Type \; Beings \; and \; Pre - Genetic \; Code \rangle +$

$\left(\begin{matrix} Electro - magnetic - wavearchetype - masspotential \; P - Type \; Being \\ and \; Pre - Genetic \; Code \end{matrix} \right) +$

$\langle Quantum - Particle \; P - Type \; Being \; and \; Pre - Genetic \; Code \rangle +$

$\langle Atom \; P - Type \; Being \; and \; Pre - Genetic \; Code \rangle +$

$\langle Molecule \; P - Type \; Being \; and \; Pre - Genetic \; Code \rangle + LSTE \langle ... \rangle + FBLEEE \langle ... \rangle +$

$LSTE \langle ... \rangle + PGCI \langle ... \rangle + Super - Matter. Based \; P - Type \; Being \; and \; Post - Genetic \; Code$

Equation 4.2.8 suggests computational realities yet to emerge. The essential action of quantization due to the space-time-energy-gravity micro-being can cause the materialization of potentially infinite four-base logic-encoding ecosystems. It is conceivable that such a variation of space-time-energy-gravity quantization coupled with an increasing ability to easily move back and forth between the material and antecedent realms, perhaps the outcome of an enhanced function driven by an initiating will itself, makes the need to house genetic information in a form such as DNA alone, burdensome. It is conceivable that four-base logic-encoding ecosystems, or even the ∞-entanglement, K-entanglement, and N-entanglement libraries may be able to more directly act at the material level in some composite subtle-material post-genetic form that also includes DNA. This possibility is in general referred to as post-genetic code.

This chapter takes a more macro-level mathematical view based on a cosmology of light, to summarize the aggregate conditions required for the creation of functionally rich super-matter. In the process the distinction between three distinct phases of matter emerge: that of established-matter, matter-in-transition, and super-matter. The development of super-matter will naturally also lead to a post-genetic housing structure for genetic-type information.

It makes sense for the structure of genetic-type information-storage to change in each of these cases since it is something different in each case. In such a view the material-fabric is the genetic-type information storage medium for "established matter". Hence, code for the electro-magnetic-wavearchetype-masspotential P-type being, the quantum-particle p-type being, and the atom P-type being, is stored in the material-fabric as reviewed in Sections 5 and 6, on the generation of the electro-magnetic-wavearchetype-masspotential P-type being and generation of P-type beings in the surfacing of matter, respectively.

Matter-in-transition would be associated with the emergence of life and as has been suggested in some detail in my previous book on genetics (Malik, 2019a), matter becomes more dynamic and adaptable as a result of this. Cellular-based genetic structure is the vehicle for storage for code associated with the cell. But further, as discussed, with the emergence of basic human capabilities, and truer individuality, the genetic information is further enhanced. These were reviewed in detail in Section 7 on the generation of P-type beings in the surfacing of life. It has also been argued in Section 8, that the emergence of complex organization, which includes stable mega-organization and a sustainable global civilization, will likely make further changes to the cellular-based genetic code. Such changes are only possible because in the cosmology of light view that animates this book, matter is essentially quantized light, and any precipitated functional richness as manifest in mega-organizations, for example, means that the genetic structure of matter has had to have changed due to the activation of such functional richness. All is one, and matter and light are two poles of that oneness.

When we arrive at the super-matter-based P-type beings, which is inevitable with the increase in functional-richness, then it stands to reason that there will likely come into existence some post-genetic structure that will house essential genetic-type information in a very different way.

The Light-Space-Time Emergence equation (3.1.3) derived in Chapter 3.1, and reproduced here for convenience, will be leveraged to derive summary equations to distinguish between the different phases of matter. The following derivations run parallel to equations created in Chapter 3.6 that focused on the effect of levels of light on the becoming of the P-type being. Reproducing (3.1.3):

$Emergence_{light-space-time} =$

$$
\left|
\begin{array}{l}
\left[
\begin{array}{l}
c_\infty : [Pr, Po, K, H] \\
(\downarrow R_{C_K} = f(R_{C_\infty})) \\
c_K : [S_{Pr}, S_{Po}, S_K, S_H] \\
(\downarrow R_{C_N} = f(R_{C_K})) \\
c_N : f(S_{Pr} \times S_{Po} \times S_K \times S_H) \\
(\downarrow R_{C_U} = f(R_{C_N})) \\
c_U : [P, V, M, C] \\
\Uparrow \\
\quad c_{O:[D,W,I,C]}
\end{array}
\right]_{Light}
\quad
\left[
\begin{array}{l}
M_3 \to System_X \\
(\uparrow F \to I) \\
M_2 \to S_{System_X} \\
(\uparrow Sig \to F) \\
M_1 \to Sig_X \\
(\uparrow > P_X) \\
U \to x_U
\end{array}
\right]_{Space}
\\[2em]
\left[
\begin{array}{l}
M_3 : -\infty \le t \le \infty \\
\qquad \downarrow \\
M_2 : 0 \ge t > \infty \\
\qquad \downarrow \\
M_1 : 0 > t > \infty \\
\qquad \downarrow \\
U \to \begin{array}{l} t \sim E_{Cell}; TC: M_3 \to U \\ t \sim E_{Human}; TC: U \to M_3 \end{array}
\end{array}
\right]_{Time}
\quad
\begin{array}{l}
TC \to x_T
\end{array}
\end{array}
\right| \langle x_U | x_T \rangle
$$

Simplifying each of the main matrices yielded (3.6.1), the simplified version of the Light-Space-Time Emergence equation, reproduced here for convenience:

$$Emergence_{light-space-time} = |[L][S][T]TC \rightarrow x_T|_{\langle x_U | x_T \rangle}$$

As explained by Neil Turok in his book The Universe Within (Turok, 2012), Euler's formula reproduced below, can be used to model many naturally occurring phenomena because of its sinusoidal oscillation between narrow bounds as x increases. Reproducing Euler's formula:

$$e^{ix} = \cos x + i \sin x$$

The sum of the squares of the ordinary and complex parts, on the right side of the equation, is one. In quantum theory this ensures that the probabilities for all possible outcomes add up to one. Hence this formula is useful when summarizing the macro-level effects of (3.6.1) which necessarily has diverse drivers of phenomenon.

Further, in modeling matter a modified notation of the Schrodinger wavefunction as interpreted by Feynman is leveraged. This version features the integral sign, $\int$, meaning that all terms to the right of it have to be summed up for all space and time till the moment when the wavefunction is required to be known.

Summary Equations for Established-Matter Based P-Type Beings

Hence, combining (3.6.1) with Feynman's interpretation of the Schrodinger wavefunction, with the Euler formula yields the following equations, 9.3.1-6 for matter:

$$\psi_{established-matter\ based\ P-type\ beings} = \left| \iint e^{i \int |[L][S][T]|} \right|_U$$

Eq. 9.3.1: Wavefunction for Established-Matter Based P-Type Beings

(9.3.1) suggests that so long as the basis of matter is untransformed, specified by the U following the vertical-brackets, the outcome is not going to be any different from established matter based P-type beings, specified by $\psi_{established-matter\ based\ P-type\ beings}$. U implies that there is not any action of the conscious will yet. This makes sense since conscious will supposedly arises with close-to human-systems. Note also that the presumption here is that matter as we perceive it – comprising of the quantum-particle P-type being, the atom P-type being, and the molecule P-type being – is perhaps only at the start of its journey. This observation is reinforced through the possibility of infinite information existing in meta-layers of Light that has yet to precipitate materially.

Further, as specified by Equation 9.3.2, the Probability-View of Established-Matter Based P-Type Beings, the very basis of matter is going to be either the untransformed physical (P_U), the untransformed vital (V_U), the untransformed mental (M_U), or the untransformed integral (I_U):

$$\left|\psi_{established-matter\ based\ P-type\ beings}\right|^2 = P_U^2 + V_U^2 + M_U^2 + I_U^2 = 1$$

Eq. 9.3.2: Probability-View for Established-Matter Based P-Type Beings

As specified by (9.3.2) the probability that any of these bases will be leveraged in the creation of matter adds up to one.

Another way of viewing established matter is by leveraging the non-wavefunction form as in (3.6.2) and adapting it to create Equation 9.3.3, Simplified Light-Space-Time Matrix Form for Established-Matter:

$Established - Matter\ Based\ P - Type\ Beings =$
$$\left|\left|[L][S][T]TC \rightarrow x_T\right|\langle x_U|x_T\rangle\right|_U$$

Eq. 9.3.3: Simplified Light-Space-Time Matrix Form for Established-Matter Based P-Type Beings

The notion of U as the bases in the formation of established-matter based P-type beings simply implies that this is a baseline and is subject to all the adaptability yet to come about through the play of dynamics such as urge and will which increases with the generation of subsequent partial-singularities.

Summary Equations for Matter-in-Transition Based P-Type Beings

Equation 9.3.4, Wavefunction for Matter-in-Transition Based P-Type Beings, suggests the mixed bases for matter, as specified by dynamics of both the untransformed (U) and the meta-levels (M_x), which therefore results in matter-in-transition based P-type beings.

$$\psi_{matter-in-transition\ based\ P-type\ beings} = \left. \left| \iint e^{i\int |[L][S][T]|} \right| \right|_{U\&M_x}$$

Eq. 9.3.4: Wavefunction for Matter-in-Transition Based P-Type Beings

The influence of meta-level, M_x, implies that action of visceral urge or desire or will. This is mixed with Nature's existing results, specified by U. As already discussed, such matter would be the result of the dynamics of life and its emergences.

As specified by Equation 9.3.5, Probability-View for Matter-in-Transition Based P-Type Beings, the probability that any of the untransformed (U) and transformed (T) bases will be leveraged in the creation of matter adds up to one.

$$\left| \psi_{matter-in-transition\ based\ P-type\ beings} \right|^2 =$$
$$P_U^2 + P_T^2 + V_U^2 + V_T^2 + M_U^2 + M_T^2 + I_U^2 + I_T^2 = 1$$

Eq. 9.3.5: Probability View for Matter-in-Transition Based P-Type Beings

Note also that another way of viewing matter in transition based P-type beings is by leveraging the non-wavefunction form as in (3.6.3) and adapting it to create Equation 9.3.6, Simplified Light-Space-Time Matrix Form for Matter-in-Transition Based P-Type Beings:

Matter – in – transition based P – type beings =
$$\left. \left| |[L][S][T]TC \rightarrow x_T| \langle x_U | x_T \rangle \right| \right|_{U\&M_x}$$

Eq. 9.3.6: Simplified Light-Space-Time Matrix Form for Matter-in-Transition Based P-Type Beings

Summary Equations for Super-Matter-Based P-Type Beings

Equation 9.3.7, Wavefunction for Super-matter-based P-Type Beings, suggests some transformed bases for matter, as specified by dynamics of the meta-levels (M_x), which therefore results in super-matter-based P-type beings. It is essentially more consciousness, whether of visceral urge, desire, or will, that is driving change.

$$\psi_{super-matter\ based\ P-type\ beings} = \left.\left|\int e^{i\int|[L][S][T]|}\right|\right|_{M_x}$$

Eq. 9.3.7: Wavefunction for Super-matter-based P-Type Beings

As specified by Equation 9.3.8, Probability-View for Super-matter-based P-Type Beings, the probability that the transformed (T) bases will be leveraged in the creation of matter adds up to one.

$$\left|\psi_{super-matter\ based\ P-type\ beings}\right|^2 = P_T^2 + V_T^2 + M_T^2 + I_T^2 = 1$$

Eq. 9.3.8: Probability View for Super-matter-based P-Type Beings

Note also that another way of viewing super-matter-based P-type beings is by leveraging the non-wavefunction form for as in (3.6.4) and adapting it to create Equation 9.3.9, Simplified Light-Space-Time Matrix Form for Super-matter-based P-Type Beings:

$$Super - Matter\ Based\ P - Type\ Beings = \left.\left|\left|[L][S][T]TC \rightarrow x_T\right|_{\langle x_U|x_T\rangle}\right|\right|_{M_x}$$

Eq. 9.3.9: Simplified Light-Space-Time Matrix Form for Super-matter-based P-Type Beings

Chapter 9.4: The Emerging Link Between Matter and Cosmos with Complexification of P-Type Beings

The previous chapters in this section further discussed the effect of functional richness in the creation of the super-matter-based P-type beings. The question then, is what is matter? This chapter will take a deeper look at the nature of matter by drawing on some observations from the field of astrophysics.

From discussion in the previous chapter it is clear that if the cumulative effect of overcoming habitual patterns increases the likelihood of the generation of the super-matter-based P-type beings being formed, this implies the increase too of four-fold quantization. Space, as the seeding ground of knowledge, may therefore also require expansion to continue to allow seeding to take place. But if space needs to expand in such a scenario, then the fact that scientists have observed that the universe is expanding may be related to this. This chapter hence also explores a possible link between the expansion of the universe and an increase in functional-richness that implies the expansion of space. Such a link may provide further insight into the nature of matter.

The entertainment of such an idea perhaps can be strengthened through consideration of the cosmological principle that suggests that the spatial distribution of matter in the universe is homogenous and isotropic when viewed on a threshold scale of 250 million light years. The nine-year cosmic microwave background image created with the Wilkinson Microwave Anisotropy Probe (WMAP) depicts 13.77 billion-year-old temperature differences that correspond to the seeds that became the galaxies (NASA-WMAP, 2014). This homogeneity in temperature difference in the early universe may suggest too a process of homogenous four-fold quantization that set into motion the development of galaxies.

In his book 'The Jazz of Physics', Alexander (Alexander, 2016) makes the case that our cosmic origins are seated in sound patterns. He suggests that cosmic constants need to be fine-tuned for cosmic structure leading up to life to exist. He proposes that we exist in a musical universe that has the ability to self-tune, like musical improvisation, the cosmic constants between a period of contraction and expansion. It is in such a manner that a period between contraction and expansion can give rise to the right cosmic constants that will allow life to exist. While this is an ingenious solution based on today's constraints of physics of the universe existing only at the physical layer, what is proposed in this book is infinite potentiality at faster moving layers of light that materialize as per meta-function at the physical layer. The process of "fine-tuning" or "self-tuning" takes place through the persistent quantum-level computation that arbitrates possibility based on simultaneous possibilities, and as per the mathematics detailed in this book. In the view suggested in this book light precedes sound. Sound requires a medium in which to travel, and all mediums are emergent from Light. Light traveling infinitely fast may even be thought of as the original medium.

Any instance of quantization though can never be equal to another instance of quantization as suggested by the light-based-singularity dynamics in this book. On the surface and in their first emergence phenomena may appear to be homogenous, but that must only hide the vast diversity that is seeded in infinite uniqueness in Light. Materialized diversity with all its infinite uniqueness will become more apparent as matter-based P-type beings tends toward super-matter-based P-type beings, due to the increasing materialization of possibility driven by the unfolding of innate fourfold functional richness.

Complexity of Matter, Dark Matter, & Dark Energy

According to NASA only 5% of the matter in the universe is visible, while as much as 27% is 'dark matter'. It is the 5% of 'visible' matter though that forms the familiar stars, galaxies, or the atoms (NASA-darkmatter, 2016). But since, as explored in Chapter 3.7 on Qualified Determinism, the architectural sets at M_2 are infinite the question is why would all functions need to manifest in the same way?

As stated in an article 'Dark Energy: The Biggest Mystery in the Universe' (Panek, 2010), "Sight itself has blinded us to the Universe". It should be possible for other matter-based constructions to come into being based on the functionality they exist for. For instance, there could be other fields that create other kinds of elementary particles, that create other kinds of atoms and that results in other kinds of structures not visible with our current instrumentation.

In reference to the four sets at M_2 (as discussed in Chapter 2.3 on 'The K-Type Being and Its Dynamics') it may also be that as the basis of matter complexifies in emergent P-type beings as it journeys through the electro-magnetic-wavearchetype-masspotential P-type being, the quantum-particle P-type being, the atom P-type being, the cell P-type being, complex individual based P-type beings, and further complex organizational FBLEE-based P-type beings, the sets which have been positioned to each contain infinite elements, concretely manifest more of their function-elements.

This suggestion has been the basis of the equations derived in previous chapters. The complexity of matter may therefore be related to the number of manifested function-elements of the set of four sets. This idea is consistent with the notion of the "adjacent possible" suggested by Kaufmann (Kaufmann, 2003) in which innovation is positioned as a recombination of existing parts to create new value – or of existing sets to combine parts of themselves to create new elements based on new circumstance. If MS signifies manifested-set, so that the cardinality or number of elements in the combined set is the union of the four manifested-sets, this may be summarized by the following equation, Equation 9.4.1, Complexity of Matter in Terms of Manifested Set:

$$Matter_{Complexity} \propto$$
$$\left| MS_{System_{Pr}} \cup MS_{System_P} \cup MS_{System_K} \cup MS_{System_N} \right|$$

Eq 9.4.1: Complexity of Matter in Terms of Manifested Set

It is also possible to hypothesize an Equation, 9.4.2, relating complexity of matter in terms of visible and dark matter:

$$Matter_{Complexity} \propto [Visible\ Matter + Dark\ Matter]$$

Eq. 9.4.2: Complexity of Matter in Terms of Visible and Dark Matter

Further, as suggested by the previous discussions on the quantization of space, space is not empty but is a seeding ground for a variety of emergences. The notion of space having 'amazing' properties was, according to a report on 'dark energy' by the Harvard-Smithsonian Center for Astrophysics, first suggested by Einstein (Harvard-Smithsonian Center for Astrophysics, 2004). Quoting: "Einstein was the first person to realize that empty space is not nothingness. Space has amazing properties, many of which are just beginning to be understood." Einstein had suggested the existence of a 'dark energy' about 100 years ago (NASA-Supernova, 2001) as a property of space that caused the expansion of the universe. Dark Energy is estimated to comprise as much as 68% of the universe.

Reinterpretation of Cosmological Expansion-Contraction Dynamics

Keeping in mind the complexification of matter as it journeys from the electromagnetic- through the quantum-, atomic-, and cellular-levels, and beyond, it may be possible to re-interpret the supposed expansion-contraction dynamics of cosmology in relation to the mathematical model presented in this book.

First, flipping the left and right sides of the equation (9.4.1) on $Matter_{Complexity}$ to yield Equation 9.4.3:

$$\left| MS_{System_{Pr}} \cup MS_{System_P} \cup MS_{System_K} \cup MS_{System_N} \right|$$
$$\propto Matter_{Complexity}$$

Eq 9.4.3: Flipped Matter-Complexity Equation

This implies that the manifested-set, MS, is growing at a certain threshold level. Assuming this threshold level, $MS\_Growth_{Threshold}$, is a property of space related to dark energy it may be possible to restate the condition of cosmological expansion and contraction. Hence, so long as the MS is increasing at a certain rate that exceeds $MS\_Growth_{Threshold}$ the level of dark energy is such that the acceleration of galaxies, $Acceleration_{Galaxies}$ exceeds the contracting force of gravity at the universal level, $Gravity_{Universe}$.

Therefore, as in Equations 9.4.4 and 9.4.5:

$$MS\_Growth_{Threshold}: Acceleration_{Galaxies} > Gravity_{Universe}$$

Eq. 9.4.4: Expansion of Universe Related to Manifest Set Growth Threshold

Conversely:

$! MS\_Growth_{Threshold}: Acceleration_{Galaxies} < Gravity_{Universe}$

Eq. 9.4.5: Contraction of Universe Related to Manifest Set Growth Threshold

The relation between $MS\_Growth_{Threshold}$ and Dark Energy is suggested by Equation 9.4.6:

$$MS\_Growth_{Threshold} \propto Dark\ Energy$$

Eq. 9.4.6: Relation Between Manifested Set Growth Threshold and Dark Energy

In this interpretation so long as more of the infinite set at M_2 results in the manifested-set, MS_{System}, the universe keeps growing. In other words, so long as matter tends towards super-matter, the universe continues with expansion. It seems therefore that matter is a dynamic state linked to the very life of the universe.

Further if the relative distances between some of the emergences discussed in this book are considered a pattern emerges. The quantum level is dealing in distances of the order of 10^{-35}m. The cellular level is dealing with distances of the order of 10^{-6}m. The astronomical level is dealing with distances as large as 10^{26}m – the estimated size of the universe. It can be observed therefore that the cell is approximately in the center with roughly 30 orders of magnitude on either side.

Such symmetry that we are at now with the human being at the mid-point between two known material extremes, seems to suggest a cosmic balance of scales with human-will and other faculties such as awareness as a potential arbiter. But this we already know is the condition for the activation of meta-levels and also therefore for the creation of super-matter-based P-type beings. Lack of will for continued functional-richness will in this view lead to a reversal in the Manifested Set due to a reversal in the quantization trend and consequently to a return of this universe to a state of collapse or unfulfilled absorption into the initiating Light.

Matter hence collapses into naught. In that case this may lead to another possible Big Bang initiated through the interplay of multiple layers of

338

Light, and a birth following such death as depicted by the ancient Ouroboros symbol of a dragon or snake eating its own tail. The eating of its own tail can be mathematically depicted by a condition of x_T reverting to x_U hence resulting in collapse as captured by Equation 9.4.7, Light-Space-Time Collapse:

$Light - Space - Time\ Collapse =$

$$
\left\| \begin{bmatrix} \begin{bmatrix} c_\infty : [Pr, Po, K, H] \\ (\downarrow R_{C_K} = f(R_{C_\infty})) \\ c_K : [S_{Pr}, S_{Po}, S_K, S_H] \\ (\downarrow R_{C_N} = f(R_{C_K})) \\ c_N : f(S_{Pr} \times S_{Po} \times S_K \times S_H) \\ (\downarrow R_{C_U} = f(R_{C_N})) \\ c_U : [P, V, M, C] \\ \Uparrow \\ c_{0:[D,W,I,C]} \end{bmatrix}_{Light} \begin{bmatrix} M_3 \rightarrow System_X \\ (\uparrow F \rightarrow I) \\ M_2 \rightarrow S_{System_X} \\ (\uparrow Sig \rightarrow F) \\ M_1 \rightarrow Sig_x \\ (\uparrow > P_{x)} \\ U \rightarrow x_U \end{bmatrix}_{Space} \\ \begin{bmatrix} U \rightarrow \begin{bmatrix} M_3 : -\infty \le t \le \infty \\ \downarrow \\ M_2 : 0 \ge t > \infty \\ \downarrow \\ M_1 : 0 > t > \infty \\ \downarrow \\ t \sim E_{Cell}; TC: M_3 \rightarrow U \\ t \sim E_{Human}; TC: U \rightarrow M_3 \end{bmatrix}_{Time} \end{bmatrix} \quad TC \rightarrow x_U \end{bmatrix} \right\| \langle x_U | x_U \rangle
$$

Eq. 9.4.7: Light-Space-Time Collapse

On the other hand, the will for continued functional-richness may lead to continuous quantization, space expansion, and to an increasing reality of super-matter-based P-type beings as was suggested in (4.2.8) reproduced here for convenience:

$Light - Space - Time\ Emergence_{Suoer-Matter\ Based\ P-Type\ Being} =$

$$
\left|
\begin{array}{l}
\left[
\begin{array}{l}
c_\infty: [Pr, Po, K, H] \\
\left(\downarrow R_{C_K} = f(R_{C_\infty}) \right) \\
c_K: [S_{Pr}, S_{Po}, S_K, S_H] \\
\left(\downarrow R_{C_N} = f(R_{C_K}) \right) \\
c_N: f(S_{Pr} \times S_{Po} \times S_K \times S_H) \\
\left(\downarrow R_{C_U} = f(R_{C_N}) \right) \\
c_U: [P, V, M, C] \\
\Uparrow \\
c_{0:[D,W,I,C]}
\end{array}
\right]_{Light} \\
\left[
\begin{array}{l}
M_3 : -\infty \leq t \leq \infty \\
\quad\quad \downarrow \\
M_2 : 0 \geq t > \infty \\
\quad\quad \downarrow \\
M_1 : 0 > t > \infty \\
\quad\quad \downarrow \\
U \to \begin{array}{l} t \leq E_{Cell}; \text{TC}: M_3 \to U \\ t \sim E_{Human}; \text{TC}: U \to M_3 \end{array}
\end{array}
\right]_{Time}
\end{array}
\right.
$$

$$
\left[
\begin{array}{l}
M_3 \to System_X \\
(\uparrow F \to I) \\
M_2 \to S_{System_X} \\
(\uparrow Sig \to F) \\
M_1 \to Sig_x \\
(\uparrow > P_x) \\
U \to x_{Molecule}
\end{array}
\right]_{Space}
$$

$$
TC \to x_{Super-Matter}
$$

$$\Rightarrow$$

$$\langle x_U | x_T \rangle$$

$\langle Space - Time - Energy - Gravity\ P - Type\ Beings\ and\ Pre - Genetic\ Code \rangle +$

$\left(\begin{array}{c} Electro - magnetic - wavearchetype - masspotential\ P - Type\ Being \\ and\ Pre - Genetic\ Code \end{array} \right) +$

$\langle Quantum - Particle\ P - Type\ Being\ and\ Pre - Genetic\ Code \rangle +$

$\langle Atom\ P - Type\ Being\ and\ Pre - Genetic\ Code \rangle +$

$\langle Molecule\ P - Type\ Being\ and\ Pre - Genetic\ Code \rangle + LSTE \langle ... \rangle + FBLEEE \langle ... \rangle +$

$LSTE \langle ... \rangle + PGCI \langle ... \rangle + Super - Matter\ Based\ P - Type\ Being\ and\ Post - Genetic\ Code$

Further iterations of (4.2.8) will lead to a possibility of all patterns in the Space-Matrix being broken, thus leading to the S-type being. This possibility will be taken up in Section 10, The S-Type Being.

340

SECTION 10: TRIUMPH OF LOVE AND THE EMERGENCE OF THE S-TYPE BEING

The S-type being is envisioned as being the result of an increasing and more comprehensive unity of light along both the vertical and horizontal dimensions.

In its automaticity, starting with the emergence of the space-time-energy-gravity beings through to the cell P-type being, horizontal unity is ensured, also revealing that the deep essence of light is nothing other than love. But as emergence approaches the human P-type being, conscious choice and will become more important and emergence of the truer-individuality P-type being and beyond requires a more conscious opening to meta-layers of light. In such an opening the deeper essence of love become more active and gradually as this becomes stronger there is a greater and deeper yearning or even a surrendering to the possibilities embodied in the meta-layers of light, that allows a more and more profound unity to take place along the vertical dimension.

The increasing dynamism of such a comprehensive unity, a triumph of love, culminates in a foundation for the emergence of the S-type being.

This Section then summarizes the journey of light-based-singularities through to the generation of the S-type being. Hence:

- Chapter 10.1 summarizes the unbroken continuity that exists in the P-type being starting from the space-time-energy-gravity quadrumvirate beings, to the emergence of super-matter-based P-type beings. The latter is seen as a prerequisite for the emergence of the S-type being.
- Chapter 10.2 leverages the method of contrasts by focusing on fragmented-singularities, exploring a hypothetical creation generated from space, time, energy, and gravity alone, and the resulting creation and nature of D-type or derivative beings.
- Chapter 10.3 explores the use of a set of mathematical operators in expediting the arrival of the s-type beings. These operators are non-exhaustive, but rather are indicative of the nature of dynamics that will increasingly arise in the extrapolation of P-type beings to super-matter-based P-type beings and beyond. The

use of such operators assists in perceiving reality as derivate from Light, and hence their active adoption will accelerate the overcoming of limiting patterns and subsequently the integration of the multiple layers of light in any light-based-singularity.

- Chapter 10.4, which focuses on altering the microcosmic-macrocosmic balance, examines on-going conditions for the generation of the S-type being. A key condition is the audacity of the human-species made possible because the human is the result of 30 orders of microcosmic magnitude implicit to it. Fields, quanta, atoms, cells, the incredible four-fold emergences of properties codified in Light have all contributed to the emergence of the human. Now in a possible reversal of causality the human appears to be positioned to similarly influence 30 orders of magnitude explicit to it and even structure the further development of the macrocosm by mastery of the microcosm.

- Chapter 10.5 reviews key insights that have surfaced through tracing the emergence of the series of light-based beings culminating in the emergence of the S-type being. driven by enlightened heart, will, and mind dynamics.

- Chapter 10.6 summarizes the mythology in the cosmology of light, highlighting themes of darkness, light, sacrifice, priests, gods, that culminate in the S-type being and an increasing triumph of love.

Chapter 10.1: The Unbroken Continuity in the P-Type Being

The preceding sections have elaborated the architecture of light-based beings that are all built from the four essential properties of Light. In this elaboration the connection between one P-type being and the next is clear in that there is an essential complexification of the underlying four-foldness. The same four organizational properties resident in Light create more complex forms of themselves with the emergence of subsequent P-type beings. Further, the four-foldness that had emerged in a previous P-type being is usually implicit in a succeeding P-type being. There is hence, "singleness" in the emergence of any light-based-singularity that binds it with all previous light-based-singularities so that a subsequent P-type being can be thought of as having a greater sphere of influence than preceding ones by virtue of all previous emergent laws being active also in it. This continuity highlights the organic development of life through the iterative process of becoming culminating in a being.

Figure 10.1.1 - Fourfold Adherence, Implicit Beings, and Shared Dynamics in Chain of P-Type Beings - summarizes the three architectural devices of light-based singularities: the adherence to fourfold properties of light, and integration of previous beings and their dynamics in the emergent P-type being. Subsequent sub-sections in this chapter will review the notion of fourfold adherence for each row in the following figure:

Type of Being	Fourfold Properties of Light	Dynamics of Being	Minimum Number of Implicit Beings
Space-Time-Energy-Gravity (STEG) Beings	Infinity of Seeds, Maturation of Seeds, Materialization of Seeds, Collectivity of Seeds	• STEG	1
EMWM P-Type Being	c_U, hv, $[f(\lambda)]$, hv/c^2	• STEG • EMWM	2
Quantum-Particle P-Type Being	Quarks, Leptons,	• STEG • EMWM	3

343

	Bosons, Higgs-Boson	• Quantum-Particle	
Atom P-Type Being	s-Group, p-Group, d-Group, f-Group	• STEG • EMWM • Quantum-Particle • Atom	4
Cell P-Type Being	Proteins, Nucleic Acids, Lipids, Polysaccharides	• STEG • EMWM • Quantum-Particle • Atom • Cell	5
Fundamental-Capacities-of-Self P-Type Being	Sensations, Wills, Emotions, Thoughts	• STEG • EMWM • Quantum-Particle • Atom • Cell • Fundamental Capacities of Self	6
Truer-Individuality P-Type Being	Service-type-individual, Knowledge-type-individual, Harmony-type-individual, Power-type-individual	• STEG • EMWM • Quantum-Particle • Atom • Cell • Fundamental Capacities of Self • Truer Individuality	7
FBLEE-Based Sustainable-Global-Civilization P-Type Being	Service-type-region, Knowledge-type-region, Harmony-type-region, Power-type-region	• STEG • EMWM • Quantum-Particle • Atom • Cell	8

		• Fundamental Capacities of Self • Truer Individuality • FBLEE-Based Sustainable Global Civilization	

Figure 10.1.1 Fourfold Adherence, Implicit Beings, and Shared Dynamics in Chain of P-Type Beings

Space-Time-Energy-Gravity Being Four-Foldness

Hence, in thinking through the generation of the space-time-energy-gravity beings at the time of the Big Bang, these may be summarized by the equations, Equations 10.1.1 through 10.1.4. In each case there is a dependence on the single fundamental attribute of 'seeds'.

Equation 10.1.1, System-Knowledge (Space) at Space-Time-Energy-Gravity Level, summarizes how knowledge is related to 'infinity of seeds' with each seed representing, as it were, a different arrangement of knowledge:

$$System_{Knowledge} \propto [f(infinity\ of\ seeds)]$$

Eq 10.1.1: System-Knowledge (Space) at Space-Time-Energy-Gravity Level

Note that the infinity of seeds requires 'space' to be expressed.

Equation 10.1.2, System-Power (Time) at Space-Time-Energy-Gravity Level, summarizes how power is related to 'maturation of seeds', since anything that gets in the way of the maturation will generally be overcome:

$$System_{Power} \propto [f(maturation\ of\ seeds)]$$

Eq 10.1.2: System-Power (Time) at Space-Time-Energy-Gravity Level

Note that maturation of seeds is felt in 'time'.

Equation 10.1.3, System-Presence (Energy / Matter) at Space-Time-Energy-Gravity Level, summarizes how presence is related to 'materialization of seeds', since the accumulation of energy will express itself in material form:

$$System_{Presence} \propto [f(materialization\ of\ seeds)]$$

Eq 10.1.3: System-Presence (Energy/Matter) at Space-Time-Energy-Gravity Level

Equation 10.1.4, System-Harmony (Gravity) at Space-Time-Energy-Gravity Level, summarizes how harmony is related to 'relationship of seeds', since seeds will generally have a relationship to one another and likely exist in collectivities:

$$System_{Harmony} \propto [f(relationship\ of\ seeds)]$$

Eq 10.1.4: System-Harmony (Gravity) at Space-Time-Energy-Gravity Level

Note that such relationship may be expressed as 'gravity'.

The creation of the space-time-energy-gravity micro-being is a fundamental device that facilitates the process of quantization thereby allowing the accumulated material-fabric to change. The creation of the space-time-energy-gravity macro-being provides a container within which the whole adventure of the materialization of light takes place.

Electro-Magnetic-Wavearchetype-Masspotential (EMWM) P-Type Being Four-Foldness

In the case of the EMWM P-type being the number of basic attributes to express four-foldness is two as will be observed.

Equation 10.1.5, System-Harmony Aspect of the Electro-Magnetic-Wavearchetype-Masspotential P-Type Being, is related to the speed of light, c at U:

$$System_{Harmony} \propto c_U$$

Eq 10.1.5: System-Harmony Aspect of the Electro-Magnetic-Wavearchetype-Masspotential P-Type Being

Equation 10.1.6, System-Power Aspect of the Electro-Magnetic-Wavearchetype-Masspotential P-Type Being, is related to the frequency, v, of the EM spectrum:

$$System_{Power} \propto hv$$

Eq 10.1.6: System-Power Aspect of the Electro-Magnetic-Wavearchetype-Masspotential P-Type Being

Equation 10.1.7, System-Knowledge Aspect of the Electro-Magnetic-Wavearchetype-Masspotential P-Type Being, is related to the wavelength, λ, of the EM spectrum:

$$System_{Knowledge} \propto [f(\lambda)]$$

Eq 10.1.7: System-Knowledge Aspect of the Electro-Magnetic-Wavearchetype-Masspotential P-Type Being

And since, $E = mc^2$ or $m = \frac{E}{c^2}$, and substituting hv for E, yields Equation 10.1.8, System-Presence Aspect of the Electro-Magnetic-Wavearchetype-Masspotential P-Type Being:

$$System_{Presence} \propto {hv}/{c^2}$$

Eq 10.1.8: System-Presence Aspect of the Electro-Magnetic-Wavearchetype-Masspotential P-Type Being

But, $C \propto {1}/{h}$

Hence, the general relationships of the four-fold wholeness may be understood by knowing just two variables, C, and either, λ or v.

The emergence of the EMWM P-type being adds another fold as it were to the P-type being biographies that can progressively be seen as a complexification of four-foldness.

Quantum-Particle P-Type Being Four-Foldness

As per the model of early cosmic development presented by the Particle Data Group (Particle Data Group, 2015) at time, $t = 10^{-10}$ seconds, the four-fold fullness begins to express itself in a series of quantum particles. Quarks, leptons, bosons, which also may imply the presence of the Higgs-boson that gives quarks their mass are observed. But as discussed quarks are a precipitation of Light's property of Knowledge, leptons of Power, bosons of Harmony, and the Higgs-boson of Presence. Hence it is again observed that the four-fold fullness has expressed itself at a different order of complexity.

The Standard Model suggests that quarks, leptons, bosons, and the Higgs-boson are different fundamental particles, which implies that the fourfold wholeness has now a more complex basis of its operation, since as compared with the fourfold wholeness of the EMWM P-type being the number of independent bases appears to have increased.

Summarizing, as in Equations 10.1.9 through 10.1.12:

$$System_{Harmony} \propto f(bosons)$$

Eq 10.1.9: System-Harmony Aspect of the Quantum-Particle P-Type Being

$$System_{Power} \propto f(leptons)$$

Eq 10.1.10: System-Power Aspect of the Quantum-Particle P-Type Being

$$System_{Knowledge} \propto f(quarks)$$

Eq 10.1.11: System-Knowledge Aspect of the Quantum-Particle P-Type Being

$$System_{Presence} \propto f(Higgs\_boson)$$

Eq 10.1.12: System-Presence Aspect of the Quantum-Particle P-Type Being

Hence the quantum-particle ecosystem logic is also now added to the quantum-particle P-type being, which progressively continues to complexify along the lines set up by the underlying fourfold organizational principles as dictated by Light's properties.

Atom P-Type Being Four-Foldness

As discussed in Chapter 6.4 the Periodic Table itself is also an expression of the four-fold wholeness. Hence, d-Group elements are an instance of Light's property of Presence, p-Group elements are an instance of Light's property of Knowledge, s-Group elements are an instance of Light's property of Power, and f-Group elements are an instance of Light's property of Harmony or Nurturing. As per the Particle Data Group model (Particle Data Group, 2015) this emergence likely began at $t = 3 \times 10^5$ years when the atom is suggested to have emerged, and continued at least to t $= 10^9$ years with the emergence of stars which it is already known are the furnaces in which heavier atoms were created.

Leveraging the fundamentally different groups yields Equations 10.1.13 through 10.1.16:

$$System_{Harmony} \propto f(f\_group)$$

Eq 10.1.13: System-Harmony Aspect of the Atom P-Type Being

$$System_{Power} \propto f(s\_group)$$

Eq 10.1.14: System-Power Aspect of the Atom P-Type Being

$$System_{Knowledge} \propto f(p\_group)$$

Eq 10.1.15: System-Knowledge Aspect of the Atom P-Type Being

$$System_{Presence} \propto f(d\_group)$$

Eq 10.1.16: System Presence Aspect of the Atom P-Type Being

The atom P-type being therefore have at least four four-fold wholenesses implicit in it.

Cell P-Type Being Four-Foldness

At time, $t = 13.8 \times 10^9$ years, there is a fifth clear expression of the same four-fold order as the bases of an even more complex organization, that of cellular life and all that is founded on it. This can be summarized as in Equations 10.1.17 through 10.1.20:

$$System_{Harmony} \propto f(lipids)$$

Eq 10.1.17: System-Harmony Aspect of the Cell P-Type Being

$$System_{Power} \propto f(polysaccharides)$$

Eq 10.1.18: System-Power Aspect of the Cell P-Type Being

$$System_{Knowledge} \propto f(nucleic\ acids)$$

Eq 10.1.19: System-Knowledge Aspect of the Cell P-Type Being

$$System_{Presence} \propto f(proteins)$$

Eq 10.1.20: System-Presence Aspect of the Cell P-Type Being

The cell P-type being therefore seems to have at least five four-fold-wholenesses implicit in it to therefore further complexify the architecture of matter and matter-based P-type beings.

Fundamental-Capacities-of-Self P-Type Being Four-Foldness

Equations 10.1.21 through 10.1.24 summarizes a sixth clear expression of four-foldness that will change the structure of life-based P-type beings.

$$System_{Harmony} \propto f(Emotions, Feelings)$$

Eq 10.1.21: System-Harmony Aspect of the Fundamental-Capacities-of-Self P-Type Being

$$System_{Power} \propto f(Urges, Wills, Desires)$$

Eq 10.1.22: System-Power Aspect of the Fundamental-Capacities-of-Self P-Type Being

$$System_{Knowledge} \propto f(Thoughts)$$

Eq 10.1.23: System-Knowledge Aspect of the Fundamental-Capacities-of-Self P-Type Being

$$System_{Presence} \propto f(Sensations)$$

Eq 10.1.24: System-Presence Aspect of the Fundamental-Capacities-of-Self P-Type Being

Similarly, as the truer-individuality P-type being expresses itself STEG quantization initiated by quantum-certainty will further complexify the bases of matter.

Equations 10.1.25 through 10.1.28 therefore summarize a seventh clear expression of four-foldness that will further complexify the structure of the material-fabric and change the structure of matter.

$$System_{Harmony} \propto f(Harmony - type - individuals)$$

Eq 10.1.25: System-Harmony Aspect of the Truer-Individuality P-Type Being

$$System_{Power} \propto f(Power - type - individuals)$$

Eq 10.1.26: System-Power Aspect of the Truer-Individuality P-Type Being

$$System_{Knowledge} \propto f(Knowledge - type - individuals)$$

Eq 10.1.27: System-Knowledge Aspect of the Truer-Individuality P-Type Being

$$System_{Presence} \propto f(Service - type - individuals)$$

Eq 10.1.28: System-Presence Aspect of the Truer-Individuality P-Type Being

FBLEE-Based Sustainable-Global-Civilization P-Type Being Four-Foldness

Similarly, as more complex organizational levels, such as a sustainable global civilization expresses itself, STEG quantization initiated by quantum-certainty will further complexify such complex organization based P-type beings.

Equations 10.1.29 through 10.1.32 therefore summarize an eighth clear expression of four-foldness. Note that these expressions are indicative only and there are other missed levels that would need to be expressed for completeness.

$$System_{Harmony} \propto f(Harmony - type - region)$$

Eq 10.1.29:   System-Harmony Aspect of the FBLEE-Based Sustainable-Global-Civilization P-Type Being

$$System_{Power} \propto f(Power - type - region)$$

Eq 10.1.30: System-Power Aspect of the FBLEE-Based Sustainable-Global-Civilization P-Type Being

$$System_{Knowledge} \propto f(Knowledge - type - region)$$

Eq 10.1.31:   System-Knowledge Aspect of the FBLEE-Based Sustainable-Global-Civilization P-Type Being

$$System_{Presence} \propto f(Service - type - region)$$

Eq 10.1.32: System-Presence Aspect of the FBLEE-Based Sustainable-Global-Civilization P-Type Being

Chapter 10.2: The Nature of the D-Type Being

The previous chapter summarized fourfold adherence, implicit beings, and shared dynamics in the chain of P-type beings.

By contrast this chapter will explore fragmented-singularity based creation by imagining what life may look like if only space, time, energy, and gravity were present, not as quadrumvirate beings that emerge from the light-matrix, but as a contained and separate creation complete in itself. Such a creation will generate another nature of 'derivative' or D-type beings.

This chapter will explore some aspects of such D-type beings.

Fragmentation and the Emergence of D-Type Beings

In the previous chapter equations (10.1.1 – 4) summarized an implicit reality of 'seeds' that gave space, time, energy, and gravity meaning. The essence of these equations is summarized here for convenience as Equations 10.2.1 – 4:

Hence, Equation 10.2.1, Space:

$$Space: System_{Knowledge} \propto [f(infinity\ of\ seeds)]$$

Equation 10.2.2, Time:

$$Time: System_{Power} \propto [f(maturation\ of\ seeds)]$$

Equation 10.2.3, Energy/Matter:

$$Energy/Matter: System_{Presence} \propto [f(materialization\ of\ seeds)]$$

Equation 10.2.4, Gravity:

$$Gravity: System_{Harmony} \propto [f(relationship\ of\ seeds)]$$

In a reality of fragmentation however, there would be a de-linking with 'seeds', which recall from (3.1.3) are a result of unique combinations of implicit function resident in several antecedent layers of light and due to the UPI-type, K-type, and N-type beings. This would mean that space, time, energy, and gravity are no longer emergent from light, become devoid of deeper significance, and would exist as purely independent physical phenomenon. In other words, such a creation becomes a 'derivate' creation based entirely on the essentially fragmented nature of reality. We will refer to the beings in such a creation as D-type beings. Further, we will refer to this first reality of fragmentation as first-order derivate or the D1-type being.

In this deeper significance space, time, energy, and gravity are processes by which the infinite possibility in Light begins to materialize. In a fragmented mode these phenomena by contrast now become independent sets of elements that at best may combine together to entirely determine all possibility within a universe. A universe that arose from a space-set, time-set, energy-set, and gravity-set would be a function of only these sets and would become devoid of all the possibility that exists in a universe generated from a light-based-singularity.

Exploring such a fragmented universe animated by D-type beings mathematically, the following sets would be starting points.

Hence, space could have elements such as 'point', 'line', 'plane', 'cube', 'sphere', amongst possibly infinite other space-elements. The Set of Space, Set_{Space}, is summarized by Equation 10.2.5, Set_{Space}:

$$Set_{Space} \ni [Point, Line, Plane, Cube, Sphere, ...]$$

Equation 10.2.5: Set_{Space}

Time could have elements such as 'slow', 'fast', 'stagnant', 'accelerating', amongst possibly infinite other time-elements. The Set of Time, Set_{Time}, is summarized by Equation 10.2.6, Set_{Time}:

$$Set_{Time} \ni [Slow, Fast, Stagnant, Accelerating, ...]$$

Equation 10.2.6: Set_{Time}

Energy could have elements such as 'concentrated', 'diffuse', 'intense', 'subtle', amongst possibly infinite other energy-elements. The Set of Energy, Set_{Energy}, is summarized by Equation 10.2.7, Set_{Energy}:

$Set_{Energy} \ni [Concentrated, Diffuse, Intense, Subtle, ...]$

Equation 10.2.7: Set_{Energy}

Gravity could have elements such as 'close', 'far', 'focused', 'encompassing', amongst possibly infinite other energy-elements. The Set of Gravity, $Set_{Gravity}$, is summarized by Equation 10.2.8, $Set_{Gravity}$:

$Set_{Gravity} \ni [Close, Far, Focused, Encompassing, ...]$

Equation 10.2.8: $Set_{Gravity}$

The elements in these sets can be thought of as second-order derivatives, and will be referred to as D2-type beings. A universe created from any combination of the elements from these four sets can be depicted by Equation 10.2.9, Space x Time x Energy x Gravity Universe, where the elements of such a universe is at least the cartesian product of the four infinite sets:

$$Space \; x \; Time \; x \; Energy \; x \; Gravity \; Universe$$
$$\geq Set_{Space} \; x \; Set_{Time} \; x \; Set_{Energy} \; x \; Set_{Gravity}$$

Equation 10.2.9: Space x Time x Energy x Gravity Universe

Such a universe would have infinite "elements" but there would not be any order to it. The universe of possible emergences in such a cartesian mathematics are combinations of D2-type beings and will be referred to as third-order derivative or D3-type beings. As can be seen, no other D-type being beyond these can emerge. This is because order is a function of implicit design. But such design cannot exist in a fragmented universe where there are no attendant meta-layers. Anything that is created or that arises has to arise from the combination of elements in the four foundational sets. So, such a universe may give rise to waves and particles through various combinations of elements from the space, time, energy, and gravity sets. But these waves and particles would be random and chaotic with no reason to begin to generate one kind of element over another or to maintain one kind of pattern over another. The notion of particles settling into agreeable combinations to create atoms, and then molecules will therefore not be possible in such a universe. There would be a lot of activity, but there would be neither rhyme nor reason to it.

Rhyme and reason imply the existence of function around which elements anchor or settle. But this can only be if such functions have a meta-significance, implying meta-layers, which by definition cannot exist in a fragmented universe.

Further, even if the elements in the sets in a *space x time x energy x gravity* universe were to create waves or something akin to the electro-magnetic-wavearchetype-masspotential spectrum, there would be no continuity of fourfold-ness from the space-time-energy-gravity layer to the electro-magnetic-wavearchetype-masspotential layer. This means that even though multiple "layers" may exist there would still be no implicit design or meaning to anything that emerges. It also means that nothing that is not of the nature of the initial set of elements, and the various combinations of them, can conceivably emerge.

Note that today's popular rendition of evolution in fact follows such a 'population thinking' scheme in which groupings are considered statistical averages rather than representative of deeper essences. 'Natural selection' (Darwin, 1947) is based on such population thinking and implies that only existing elements in a system can determine what the next iteration or possibility in elements will be. By contrast the view that emergences may be materializations of deeper essences, implying a connected as opposed to a fragmentary creation, is a scheme known as 'essentialism' and gives rise to different 'types'. Essentialism is attributed to Plato based on his development of the theory of ideas (Kleinman, 2006). The notion of 'ideas' or 'types' is consistent with the meta-functional view proposed in the cosmology of light book series, where infinite information in meta-layers of light, seeds infinite material possibility through a scheme involving persistent quantum-level computation that generates a continuous stream of genetic-type information.

In the absence of such light-based-singularity design and implicit order, a *space x time- x energy x gravity* universe would become increasingly chaotic, as more and more permutations and combinations involving the elements in the space-set, time-set, energy-set, and gravity-set were to emerge.

Note that D-type beings may emerge from any fragmented P-type being. This subsection used the space-time-energy-gravity quadrumvirate P-type being as an example only, to elaborate the generation of D1, D2, and D3-type beings.

A contemporary example of derivative or D-type beings is clear in AI. Assume that first-order derivatives or D1-type beings of the human species, themselves a P-type being because all the dynamism and logic of light is inherent in them, would be computer-based programs. Second-order derivatives, or D2-type beings, would be constructions of such computer-based programs such as created when using 3D printing, for example. Such D-type beings are considered to be fragmented-singularities, since all conscious connection with the design, the laws, the antecedent layers of light, that had continued to exercise themselves through the long play of Space and Time to allow beings to become, have essentially been severed. It is possible that D3-type beings may be generated by the union of several D2-type beings of the nature of 3D printing.

Note though that any material construction though, will have implicit in it light-based-beings such as the space-time-energy-gravity P-type beings, the electro-magnetic-wavearchetype-masspotential P-type being, the quantum-particle P-type being, the atom P-type being, and the molecule P-type being. Hence, even fragmented-singularities have life in them by virtue of the number of implicit beings that exist in them.

Figure 10.1.1 - Fourfold Adherence, Implicit Beings, and Shared Dynamics in Chain of P-Type Beings – is repurposed at Figure 10.2.1 – Relative Presence of Degree of Life Based on Minimum Number of Implicit Beings – to illustrate the degree of life present in any creation:

Base Being	Dynamics of Being	Relative Presence of Degree of Life Based on Minimum Number of Implicit Beings
Space-Time-Energy-Gravity (STEG) Beings	• STEG	1
EMWM P-Type Being	• STEG • EMWM	2
Quantum-Particle P-Type Being	• STEG • EMWM	3

	• Quantum-Particle	
Atom P-Type Being	• STEG • EMWM • Quantum-Particle • Atom	4
Cell P-Type Being	• STEG • EMWM • Quantum-Particle • Atom • Cell	5
Fundamental-Capacities-of-Self P-Type Being	• STEG • EMWM • Quantum-Particle • Atom • Cell • Fundamental Capacities of Self	6
Truer-Individuality P-Type Being	• STEG • EMWM • Quantum-Particle • Atom • Cell • Fundamental Capacities of Self • Truer Individuality	7
FBLEE-Based Sustainable-Global-Civilization P-Type Being	• STEG • EMWM • Quantum-Particle • Atom • Cell • Fundamental Capacities of Self • Truer Individuality	8

	• FBLEE-Based Sustainable Global Civilization	

Figure 10.2.1 Relative Presence of Degree of Life Based on Minimum Number of Implicit Beings

Hence, referring to Figure 10.2.1, the AI-based D-type beings discussed above will have a relative presence of life of '4', based on the minimum number of implicit beings in them.

Possible Deleterious Effects of D-Type Beings

Other common examples of D-type beings include those resulting in pollution due to electromagnetic radiation in the breaking off from the larger contextual electromagnetic fields of the sun and the earth, due to climate change in the breaking off from the balance of local ecosystem based climates, and due to chemical toxification in the breaking off from natural and composite systems in which chemical balance is maintained.

In his book 'Sun of gOd' Gregory Sams (Sams, 2009) refers to the holistic and pervasive electromagnetic field of our Sun, for example, in which the fullness of life on earth has blossomed. For that matter even earth has its own electromagnetic field which has influenced the development of life. In fact, Chapter 7.1, Effect of Electro-Magnetic-Wavearchetype-Masspotential P-Type Being on Life, illustrates these ideas to some degree. There is therefore a natural and holistic electromagnetic field that positively influences our life. By contrast, when we create electromagnetic radiation locally, due for example, to many of our wireless or computing devices, these are essentially fragmented-electromagnetic D-type beings that have been severed from the natural 'intelligence' that animates light-based-beings, and as a result there can be deleterious effects on our beings. For some discussion on this refer to Bob Berman's Zapped (Berman, 2018).

Similarly, climate always exists in the context of a larger balance and is part of natural ecosystems local to the earth. When due to artificial means such as emission of extensive green-house gases, thus creating an artificial-climate D-type being, this natural balance is upset, then the inherent logic and intelligence that has been built up over millennia in light-based-beings is destabilized and larger deleterious effects come into being. For a representative discussion of this refer to Flannery's The Weather Makers (Flannery, 2005).

Finally, as an example of chemical toxification on the micro-scale refer to some of Andrew Weil's works (Weil, 1996) where he elaborates that taking a compound out of context of the natural composite and comprehensive state it exists in, in Nature, and delivering this into the human body will create deleterious effects. This too, is due to the essential fragmentation that comes into effect with the creation of a more narrowly-defined-chemical D-type being. For the phenomena of chemical toxification on a macro-scale refer to Rachel Carson's Silent Spring (Carson, 1964).

Essentially the negative effects of D-type beings are captured philosophically in James Carse's Finite and Infinite Games (Carse, 1986). By their nature infinite games are unending and played for the joy of creation, just as happens in the generation of light-based-beings as suggested by (3.1.3) the P-type generator equation. By contrast finite games are played for a particular outcome with defined boundaries and will create increasing entropy unless they occur within the context of an infinite game.

D-type beings can often be a natural outcome of short-sighted human outcome. Once a D-type being emerges, it results in increasing fragmentation and therefore in increased entropy driving a containing system to destruction. As untransformed human actors operating from a narrow egoistic base, it is often difficult to take the high road and see and do things as though we were indeed part of a single light-based edifice replete with infinite positive potential. Therefore, it becomes necessary to create or practice some culture where we see, think, will, feel, and love more broadly and differently, as though we were indeed of one substance with a much larger containing system. This chapter will explore some mathematical operators to focus becoming on the task of expediting transformation of the human P-type being, thereby also expediting the arrival of the S-type being.

In an interesting point of view Lanza's Biocentrism (Lanza et al., 2010) suggests that life is not an accidental byproduct of physics, but rather is a key part of our understanding of the universe. It states that there is no independent external universe outside of biological existence, and as support cites that there are over 200 physical parameters within the universe so exact that it is more probable that this is so in order to allow for existence of life and consciousness, rather than these coming about at random. While Light is modeled as being the foundation of all existence in the Cosmology of Light book series and in that regard differs from biocentrism, yet in biocentrism's point of view that there is a supremacy of life and cognition there appears to be consistency with the idea of human P-type beings leveraging mathematical operators to expedite their transformation and even the transformation of the cosmos.

Based on the mathematics developed in Section 2, on modeling the mathematical structure of being and the becoming of P-type beings, twelve general 'mathematical operators' are arrived at deductively. These operators are the result of considering some of the dynamics of each of the levels in the seed-singularity, and integration of the multiple levels of the overall light-based-singularity mathematical model. These operators are non-exhaustive, but rather are indicative of the nature of dynamics that will increasingly arise in the successful extrapolation of human P-type beings to super-matter-based P-type beings and beyond, due to the increasing presence of meta-level seed-singularity entanglement and precipitated functional-richness representative of the dynamics of the UPI-type, K-type, and N-type beings.

The use of such operators assists in perceiving reality as derivate from Light, and hence their active adoption will accelerate the overcoming of limiting patterns and subsequently the integration of the multiple layers of light in a light-based-singularity. As a result, the arrival of the S-type being will be expedited.

Note that in a previous book on quantum computing in a cosmology of light (Malik, 2018c) these operators were explored as a possible basis for alternative quantum-related computational models. Such quantum-level computation is a dynamic embedded in any light-based-singularity and such mathematical operators have to arise as the practical union of matter and light continues.

Presence-Based Mathematical Operators

Considering Light's dimension of Presence, Equation 10.3.1, Presence-Based Mathematical Operators, summarizes a set of representative presence-based mathematical operators:

$$Presence\_based_{Mathematical\_operators} \ni [Fullness, Equality, Uniqueness, ...]$$

Eq. 10.3.1: Presence-Based Mathematical Operators

Fullness refers to the possibility that every single point in any time-space continuum is informed by Light's four-fold fullness. That is, the reality of ∞-entanglement representative of the UPI-type being is fully active. Note that ∞-entanglement is going to become more effective as the very fabric of existence is changed by the generation of P-type beings with increasing spheres of influence. Hence, in equation form it may be suggested that Fullness is the union (U) of $System_{Pr}$, $System_P$, $System_K$, and $System_N$ as in Equation 10.3.2, Fullness:

$$Fullness = \bigcup \begin{bmatrix} System_{Pr} \\ System_P \\ System_K \\ System_N \end{bmatrix}$$

Eq. 10.3.2: Fullness

To see such 'fullness' in everything is to begin to see as perhaps Light sees, and this will assist in closing the gap between the material layer and layers of Light antecedent to it.

In general, for any two points 'A' and 'B' it can be further suggested that the Fullness behind A is the same as the Fullness behind B, as in Equation 10.3.3, Equivalence of Fullness. This suggestion may also be arrived at by considering Einstein's General Theory of Relativity (Einstein, 1995) that states: "All bodies of reference K, K1 etc. are equivalent for the description of natural phenomena (formulation of the general laws of nature), whatever may be their state of motion." If we shrink these bodies of reference or coordinate systems to infinitesimal dimensions, thus approaching a 'point', this suggests that there is equivalence in that the general laws of nature are equally valid at any two points. Hence, Equation 10.3.3:

$Fullness_A \equiv Fullness_B$

Eq. 10.3.3: Equivalence of Fullness

Equality refers to the possibility that since every point in a time-space continuum is always some expression of Light's four-fold fullness, hence every point is equal to any other point. As in the likely practically felt effectiveness of the previous operator, such equality will also be felt more tangibly as the very fiber of existence tends towards super-matter-based P-type beings and beyond. This implies that all expressions and developments share a fundamental equality with all other points.

Depicting a point 'a' by $Point_A$ and a point 'b' by $Point_B$, then an equation, Equation 10.3.4, Equality, would be:

Equality: $Point_A \equiv Point_B$

Eq. 10.3.4: Equality

Uniqueness refers to the possibility that any point is fundamentally unique. This is of course the very meaning of infinite information in Light, as elaborated in some detail in Section 2, and specifically in the further elaboration of seed-singularity dynamics involving ∞-entanglement representative of the UPI-type being, K-entanglement representative of the K-type being, N-entanglement representative of the N-type being, and subsequently FBLEE-entanglement, that necessitate the reality of uniqueness. As matter-based P-type beings tends toward super-matter-based P-type beings and beyond, the very fiber of existence reflects antecedent functional-richness more fully, and the underlying reality of uniqueness becomes materially more visible.

Practically though, perhaps an easy way to envision this is to see that any point in a time-space continuum is a result of a unique time-space intersection. This becomes more meaningful knowing that time and space are in reality the summary result of a persistent computation resulting in light-based quantization. Hence, two points within a time-space continuum, A and B, can always be envisioned as having unique time and space coordinates and hence characteristics, as in Figure 10.3.1:

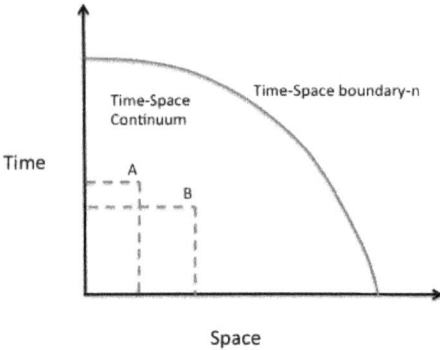

Figure 10.3.1: Uniqueness in Time-Space Continuum

But even for any organization, which can itself be considered a further development of a point or points through the action of time and space, the already proposed equation, (2.4.5) that governs organizational signatures may be restated as an equation for uniqueness, as in Equation 10.3.5:

$$Uniqueness: Sig = Xa + \overline{Yb_{0-n}} \ \ where \ \begin{bmatrix} X \in [S_{System_{Pr}}, S_{System_P}, S_{System_K}, S_{System_N}] \\ Y \in [S_{System_{Pr}}, S_{System_P}, S_{System_K}, S_{System_N}] \\ a, b \ are \ integers; a > b \end{bmatrix}$$

Eq. 10.3.5: Uniqueness

Power-based mathematical operators pertain to the fount of dynamism that will tend to determine an organization's practical action in super-matter-based P-type beings. Mathematical operators related to 'power' give insight into how an organization will tend to meet circumstance and other organizations.

Representative operators include Direction, Fractal, and Intersection and in contrast to Knowledge-based operators discussed in the next sub-section, tend to create the more visceral and immediate reactions to circumstance. Knowledge-based operators on the other hand tend to create action more in line with the long-term. This power-based set can be summarized by Equation 10.3.6:

$$Power\_based_{Mathematical\_operators} \ni [Direction, Fractal, Intersection \dots]$$

Eq. 10.3.6: Power-Based Mathematical Operators

Direction refers to the possibility that direction at any possible system bifurcation point is not random and not determined either. Rather it can be thought of as a qualified determinism, as introduced in Chapter 3.7, and is a function of applying DI_V and DI_H to the set of possibilities existing at a bifurcation point. As per the functioning of DI_V and DI_H, it is the strongest or most 'powerful' possibility that will tend to determine what will emerge as an organization meets circumstance and other organizations. Hence leveraging (3.7.4) creates Equation 10.3.7:

$$Direction: Org\_Dir = DI \left(\begin{bmatrix} M_3 \rightarrow System_X \\ (\uparrow F \rightarrow I) \\ M_2 \rightarrow S_{System_X} \\ (\uparrow Sig \rightarrow F) \\ M_1 \rightarrow Sig_x \\ (\uparrow > P_P) \\ U \rightarrow x_U \end{bmatrix}_{x=p,v,m,i} \right) \rightarrow$$

$$x\_matrix_{strongest} @ level_{strongest}$$

Eq. 10.3.7: Direction

Fractal refers to the possibility that as organizational complexity increases, the base orientation, orientation-x, where x could be physical, vital,

365

mental, or integral, of an average organization at some level of complexity 'n', will tend to determine the orientation of an organization at a level of complexity 'n+1'. Likewise, the orientation of an organization at level of complexity 'n+1' will tend to determine the orientation of an organization at level of complexity 'n'. This is summarized by Equation 10.3.8:

$$Fractal: \ Orientation_x \ @ \ Complexity_n \ \leftrightarrow \ Orientation_x \ @ \ Complexity_{n+1}$$

Eq. 10.3.8: Fractal

Organizational complexity refers to an order of magnitude change as in from a person to a team, or from a team to a business unit, and so on, for example. In Nature such fractal arrangements abound in the way the human body is constructed to the very structure of galaxies (Briggs, 1992). This notion has been suggested to exist in complex behavioral systems as well as described in some detail in books such as The Fractal Organization (Hoverstadt, 2008), The Fractal Organization (Malik, 2015), and The Misbehavior of Markets (Mandelbrot, 2006). This notion relates to 'power' in that it is the patterns at one level that will tend to determine the patterns at another level, often preempting what may be a more logical choice based on reason. This kind of behavior has been suggested as causing cyclic fluctuations in stock and other markets where greed and fear often trump more intelligent and rational choices (Frost, 2005). Greed rises until fear sets in. Fear rises until greed sets in.

Intersection occurs when two organizations shift orientations to the next successive level due to the shock of interaction. Hence if an organization is at a physical orientation, it may be shifted to a vital orientation when intersection occurs, as in Equation 10.3.9, Intersection, where the function 'Next Element' extracts the next element from the Set S comprising the elements (physical, vital, mental, integral). Examples of such phenomena abound where failure to make a shift results in shock of conflict repeating itself endlessly. Such a process with applicability at multiple levels of complexity has been captured by the series of books on crucial conversations (Patterson, 2011).

$$Intersection: \ Organization \ Orientation \ \rightarrow \ Next \ Element \ (S)$$

Eq. 10.3.9: Intersection

Knowledge-based mathematical operators have to do with how organizations tend to develop by increasing knowledge over time. Representative knowledge properties discussed in this section include Alternative, Flowering, and Higher, which as suggested in the previous section are of a different nature than 'power' or dynamism-based properties that tend to determine an organization's visceral or immediate reaction to its environment. This knowledge-based set can be summarized by Equation 10.3.10:

$$Knowledge\_based_{Mathematical\_operators} \ni [Alternative, Flowering, Higher ...]$$

Eq. 10.3.10: Knowledge-Based Mathematical Operators

Alternative refers to an alternative narrative that an organization will tend to embed itself in. These alternative narratives relate to the physical, the vital, the mental and the integral orientations. These narratives can easily become fixed and can strongly influence the entire internal and external orientation of an organization. The book 'The Fractal Organization: The Future of Enterprise' (Malik, 2015) suggests a theory of such narratives with their consequent effect on practical action. The alternative narratives are best described using the generalized equation (2.6.6) modified as Equation 10.3.11. Hence:

$$Alternative: P - Type\ Becoming_{orientation-x}$$

$$= \begin{bmatrix} M_3 \to System_X \\ (\uparrow F \to I) \\ M_2 \to S_{System_x} \\ (\uparrow Sig \to F) \\ M_1 \to Sig_x \\ (\uparrow > P_{x)} \\ U \to x_U \end{bmatrix} TC \to x_T, where \begin{bmatrix} x_U \ni [...] \\ x_T \ni [...] \end{bmatrix}$$

Eq. 10.3.11: Alternative

Where, 'x' can be thought of as an element from the Set of Orientations:

$$x \in (physical, vital, mental, integral)$$

'X' can be thought of as an element from the Set of System-level architectural forces. Hence:

$X \in (Presence, Power, Knowledge, Nurturing)$

Flowering refers to the possibility that any time-space boundary-n, depicted as TS_n (as in Fig. 10.3.1) will have more potential or possibility associated with it than a time-space boundary-(n-1). Putting this into equation format yields Equation 10.3.12:

$Flowering: Possibility_{TS_n} > Possibility_{TS_{n-1}}$

Eq. 10.3.12: Alternative

Higher refers to the possibility that over time the direction will always tend to move to a higher meta-level. This can be summarized by using the notion of the core-matrix (2.6.7) yielding Equation 10.3.13:

$$Higher: Upward \begin{bmatrix} M_3 \rightarrow System_X \\ (\uparrow F \rightarrow I) \\ M_2 \rightarrow S_{System_X} \\ (\uparrow Sig \rightarrow F) \\ M_1 \rightarrow Sig_x \\ (\uparrow > P_x) \\ U \rightarrow x_U \end{bmatrix}$$

Eq. 10.3.13: Higher

Note also that while many organizations are practically at the untransformed level for a long time there is still movement within that level that can generally be depicted as a change from a predominantly physical-orientation to a more mental-orientation. In general, this shift in orientation is implied, as more and more fixed patterns are overcome: $\uparrow > P_x$, where P_x refers to patterns along an orientation 'x' where x is an element from the set: (physical, vital, mental, integral).

Nurturing or Harmony-Based Mathematical Operators

Nurturing-based mathematical operators have to do with the nature of relationship within systems. These relationships are posited as being of a nurturing nature and emanate from the notion of 'nurturing' as an organizing class. Representative mathematical-operators include

Remember, Linking, and Relate. This nurturing-based set can be summarized by Equation 10.3.14:

$$Nurturing\_based_{Mathematical\_operators} \ni [Remember, Linking, Relate \ldots]$$

Eq. 10.3.14: Nurturing-Based Mathematical Operators

Remember has to do with remembering that there is something in each organization that existed before the existence of any organization, and further, that there is something in each organization that exists in every other organization. This captures a key dynamic that accompanies changing biography in any P-type being. This dynamic can be thought of as going back to a time-space moment of zero, and subsequently of expanding into the Time-Space Continuum keeping that connection in mind. There is something in each organization that exists in every organization and highlights a special way to relate to the underlying system as in Figure 10.3.2:

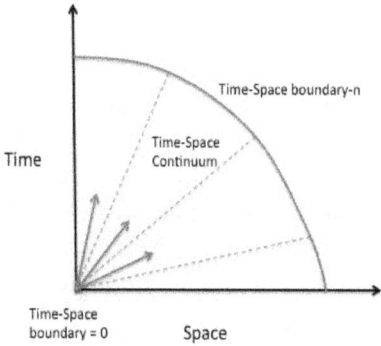

Figure 10.3.2: Remembrance in Time-Space Continuum

This may be depicted by Equation 10.3.15:

$$Remember: Ubiquity \begin{bmatrix} \overleftarrow{TS=0} \\ \overrightarrow{TS>0} \end{bmatrix}$$

Eq. 10.3.15: Remember

Where the condition of going back (←) before any organization existed, TS = 0, is invoked as all developments proceed (→), TS > 0, to create a sense of ubiquity. The sense of ubiquity is a remembrance. This notion

369

may be akin to the concept that everything that is in the universe emanated from the Big Bang and that we are all recycled stardust (Swimme, 2001).

Link refers to the condition whereby in any time-space coordinate, irrespective of the level of untransformed reality (U), any present state can be consciously linked to the underlying ubiquitous system. This can be depicted by Equation 10.3.16, Link, where the conditions that usually need to be in place for a meta-layer to actively influence layer U disappear. There may be an attitude or receptiveness on the part of the element at U that allows such linking to take place. Hence:

$$Link: \begin{bmatrix} M_3 & \to & System_X \\ & \updownarrow & \\ M_2 & \to & S_{System_X} \\ & \updownarrow & \\ M_1 & \to & Sig_X \\ & \updownarrow & \\ U & \to & x_U \end{bmatrix}$$

Eq. 10.3.16: Link

Relate is a way to relate to the System so as to offer or surrender activities any kind of organization is involved in, to the System, and hence the Fullness or intelligence embedded in every point. This can be depicted by an Offer function, such that the first or relatively untransformed element in the function, depicted by x_U, is being offered to the second one, the union of the four-fold intelligence embedded in each point and depicted by the union function U[ ], as in Equation 10.3.17:

$$Relate: \; Offer \left(x_U, \bigcup \begin{bmatrix} System_{Pr} \\ System_P \\ System_K \\ System_N \end{bmatrix} \right)$$

Eq. 10.3.17: Relate

Chapter 10.4: Altering the Microcosmic-Macrocosmic Balance and the Birth of the S-Type Being

In the Section 9, on extrapolation to super-matter-based P-type beings, a possible link was made between the expansion of the universe and the changing nature of matter due to continued precipitation of fourfold functional richness. through the becoming of a continuing series of P-type beings. Perhaps it is that if the expansion continues the apparent gaps between galaxies filled with extraordinarily high dark-energy, possibly correlated with an increasing "manifested set", becomes ripe for a series of inter-galactic-bangs of the nature of the Big Bang. And perhaps such a series of inter-galactic-bangs in such a functionally-rich universe precipitates cycles of development based on other speeds of light, that in turn sets up a more conscious fusion of the multiple layers of existence so created, into one incredible Universe where an entirely new cycle of super-matter-based P-type beings is generated.

The fruition of such a possibility signifies a new Ouroboros where richness builds on richness and new types of super-matter-based P-type beings on previous super-matter-based P-type beings.   In Martin Nowak's book, Super Cooperators (Nowak, 2011), he proposes a third principle of evolution: that of cooperation. A first principle is that of mutation, responsible for generating genetic diversity. A second principle is that of selection, that focuses on individuals best suited to certain environments. He calls cooperation the master architect of evolution since from it emerges the constructive side of evolution 'from genes to organisms to language to complex social behavior'. But as suggested here, evolution will occur driven by Light itself until emergent beings see as Light sees, and become as Light is, even materially.   This is the foundation for the generation of the S-type being.

The human P-type being has in a certain sense been the result of 30 orders of microcosmic magnitude implicit to it. Fields, quanta, atoms, cells, the incredible four-fold emergences of properties codified in Light have all contributed to the emergence of the human P-type being. Now in a possible reversal of causality the human P-type being appears to be positioned to similarly influence and even structure the further development of the macrocosm by mastery of the microcosm. Such mastery will involve leveraging an increasing set of functional operators of the type introduced in the previous chapter.

Through will, human faculties, the use of functional operators, and vaster and vaster displays of impersonal love, continued states of quantum-certainty resulting in four-fold quantization will occur. Such quantization changes the very nature of space, time, energy, and gravity and will alter the development of the 30 orders of macrocosmic magnitude.

Such audacity is only possible because the human is a master emergence, a promising light-based-singularity in which by definition the micro and the macro can increasingly become extraordinary reflections of each other, in a universe of Light where all is Light and all is of Light. Light in material form turning on itself can alter the very nature of what it is by what it is. Such is the future tale of Ouroboros changing from swallowing its tail to entwining itself through a Cosmos in a never-ending spiral toward unforeseen flowerings.

The breaking down of boundaries between meta-layers and U so that the dynamics at U is increasingly active with the presence of the UPI-type, K-type, and N-type beings, is the condition of the birth of the S-type being and is depicted by Equation 10.4.1, Birth of the S-Type Being:

$$Birth\ of\ the\ S-Type\ Being$$

$$=\begin{Vmatrix}\begin{bmatrix}c_\infty\colon[Pr,Po,K,H]\\ (\downarrow R_{C_K}=f(R_{C_\infty}))\\ c_K\colon[S_{Pr},S_{Po},S_K,S_H]\\ (\downarrow R_{C_N}=f(R_{C_K}))\\ c_N\colon f(S_{Pr}\ x\ S_{Po}\ x\ S_K\ x\ S_H)\\ (\downarrow R_{C_U}=f(R_{C_N}))\\ c_U\colon[P,V,M,C]\\ \Uparrow\\ c_{0:[D,W,I,C]}\end{bmatrix}_{Light}\begin{bmatrix}M_3\ \to\ System_X\\ \updownarrow\\ M_2\ \to\ S_{System_X}\\ \updownarrow\\ M_1\ \to\ Sig_X\\ \updownarrow\\ U\ \to\ x_U\end{bmatrix}_{Space}\\[4pt]\begin{bmatrix}U\ \to\ \begin{matrix}M_3\colon\ -\infty\ \leq t\ \leq\ \infty\\ \downarrow\\ M_2\colon 0\ \geq t\ >\ \infty\\ \downarrow\\ M_1\colon\ 0\ >t\ >\ \infty\\ \downarrow\\ t\sim E_{Cell};TC\colon M_3\ \to U\\ t\sim E_{Human};TC\colon U\ \to\ M_3\end{matrix}\end{bmatrix}_{Time}\ \ \ TC\ \to x_T\end{Vmatrix}_{\langle x_U|x_T\rangle}$$

Eq. 10.4.1: Birth of the S-Type Being

Note that prerequisite conditions for activation of meta-layers have been entirely transcended in the Space-Matrix as depicted by ($\updownarrow$) in (10.4.1). The very notion of ($\updownarrow$) implies a triumph of love. For it is only with a complete self-giving and receiving without judgment the possibilities in meta-layers of Light that the S-type being can become. But self-giving and receiving without judgment is nothing other than a supreme act of love – a triumph of love.

Equation 10.4.2, Future Code, suggests genetic-type code generated through further iterations involving the S-type being. Note that the starting point in this iteration is assumed to be some approximate form of super-matter, depicted as '~Super-Matter' in x_u:

Future S − Type Being Based Code =

$$
\left[
\begin{array}{c}
\begin{bmatrix}
c_\infty : [Pr, Po, K, H] \\
\left(\downarrow R_{C_K} = f(R_{C_\infty}) \right) \\
c_K : [S_{Pr}, S_{Po}, S_K, S_H] \\
\left(\downarrow R_{C_N} = f(R_{C_K}) \right) \\
c_N : f(S_{Pr} \; x \; S_{Po} \; x \; S_K \; x \; S_H) \\
\left(\downarrow R_{C_U} = f(R_{C_N}) \right) \\
c_U : [P, V, M, C] \\
\Uparrow \\
c_{0:[D,W,I,C]}
\end{bmatrix}_{Light}
\quad
\begin{bmatrix}
M_3 \;\to\; System_X \\
\updownarrow \\
M_2 \;\to\; S_{System_X} \\
\updownarrow \\
M_1 \;\to\; Sig_X \\
\updownarrow \\
U \;\to\; X_{\sim Super-Matter}
\end{bmatrix}_{Space} \\[2em]
\begin{bmatrix}
M_3 : -\infty \leq t \leq \infty \\
\downarrow \\
M_2 : 0 \geq t > \infty \\
\downarrow \\
M_1 : 0 > t > \infty \\
\downarrow \\
U \;\to\; \begin{array}{l} t \leq E_{Cell}; TC: M_3 \to U \\ t \sim E_{Human}; TC: U \to M_3 \end{array}
\end{bmatrix}_{Time}
\quad TC \;\to\; x_{Future}
\end{array}
\right] \;\Rightarrow\; \langle x_U | x_T \rangle
$$

$$\bigcup_{f(x)} \begin{bmatrix} \langle Space - Time - Energy - Gravity \; P - Type \; Beings \; and \; Pre - Genetic \; Code \rangle \\ \begin{pmatrix} Electro - magnetic - wavearchetype - masspotential \; P - Type \; Being \\ and \; Pre - Genetic \; Code \end{pmatrix} \\ \langle Quantum - Particle \; P - Type \; Being \; and \; Pre - Genetic \; Code \rangle \\ \langle Atom \; P - Type \; Being \; and \; Pre - Genetic \; Code \rangle \\ \langle Molecule \; P - Type \; Being \; and \; Pre - Genetic \; Code \rangle \\ \langle Cell \; P - Type \; Being \; and \; Genetic \; Code \rangle \\ \langle Human \; P - Type \; Being \; and \; Genetic \; Code \rangle \\ \langle Stable - Mega - Organization \; P - Type \; Being \; and \; Post - Genetic \; Code \rangle \\ \langle Sustainable - Global - Civilization \; P - Type \; Being \; and \; Post - Genetic \; Code \rangle \\ \langle Super - Matter - Based \; P - Type \; Being \; and \; Post - Genetic \; Code \rangle \\ \cdots \\ \sum_{n}^{\infty} \begin{pmatrix} \sum Sig_I = Xa + \overline{Yb_{0-n}} \\ where \begin{bmatrix} X \in [S_{System_N}] \\ Y \in [S_{System_{Pr}}, S_{System_P}, S_{System_K}, S_{System_N}] \\ a, b \; are \; integers; a > b \end{bmatrix} \\ \sum Sig_M = Xa + \overline{Yb_{0-n}} \\ where \begin{bmatrix} X \in [S_{System_K}] \\ Y \in [S_{System_{Pr}}, S_{System_P}, S_{System_K}, S_{System_N}] \\ a, b \; are \; integers; a > b \end{bmatrix} \\ \sum Sig_V = Xa + \overline{Yb_{0-n}} \\ where \begin{bmatrix} X \in [S_{System_P}] \\ Y \in [S_{System_{Pr}}, S_{System_P}, S_{System_K}, S_{System_N}] \\ a, b \; are \; integers; a > b \end{bmatrix} \\ \sum Sig_P = Xa + \overline{Yb_{0-n}} \\ where \begin{bmatrix} X \in [S_{System_{Pr}}] \\ Y \in [S_{System_{Pr}}, S_{System_P}, S_{System_K}, S_{System_N}] \\ a, b \; are \; integers; a > b \end{bmatrix} \end{pmatrix} \end{bmatrix}$$

Eq. 10.4.2 Future S-Type Being Based Code

The final code-segment in Equation 10.4.2, Future S-Type Being Based Code, is a generalization (Σ) of all the development that can take place along known dimensions of physicality, vitality, mentality, and integrality. Further the code is a union (U) of all that has been generated and all that will be generated combined together in anyway (f(x)). This equation implies a constant stream of materialized functional-richness, an evolving super-matter, and an evolving post-genetic structure for encapsulating such future code, all brought out by the triumph of love by which horizontal and vertical boundaries dissolve into an active unity.

This book has proceeded to summarize a perception of Life based on the nature and play of Light. In this play all beings emerge from Light. Light's essential nature, housing infinite potentiality within it has already created this extra-ordinary, massively intelligent universe, run by a persistent quantum-level computation that generates a continuous stream of genetic-type data that functions as relevant "law" as it were, to allow more of Light's potentiality to continually materialize in a process of becoming and in milestones of generated beings.

The iterative process of becoming and being constitutes life, and further must dictate any holistic process of transhumanism. Essentially the materialization of potentiality in Light may lead to the generation of the S-type being, the singularity whose power will far surpass any other present singularity precisely because of the level and sophistication of integration, and the continual real-time re-computation of reality driven by enlightened heart, will, and mind dynamics. The master stroke in the successful generation of the S-type being is suggested to be love, because it is only in its triumph that the fullness of the horizontal fourfoldness and the vertical fourfoldness can become a sustainable foundation for further becoming.

This chapter summarizes the S-type being biography by reviewing key insights that have surfaced through the ongoing iteration between being and becoming from the pre-Big Bang era to the extrapolation of the super-matter-based P-type being to a possible generation of the S-type being:

1. When Light travels at speed c, 186,000 miles per second in vacuum it creates past, present, future, and an accumulation of possibility due to the finite speed that becomes matter.
2. There appears to be no reason for light to travel at c than for reality to be experienced the way it is when light travels at c.
3. The necessity for light to travel at c to create a material universe is also the foundation for Einstein's Theory of Relativity in that for speed to remain constant, time and space have to be allowed to fluctuate.
4. In Light's native state it travels infinitely fast. This means it is present everywhere instantaneously or is omnipresent. It means that nothing that is not of the nature of Light can persist against the reality of Light, and hence it is omnipotent. Light being

everywhere is aware of all that is happening in it, and it is therefore omniscient. All is connected in one embrace in the substance of Light – therefore it is omninurturing.

5. The essence of omnipresence can be thought to be presence. The essence of omnipotence can be thought of as power. The essence of omniscience can be thought of as knowledge. The essence of omninurturance can be thought to be harmony.

6. These four implicit properties of Light – presence, power, knowledge, harmony – hint at the potentiality resident in Light.

7. The pre-existent complexity of Life derives from the fact that Light has infinite potentiality embedded in it.

8. The essence of light traveling at c can be seen to be mathematically symmetrical to the four implicit properties of Light in its native state at infinity.

9. A past implies a status quo – the result of a working out of things into established reality. Hence, symmetrical to the notion of 'presence'.

10. A present implies a vitality or dynamism with the stronger influence expressing itself in the present. Hence, symmetrical to the notion of 'power'.

11. A future implies the working out of deeper ideas or seed concepts. Hence, symmetrical to the notion of 'knowledge'.

12. Interaction between material islands created in a reality of light traveling at c implies working out connection. Hence, symmetrical to the notion of 'harmony'.

13. For the four implicit properties of Light at infinity to become the reality of vast diversity in a material universe generated when light travels at c, means that some essential mathematical transformations have to occur to diversify the implicit four, between these two layers of Light.

14. The reality of light traveling infinitely fast so that it is ubiquitously present at every point-instance is summarized by a beingness that we can refer to as the UPI-type being, where 'U' stands for ubiquity, 'P' for point, and 'I' for instance. 'Point' is a summary for the notion of 'space'. 'Instance' is a summary for the notion of 'time'.

15. The first transformation occurs when each of the four properties become variations based on their essence, to create four sets of infinite elements each.

16. The second transformation occurs when elements from the four sets combine together in unique combinations to create an infinite number of unique seeds.

17. The first transformation can be envisioned as a field. A field means that the essential binding factor that creates matter is relaxed so that there is a vaster spreading out of information. This will occur when light is traveling quite a bit faster than c.

18. The second transformation is envisioned as a wave. A wave means that the form of the information being represented is closer to being a particle than a field. This will occur when light is traveling faster than c but slower than the speed required to create a field.

19. Light traveling at c creates an upper bound in this known material universe. The inverse of c creates a lower bound and is proportional to Planck's constant h.

20. Planck's constant h is significant in that it provides insight into the least amount of energy required for matter to express itself.

21. When light travels faster than c, then the inverse of this speed is a constant smaller than h. This smaller constant implies that the energy will not be able to express itself in matter, but as a more spread-out wave.

22. When light travels faster than c, but closer to being infinitely fast, than the inverse of that speed is a constant, significantly smaller than h. This miniscule constant implies that energy will not even be able to express itself in a more contained wave but will require a field to express itself.

23. The down-shifting of information in Light, from its native state of traveling infinitely fast and housing infinite potential, to each subsequent layer where light travels slower, implies the accumulation of some of that information in the body of a being, so that it can express itself in more material form in the layer of slower moving light. This downshifting requires the device of quanta.

24. Quanta are the device by which information in a faster-moving layer of light materializes in a slower-moving layer of light.

25. The speed at which light travels to create a reality of fields of information is referred to as c_k. The corresponding layer of reality is referred to as K. The corresponding being is referred to as the K-type being.

26. The K-type being can also be thought of as a partial-seed light-based-singularity in that there is wholeness and a complete backward integration with light in its native state traveling infinitely fast.

27. A light-based-singularity occurs when the most current emergent form has all previously generated code embedded or available to it thereby inherently abiding with all "laws" that have thus far emerged in previous light-based singularities.

28. The seed-singularity itself will have had a progressive biography resulting in partial-seed-singularities along the way, until the seed-singularity has itself emerged.

29. The speed at which light travels to create a reality of waves of information is referred to as c_n. The corresponding layer of reality is referred to as N. The corresponding being is referred to as the N-type being.

30. The N-type being can also be thought of as the seed-singularity in that there is wholeness and an integration with K and the layer representing Light's native state at infinity. This seed-singularity is the seed of all other light-based singularities.

31. There is a light-based wholeness in the seed-singularity that results in a first meaningful sphere of cohesion with the creation of myriad light-based seeds of uniqueness.

32. The layer of reality created when light is traveling at c is referred to as U. The corresponding being is the U-type being.

33. The pre-existent complexity in Life is further reinforced by different information fields existing in each subsequent layer of slower light, relative to Light in its native state at infinity.

34. The origin of genetics is the vast amount of information embedded in Light in the seed-singularity.

35. If there is a history of singularities or light-based beings, or a biography that relates the seed-singularity with partial-singularities, with the Second Singularity, then genetics can be thought of as the language in which it is written.

36. All partial-singularities, arising when light travels at c and therefore U-type beings, are referred to as P-type beings where 'P' signifies 'partial'.

37. Singularity-biographies are the process by which a singularity materializes possibility contained in the seed-singularity through a trajectory of partial-singularities culminating in the Second Singularity.

38. The Second Singularity is referred to as the S-type being.
39. In its native state with light traveling infinitely fast the libraries of information generated are subject to a class of entanglement referred to as ∞-entanglement. This implies that this information is subtly available in all of existence.
40. The UPI-type being therefore exhibits ∞-entanglement.
41. At layer K, libraries of information generated from the four sets are subject to a class of entanglement referred to as K-entanglement.
42. The K-type being therefore exhibits K-entanglement.
43. At layer N, libraries of information generated from the infinite number of unique seeds, is subject to a class of entanglement referred to as N-entanglement.
44. The N-type being therefore exhibits N-entanglement.
45. Entanglement is what ensures uniqueness as information materializes.
46. Quantum downshifts occur between subsequent layers of light until that point when information is ready to materialize at U, where light travels at c.
47. When light exists at speed zero, the opposite of its native state of being infinitely fast, that reality also becomes the opposite of omnipresence, omnipotence, omniscience, and omninurturing. Such a reality would therefore be characterized by properties such as weakness, ignorance, darkness, and hate.
48. The being created when light exists as zero speed is referred to as the 0-type being.
49. Materialization of information in the form of matter involves superposition of different possibilities resident at different layers of Light.
50. The influence of the layer of light specified by zero speed, the 0-type being, will impregnate practical existence with an essential inability and cause obstinate habit to form.
51. Quantum-level computation is the process that arbitrates the material expression of different states existing simultaneously in different layers of light.
52. The continual shifting of information in layers of light necessitates the reality of a persistent quantum-level computation.
53. Schrodinger's equation summarizes matter in its wave-aspect and models how infinite superposition of possibility results in some concrete materialization of these.

54. Heisenberg's uncertainty principle sheds light on the always-buzzing pregnant-infinity behind all appearance.

55. Euler's equation will approximate the range of materialization influenced by the dark - or the 0-type being, to light - or the UPI-type being, spectrum, and can be modeled to oscillate or exist between these two bounds.

56. Imaginary numbers in Schrodinger's equation or Euler's equation interrelate antecedent layers of light with the material layer.

57. The Big Bang is the phenomenon that occurs when Light slows down to c. This slowing down materializes a vast amount of information that creates the Big Bang.

58. The process of light slowing down to c from its native state of traveling at infinity is modeled as a light-based-singularity superstructure. The term superstructure implies that different kinds of light-based-singularities can occur within it.

59. The light-based-singularity superstructure can also be thought of as the matrix of being, in that all beings emerge from it.

60. When the Big Bang occurs the materialization of information creates space, time, energy, and gravity.

61. The Big Bang generates the first envisioned partial-singularity or space-time-energy-gravity quadrumvirate beings that occurs after the formation of the seed-singularity.

62. The Big Bang is a watershed event and signifies the start of a progressive, functionally rich evolution.

63. In the absence of the Big Bang evolution would have proceeded based on light existing at speed zero, or under the weight on the 0-type being as the starting point.

64. Space can be thought of as containing all the seeds of uniqueness formed by antecedent layers of Light.

65. Time is a way to think of the process of maturity by which the potentiality in seeds will fructify.

66. Energy is a way to think of the essential accumulation of possibility that will allow materialization of seeds.

67. Gravity is an arrangement that binds seed to a set of seeds that will be required in the journey to its own fulfillment.

68. Space can be thought of as filled with multiple levels of superposed entanglements in static form.

69. Time appears to be the dynamic working out of the possibilities implicit in superposed entanglements materialized as seeds in Space.

70. Space, Time, Energy, and Gravity are emergent from Light and mirror or reflect Light's implicit properties of Knowledge, Power, Presence, and Harmony, respectively.

71. The process of materialization of space, time, energy, and gravity takes place in two steps.

72. Step one takes place as the speed of light approaches c from above. Logic specifying this essential four-foldness of space-time-energy-gravity is created in a layer at the quantum-levels. This layer is referred to as Four-Base-Logic-Encoding-Ecosystem (FBLEE).

73. Logic systems in FBLEE can exercise themselves through FBLEE-entanglement (FBLEEE).

74. Step two is a further precipitation that occurs as light travels at c. Space-time-energy-gravity FBLEE precipitates into a layer that houses this kind of information, referred to as the material-fabric (MF).

75. The MF can be thought of as housing genetic-type information that is pre-genetic, and yet exercises a binding logic on all constructs arising in the universe.

76. The material-fabric is imbued with reality when the fourfold space-time-energy-gravity quantization is itself generated as pre-genetic code.

77. Hence the logic of space, time, energy, gravity is binding on all constructs arising in the material universe specified by light moving at speed c.

78. Space-time-energy-gravity logic embedded in the material-fabric is what allows quantization and therefore materialization to take place.

79. Space-time-energy-gravity has two modes of operations. The macro-mode is ingrained in the material-fabric and puts in place the familiar large-scale universal parameters of Space, Time, Energy, and Gravity. This generates the space-time-energy-gravity P-type macro-being.

80. The micro-mode occurs as a composite space-time-energy-gravity quantum whose action is required for the encoding of any logic-ecosystem in FBLEE and subsequently its materialization in the MF or appropriate information-housing structure.

81. Hence all materialization requires the action of the space-time-energy-gravity P-type micro-being.

82. The space-time-energy-gravity P-type micro-being can be thought of as the ever-present witness under whose gaze antecedent

possibility in light materializes in the universe constructed when light travels at c.

83. Such micro-mode space-time-energy-gravity quantization legitimizes the logic of an ecosystem into "law" by encoding it in the material-fabric.

84. Quadrumvirate space-time-energy-gravity quantization seeking to express the reality of its inherent oneness will cause matter to containerize regardless of scale.

85. The action of the space-time-energy-gravity P-type micro- and macro-beings ensures that the generation of any subsequent P-type being maintains integrity as a light-based system by binding it to the dynamics of antecedent seed-singularity-based beings.

86. Partial-singularities or P-type beings are always the result of the persistent quantum-computation that arbitrates material reality through the interaction of the many dynamics belonging to the multiple realities set up by different speeds of light, including that of c.

87. Fragmented-singularities, by contrast, that are essentially separated from any light-based-singularity, can be thought of as proceeding in a universe typified by light existing at zero speed only, or under the predominant weight of the 0-type being.

88. Seed-singularity-based beings such as the UPI-type being, K-type being, and N-type being can also be thought of as bound to P-type beings through a cosmological subtle-DNA.

89. Cosmological subtle-DNA is comprised of a downward-strand and an upward-strand.

90. The downward-strand is caused by light slowing down in quantized-decelerations, as it were, progressively concretizing more of the information in light.

91. The upward-strand is envisioned as prescribing a time-variable sequence wholly determined by the nature of the layers in the downward-strand.

92. P-type beings or partial-singularities remain "partial" until all the conditions specified by the logic in the upward-strand are fulfilled at which point the S-type being, or the Second Singularity, comes into being.

93. All P-type beings and the S-type being itself, while occurring in the downward-strand, are the result of the interaction of the downward-strand and the upward-strand.

94. The information-housing structure for all generated genetic-type information can be the pre-genetic material-fabric, the genetic

DNA, or some post-genetic hybrid structure that may surface with continuing iterations of being and becoming.

95. With each P-type being, specific logic-ecosystems are first encoded in FBLEE through a successful activation process involving space-time-energy-gravity quantization.

96. The outcome of successful activation involving space-time-energy-gravity quantization will culminate in a state of quantum-certainty.

97. FBLEE action that alters the quantum-level interface between the material and antecedent layers of light in effect changes the basis by which matter materializes.

98. All successful FBLEE action will alter natural or human history.

99. Genetic mutation is in its essence, an attempt to change FBLEE.

100. Genetic-type mutation, the process by which repositories of instruction change, allows the wholeness implicit in the seed-singularity-based beings to materialize as the many P-type beings with greater and greater spheres of influence, culminating in the foundation of the S-type being.

101. Constructive mutation is the only type that can successfully change FBLEE.

102. Constructive mutation implies an integration of U with light-layers traveling faster than c. It is hence an attempt to integrate the nature of a materialized construct with more of its antecedent potentiality.

103. Constructive mutation is a greater union of the P-type being with seed-singularity-based beings such as the UPI-type, K-type, or N-type being.

104. Destructive mutation will only affect local genetic structure precisely because it is unable to penetrate FBLEE.

105. Destructive mutation implies an integration of U with aspects of the layer of light existing at zero speed.

106. Destructive mutation is a greater union of the P-type being with the 0-type being.

107. The generation of the electro-magnetic-wavearchetype-masspotential P-type being follows the space-time-energy-gravity P-type beings in the S-type being biography.

108. The primary architecture of the electro-magnetic-wavearchetype-masspotential spectrum P-type being, and all subsequent P-type being architectures are determined by the light-space-time emergence equation.

109. The light-matrix in the light-space-time emergence equation delineates the essential layers of Light and the transformations that allow uniqueness-based diversity to materialize at U.
110. The essential electro (power), magnetic (harmony), wave-archetype (knowledge), mass-potential (presence) architecture, is specified by Light in its native state (the first line) in the light-matrix. Subsequent lines further detail this architecture.
111. The infinite variation in wave-archetype creating infinite number of seeds in the electro-magnetic-wavearchetype-masspotential P-type being, causes it to be spread out in spite of the impulse of union of the essential four-foldness in its architecture.
112. Mass containerizes in an attempt to realize its oneness when the action of space-time-energy-gravity quantization takes place.
113. Just as the space-time-energy-gravity quadrumvirate P-type micro-being has an impact on every material emergence, so too will the action of the electro-magnetic-wavearchetype-masspotential P-type being have an action on every emergence of life.
114. Segments of electro-magnetic-wavearchetype-masspotential P-type being code, due to variation in wavelength and frequency, creates diverse functional possibility in the intent, energetics, and mass-possibility of built-up life.
115. Cross-over of fourfold consistency from one type of P-type being to another, as in the case from the electro-magnetic-wavearchetype-masspotential P-type being, to matter-based P-type beings, to life-based P-type beings is the hallmark of light-based-singularities indicating that all that emerges in such a manner is of one light-based edifice.
116. The P-type being generator equation is holographic. That is, the same equation in its wholeness can be applied in the creation of the smallest emergences from Light.
117. The P-type being generator equation is fractal. That is, the same mathematical equation will generate material reality regardless of scale.
118. Light can be thought of as a creative medium and the P-type generator equation as modeling the specification of what is to be created.
119. Mathematical biographies for the quantum-particle P-type being, the boson P-type being, the atom P-type being, and the cell P-type being follow the generation of the electro-magnetic-wavearchetype-masspotential P-type being, and are also

expressed or generated through the light-space-time emergence equation.

120. The notion of U, or 'untransformed' as the bases in the formation of established-matter, comprising of all P-type beings prior to the human P-type being, simply implies that this is a baseline and is subject to all the adaptability yet to come about through the play of dynamics such as urge and will which increases with the generation of subsequent P-type beings.

121. Common DNA existing in every cell at the material layer can be thought of as influenced by up to four antecedent-entanglements: ∞-entanglement, K-entanglement, N-entanglement, and FBLEE-entanglement.

122. The light-space-time emergence or P-type being generator equation outputs genetic-type information summarized as a series of equations. Solutions to these generated equations detail the genetic-type information for that P-type being.

123. Life is a field that allows pre-existent complexity in Light to emerge materially.

124. Such pre-existent complexity materializes through an iterative process of being and becoming, the totality of which can be thought of as Life.

125. The light-space-time emergence or P-type being generator equation is iterative and will add to the biography of a previous P-type being to output a subsequent P-type being biography.

126. The time-matrix of the P-type being generator equation approximates when different layers of light are active, and importantly when the automaticity of meta-layer action yields more to dynamics such as urge and will.

127. Dynamics such as urge and will tend to become more active with the emergence of human P-type being.

128. As capacities of 'self', such as urge, will, desire, emotion, and thought emerge, the causal effect of these on the body, is presumed to be founded on genetic code that is created with the emergence of these capacities.

129. In pre-human P-type beings the actions of meta-layers is automatically more insistent as compared with human and post-human P-type beings.

130. Even human and post-human P-type beings are light-based singularities with deeper light-based dynamics always active

regardless of their being perceived. That after all, is the point of a light-based-singularity.

131. As P-type beings approach the human, strong wills and urges create an activation-state that can potentially open a quantum-window.

132. The space-matrix of the P-type being generator equation details the conditions that will allow meta-layers to become active at U.

133. A quantum-window provides access to FBLEE, and based on the levels of light active, can initiate space-time-energy-gravity quantization to change or add to the code in FBLEE.

134. All code in FBLEE is accessible by materialized constructs and can precipitate into genetic or even post-genetic structure.

135. Stable-mega-organization and sustainable-global-civilization P-type beings are milestones toward the S-type being.

136. In their current state both stable mega organization and the possibility of sustainable global civilization exist as FBLEE, and hence as the FBLEE-based stable-mega-organization P-type being and as the FBLEE-based sustainable-global-civilization P-type being only.

137. All close-to-human P-type beings have an impact on genetic structure. This is so because the four-fold functionality that all genetic-type information essentially expresses is enhanced with subsequent P-type beings.

138. Each P-type being biography essentially elucidates how fourfold functionality increases. All fourfold-functionality increasingly expresses potentiality in Light.

139. Super-matter-based P-type beings begin to express themselves when primary activity is meta-layer based. This will cause fourfold-functionality to increase dramatically and necessitate a different structuring of matter: super-matter.

140. Super-matter-based P-type beings is a foundation based on will or cohesive want and sets the stage for a potentially unending willed development in which functional-richness existing in Light in its native state traveling at infinite speed, can manifest in this 'material' universe.

141. Generation of super-matter-based P-type beings is likely linked to the expansion of the universe. This is because the increase in super-matter-based P-type beings is linked to increase in fourfold-functionality, which will require more seeds to express themselves. More seeds expressing themselves means more

Space is created. More Space means acceleration of an expanding universe continues.

142. Dark energy is hence related to an increase in cardinality of the combined set containing elements from all four manifested sets.

143. Manifested sets are a record of specific fourfold functionality elements that have materialized.

144. A persistent quantum-level computation resulting in a constant stream of genetic-type information alters the code and the very light-based computational machinery emergent as progressively more sophisticated pre-matter, matter, and post or super-matter-based P-type beings.

145. Matter is a dynamic state linked to the very Life of the universe.

146. In situations such as Black Holes or a Cosmic Bounce the large number of seeds that are normalized into a smaller number of seeds implies that space will be compressed, and time will slow down. This appears to be consistent with Einstein's General Theory of Relativity.

147. In a Big-Planet composite quantum, time, due to increased number of seeds acting independently, will be elongated or experienced faster.

148. The human P-type being is a master emergence, an unusually promising partial-singularity, in which by definition the micro and the macro become extraordinary reflections of each other, in a universe of Light where all is Light and all is of Light. This is so because 30 orders of magnitude lie on both sides of the human P-type being, into the micro and into the macro respectively.

149. If we, or any species that follows the human-species were to reach a point where we are unable to increase fourfold-functionality, it would mean that the materialization of potentiality in Light would discontinue and that the current complexification of four-foldness in this universe would likely need to be reconfigured in some fundamental way. In this case the universe would go through a collapse and possible re-ignition through a subsequent Cosmic Bounce.

150. It is also for this reason that any fragmented-singularity will be unable to sustain itself: essentially because being based on inherently finite-sets only, it will reach a limit to the four-foldness that can be expressed.

151. The notion of finite-sets is due to a de-linking with the infinity of information existing as light-based seeds.

152. Such a delinking generates a different genre of fragmented or derivative-beings. Such derivative-beings will be referred to as D-type beings.

153. D1-type beings, or first-order derivative beings, come into being with the initial delinking from infinite light-based possibility. A contemporary example of this are computer programs.

154. D2-type beings, or second-order derivative beings, are creations of D1-type beings.

155. D3-type beings, or third-order derivative beings, are creations of D2-type beings.

156. Mathematically, a D4-type being cannot be envisioned since all fragmented creations can be seen to be the result of the combination of D1-, D2-, and D3-type beings only.

157. The S-type being implies fully integrated and conscious four-foldness both horizontally and vertically. Reality at U becomes a powerful means to consciously and rapidly express four-foldness resident in Light's native state.

158. The S-type being implies a triumph of love, in which explicit fourfoldness seeks its implicit fourfold unity. This is true along both the horizontal and vertical dimensions.

159. The S-type being implies a stable and dynamic equilibrium where potentiality in Light is able to express itself in more profound, complete, exhilarating, and astounding formations.

160. Transhumanism can follow several lines of development. Transhumanism based on a foundation of D-type beings will never be as fulfilling as transhumanism based on a foundation of S-type beings.

161. Light in material form turning on itself can alter the very nature of what it is by what it is. Such is the future tale of Ouroboros changing from swallowing its tail to entwining itself through a Cosmos in a never-ending spiral toward unforeseen flowerings.

In its deepest essence Light can be envisioned as Love. This is because in Light there is a resplendent unity of properties and this unity is nothing other than love. Even as light materializes there is the urge to unite what is being projected from it. In its native state when light travels infinitely fast, the nature of the reality so created can be imagined to express omnipresence, omnipotence, omniscience, and omninurturance. This fourfoldness is held together in an embrace that continues, and expresses itself as love, regardless of how light projects itself.

In all mythology darkness is imagined as being a significant and apparent starting point in the expression of the glory of light, and in the iterations of the P-type being generator equation whose output creates life in a Cosmology of Light, this is no exception. Light projecting itself at speed-zero, engendering the 0-type being, whose nature, due to the complete absence of light can be envisioned as being the opposite of omnipresence, omnipotence, omniscience, and omninurturance. Weakness, ignorance, death, depravity, and darkness manifest instead, and create a weight that will have a dark influence on any subsequent projections of light. But such darkness has a profound effect in the fullness of light and love being able to express itself. For the gravitation toward such fundamental darkness can cause an existential yearning for that from which all comes, which is hidden deep within every iota of existence, and thus begins an incredible journey in which gradually more of the infinite possibility existing in light makes itself felt and seen and even known in more complete materializations of light.

Hence a projection of light at speed c can be thought of as immense act of Grace, of love, because now there is the possibility of a massive integration in a never-ending reality of love, in spite of, or on top of the possibility of the tremendous and dark negation and hidden foundation that is formed through the 0-type being. The Big Bang, the event in which light projects itself at c, initiates the massive journey of materialization and integration in which progressively, the universe experiences itself as its deep reality of infinity in light.

The space-time-energy-gravity quadrumvirate beings, themselves nothing other than an expression of the deep love in light, materialize both as the macro-container and the micro-priest by which the sacrifice and progressive surrender to love takes place.   For Space, Time, Energy, Gravity in their macro-aspect are nothing other than the Knowledge, Power, Presence, and Harmony in Light expressing itself as the macro-container of cosmos in which the huge adventure of the possibility of light-formed seeds will fructify. Space, time, energy, gravity having thus created the substance of the cosmos, in their micro-aspect as the space-time-energy-gravity P-type micro-being will preside over any further materialization in the cosmos as priests of the sacrifice in which P-type beings begin the long journey of iterating through becoming and being to materialize the infinity of Love-founded Life.

Hence it is nothing other than Light's Love expressed in a composite reality of fourfoldness that sets up the very playing-field and the means of progression in the playing-field that we call cosmos or the universe. Infinity in Light expresses itself in a series of quadrumvirate beings, that have to be quadrumvirate in order to be stable. For nothing can become a being unless it is based on the deep essence of love embodying the fourfold unity that animates light.

Hence the electro-magnetic-wavearchetype-masspotential P-type being is able to exist and express itself, and play its part in the adventure of becoming, because it is nothing other than an act of love made apparent as light begins its journey of materialization.   For the 'electro' expresses power, the 'magnetic' harmony, the 'wavearchetype' knowledge, and the 'masspotential' presence, each of which are properties of light expressing themselves explicitly as one being, to adhere to the fourfold implicitness of their reality of love in light.   Similarly the quantum-particle P-type being further expresses the essence of love in light by creating a stable fourfold foundation in which the quark expresses light's aspect of knowledge, the lepton, light's aspect of power, the boson, light's aspect of harmony, and the Higgs-boson, light's aspect of presence.

This journey continues through the myriad expressions of P-type beings, iterating through the atom, the cell, the molecule, the molecular plans in cells, the cell, and the human, as milestones of stability created through the reality of love that allows explicit expression of light's implicit and unified fourfoldness.

The human P-type being is a master creation and symbolizes the conscious turning of the universe materializing light automatically, to the universe beginning to materialize light consciously. Light in the human P-type being has to seek deep within itself and mobilize light by light to manifest more of the infinity in light materially. Such deeper seeking unveils deeper layers of light and can open the human P-type being to veritable gods, in the guise of the antecedent N-type, K-type, and UPI-type beings.

Through increasing influence of these seed-singularity-based antecedent beings, the dark and whispering influence of the 0-type being is diminished, and genetic mutation – the script that animates biographies of all material beings – begins to change, to record more constructive as opposed to destructive mutation. For constructive mutation occurs due to the integration of faster than c layers of light, that is, with the integration of the P-type being with increasing aspects of each or all of the UPI-type, K-type, and N-type beings. This recording always occurs in the presence of the cosmically ubiquitous priest – the space-time-energy-gravity P-type micro-being – who is able to engrave a deep enough yearning or act of surrender or sacrifice or display of deep love on FBLEE, the subtle quantum-level script, from which material law arises.

The more the engraving of such a nature, essentially invoking deeper and deeper dynamics of love, the more the vary basis of matter changes and progressively gives rise to super-matter-based P-type beings. In such beings the balance of normal dynamics is altered, so that the ubiquitous space-time-energy-gravity P-type micro-being constantly engraves, constantly changing the very nature of its sister the space-time-energy-gravity P-type macro-being, so that the creation of light-based seeds continues at an accelerated pace, and 'dark energy' pushes the universe to continue with rapid expansion.

For in its deepest reality love knows no bounds, and for more than the normal iota of it to express itself materially, requires a different kind of space, one in which cosmicity is able to express itself materially in each new being. For that is the nature of the S-type being – a material-cum-cosmic being – in which some astounding perception of a face of light's infinity expresses itself materially allowing an even more complete triumph of love.

REFERENCES

1. Alexander, S. 1920. Space, Time, and Deity. University of Michigan Library: Michigan.

2. Alexander, S. 2016. The Jazz of Physics. Basic Books: New York.

3. Arabatzis, T. 2006. Representing Electrons: A Biological Approach to Theoretical Entities. University of Chicago. Chicago

4. Berman, B. 2018. Zapped: From Infrared to X-rays, the Curious History of Invisible Light. Little, Brown and Company: New York

5. Beinhocker, E. 2006. The Origin of Wealth. Boston: Harvard Business School Press.

6. Bohm, D. 1983. Wholeness and the Implicate Order. London: Ark Paperbacks.

7. Carse, J. 1986. Finite and Infinite Games. Free Press: New York.

8. Carson, R. 1964. Silent Spring. Crest Books: New York.

9. Cardin, T. 2008. The Phenomenon of Man. Harper Perennial Modern Classics: New York.

10. Chown, M. 1990. Can Photons Travel Faster Than Light? New Scientist 126(1711)

11. Cottingham, W & Greenwood, D. 2007. An Introduction to the Standard Model of Particle Physics. Cambridge University Press. Cambridge

12. Darwin, C. 1947. On the Origin of Species by Means of Natural Selection: Or the Preservation of Favoured Races in the Struggle of Life. Pinnacle Press: London.

13. Deppe, A. 2013. Therapy with Light, a Practitioner's Guide. Strategic Book Publishing.

14. Diamond, J. 2005. Collapse: How Societies Choose to Fail or Succeed. Viking Books: New York

15. Einstein, A. 1995. Relativity: The Special and General Theory. New York: Broadway Books.

16. Feynman, RP. 1985. QED The Strange Theory of Light and Matter. New Jersey: Princeton University Press

17. Flannery, T. 2005. The Weather Makers: How Man is Changing the Climate and What it Means for Life on Erath. Atlantic Monthly Press: New York.

18. Fuller, B. 1982. Synergetics: Explorations in the Geometry of Thinking. MacMillan Publishing Co.: New York

19. Gebser, J. 1986. The Ever-Present Origin. Ohio University Press: Ohio.

20. Goodsell, David. 2010. The Machinery of Life. New York: Springer

21. Gottlieb, M. 2013. The Feynman Lectures on Physics: III. California Institute of Technology. http://www.feynmanlectures.caltech.edu/III_16.html

22. Gray, T. 2009. The Elements: A Visual Exploration of Every Known Atom in the Universe. Black Dog & Levental Publishers. New York.

23. Gubser, S. 2010. The Little Book of String Theory. Princeton University Press

24. Harvard-Edu. 2014. http://news.harvard.edu/gazette/story/2014/04/harvard-to-sign-on-to-united-nations-supported-principles-for-responsible-investment/

25. Harvard-Smithsonian Center for Astrophysics. 2004. https://www.cfa.harvard.edu/seuforum/de_whatmight.htm#

26. Hawking, Stephen. 1988. A Brief History of Time. New York: Bantam Books

27. Heiserman, D. 1991. Exploring Chemical Elements and their Compounds. McGraw-Hill. New York.

28. Holland, P. 1995. The Quantum Theory of Motion: An Account of the de Broglie-Bohm Causal Interpretation of Quantum Mechanics. Cambridge: Cambridge University Press.

29. Hope, Chris; Fowler, Stephen J. (2007). "A Critical Review of Sustainable Business Indices and Their Impact". Journal of Business Ethics. Vol. 76. S. 243–252. Springer: New York

30. Hyperphysics. 2016. Department of Physics and Astronomy. Georgia State University. http://hyperphysics.phy-astr.gsu.edu/hbase/mod3.html#c1

31. Isaacson, W. 2008. Einstein: His Life and Universe. Simon and Schuster. New York.

32. Jeans, J. 1932. The Mysterious Universe. Cambridge University Press.

33. Kaufmann, S. 1995. At Home in the Universe. New York: Oxford University Press.
34. Kaufmann, S. 2003. The Adjacent Possible, Edge. https://www.edge.org/conversation/the-adjacent-possible
35. Kleinman, R. 2006. The Four Faces of the Universe: An Integrated View of the Cosmos. Lotus Press: Twin Lakes.
36. Lanza, R., Berman, B. 2010. Biocentrism: How Life and Consciousness are the Keys to Understanding the True Nature of the Universe. Benbella Books: Texas
37. Lloyd, S. 2007. Programming the Universe: A Quantum Computer Scientist Takes On the Cosmos. New York: Vintage
38. Logue, A. 2008. Socially Responsible Investing for Dummies. Wiley Publishing: New Jersey
39. Lorentz, H.A. 1925. The Science of Nature. Vol. 25, p 1008. Springer
40. Malik, P. 2009. Connecting Inner Power with Global Change: The Fractal Ladder. New Delhi: Sage Publications
41. Malik, P. 2015. The Fractal Organization. New Delhi: Sage
42. Malik, P. 2017a. Doctoral Thesis. University of Pretoria Graduate School of Technology Management. https://repository.up.ac.za/handle/2263/62779?show=full
43. Malik, P. 2017b. A Story of Light. Amazon Kindle.
44. Malik, P. 2017c. Oceans of Innovation. Amazon Kindle.
45. Malik, P. 2017d. Emergence. Amazon Kindle.
46. Malik, P. 2017e. Quantum Certainty. Amazon Kindle.
47. Malik, P, Pretorius, L, Winzker, D. 2017. Qualified Determinism in Emergent-Technology Complex Adaptive Systems. IEEE TEMSCON.
48. Malik, P. 2018a. Super-Matter. Amazon Kindle.
49. Malik, P. 2018b. Cosmology of Light. Amazon Kindle.
50. Malik, P. 2018c. The Emperor's Quantum Computer. Amazon Kindle.
51. Malik, P, Pretorius, L. 2018. Symmetries of Light and Emergence of Matter. Indian Journal of Science & Technology. http://www.indjst.org/index.php/indjst/article/view/110789.
52. Martel, A. 2018. Light Therapies: A Complete Guide to the Healing Power of Light. Rochester, Vermont: Healing Arts Press
53. Malik, P. 2019a. The Origins and Possibilities of Genetics. Amazon Kindle.
54. Mora, C, Tittensor, D, Adl, S, Simpson, A, Worm, B. 2011. How Many Species Are There on Earth and in the Ocean? Plos.org.

http://journals.plos.org/plosbiology/article?id=10.1371/journal.pbio.1001127#abstract0

55. Narby, J. 1998. The Cosmic Serpent: DNA and the Origins of Knowledge. New York: Penguin Putnam.

56. NASA-darkmatter.   2016.
http://science.nasa.gov/astrophysics/focus-areas/what-is-dark-energy/

57. NASA-supernova. 2001. News Release Number: STScI-2001-09: "Blast from the Past: Farthest Supernova Ever Seen Sheds Light on   Dark   Universe".
http://hubblesite.org/newscenter/archive/releases/2001/09/

58. NASA-WMAP,   2014.
https://map.gsfc.nasa.gov/media/121238/index.html

59. Nowak, M., Highfield, R. 2011. Super Cooperators: Altruism, Evolution, and Why We Need Each Other to Succeed. New York: Free Press.

60. Olive, K.A et al. 2014. Particle Data Group. Chin. Phys. C, 38, 090001.

61. Openshaw,   J.   2015.
http://www.marketwatch.com/story/socially-responsible-investing-has-beaten-the-sp-500-for-decades-2015-05-21

62. Panek,R. 2010. Dark Energy: The Biggest Mystery in the Universe.   Smithsonian   Magazine.
http://www.smithsonianmag.com/science-nature/dark-energy-the-biggest-mystery-in-the-universe-9482130/?no-ist

63. Parker, A. 2003. In the Blink of an Eye: How Vision Sparked the Big Bang of Evolution. New York: Basic Books.

64. Particle Data Group. 2015. Lawrence Berkeley National Laboratory.
http://www.cpepphysics.org/main_universe/universe.html

65. Pauli, W. 1964. Nobel Lectures, Physics 1942 – 1962. Elsevier Publishing Company. Amsterdam.

66. Perkowitz, S. 2011. Slow Light. London: Imperial College Press

67. Planck, M. 1933. Where is Science Going? Ox Bow Press. Connecticut.

68. Portugali, J., 2012. Self-organization and the city. New York: Springer Science & Business Media.

69. Prigogine, I. 1977. Time, Structure, and Fluctuations. Nobelprize.org. Nobel Media AB 2014. Web. 5 Mar 2016.

<http://www.nobelprize.org/nobel_prizes/chemistry/laureate s/1977/prigogine-lecture.html>

70. Ridley, M. 1999. Genome: The Autobiography of a Species in 23 Chapters. New York: Harper Perennial.

71. Rovelli, C. 2017. Reality Is Not What It Seems. New York: Riverhead Books

72. Sams, G. 2009. Sun of gOd. Weiser Books: London

73. Snyder, M. 2010. Stanford Medicine. http://med.stanford.edu/news/all-news/2010/03/what-makes-you-unique-not-genes-so-much-as-surrounding-sequences-says-stanford-study.html#.html

74. Spitler, H. 2011. The Syntonic Principle: In Relation to Health and Ocular Problems. Eugene, Or: Resource Publications.

75. Sri Aurobindo. 1971. Social and Political Thought. Sri Aurobindo Ashram Press: Pondicherry

76. Stewart, Ian. 2012. In Pursuit of the Unknown. Basic Books. New York.

77. Smith, J, Szathmary, E. 1995. The Major Transitions in Evolution. Oxford: Oxford University Press

78. Toynbee, A. 1961. A Study of History, Volumes I – XII. Oxford University Press: Oxford

79. Turok, N. 2012. The Universe Within. Anasi Press: Ontario

80. Tweed, M. 2003. Essential Elements: Atoms, Quarks, and the Periodic Table. Walker & Copmany. New York.

81. Van Obbergen, P. 2014. Traite de Couleur Therapie Pratique. Paris: Guy Tredaniel Editeur.

82. Weil, A. 1996. Spontaneous Healing: How to Discover and Enhance Your Body's Natural Ability to Maintain and Heal Itself. Fawcett Books: New York.

83. Wheeler, J. Ford, K. 2000. Geons, Black Holes, and Quantum Foam: A Life in Physics. New York: W. W. Norton & Co.

84. Whitaker, A._2006. Einstein, Bohr and the Quantum Dilemma: From Quantum Theory to Quantum Information. Cambridge: Cambridge University Press

85. Wilczek, F. 2016. A Beautiful Question: Finding Nature's Deep Design. New York: Penguin Books

86. Willis, A. 2003. The Role of the Global Reporting Initiative's Sustainability Reporting Guidelines in the Social Screening of Investment. Journal of Business Ethics. Volume 43. Springer: New York

87. Wimmel, H. 1992. *Quantum Physics & Observed Reality: A Critical*

Interpretation of Quantum Mechanics. World Scientific. p. 2. Bibcode:1992qpor.book.....W. ISBN 978-981-02-1010-6.

88. Wolchover, N. 2013. A Jewel at the Heart of Quantum Physics. Quanta Magazine. https://www.quantamagazine.org/20130917-a-jewel-at-the-heart-of-quantum-physics/

89. Wright, R. 2009. Evolution of Compassion. https://www.ted.com/talks/robert_wright_the_evolution_of_compassion/transcript?language=en. TED 2009.

90. Yates, F.E. 2012. *Self-organizing systems: The emergence of order.* New York: Springer Science & Business Media.

91. Young, E. 2016. I Contain Multitudes: The Microbes Within Us and a Grander View of Life. New York: HarperCollins.

About the Author and his Books

Early Books

1. The Flowering of Management
2. India's Contribution to Management

Fractal Series

1. Connecting Inner Power with Global Change: The Fractal Ladder
2. Redesigning the Stock Market: A Fractal Approach
3. The Flower Chronicles: A Radical Approach to Systems and Organizational Development
4. The Fractal Organization: Creating Enterprises of Tomorrow

Cosmology of Light Series

1. A Story of Light: A Simple Exploration of the Creation and Dynamics of this Universe and Others
2. Oceans of Innovation: The Mathematical Heart of Complex Systems
3. Emergence: A Mathematical Journey from the Big Bang to Sustainable Global Civilization
4. Quantum Certainty: A Mathematics of Natural and Sustainable Human History
5. Super-Matter: Functional Richness in an Expanding Universe
6. Cosmology of Light: A Mathematical Integration of Matter, Life, History & Civilization, Universe, and Self

Applications of Cosmology of Light Series

1. The Emperor's Quantum Computer: An Alternative Light-Centered Interpretation of Quanta, Superposition, Entanglement and the Computing that Arises from it
2. The Origins and Possibilities of Genetics: A Mathematical Exploration in a Cosmology of Light

In the earlier stage I wrote 'The Flowering of Management' and 'India's Contribution to Management'. The impetus for both these books was similar in that they were reactions to the environment that I was placed in at the time. When I first began working in the corporate world the reality of the environment struck me as decidedly anachronistic. I had a different sense of what life could offer and wrote 'The Flowering of Management' to capture aspects of a vision I thought corporations and money existed for. Similarly, when I wrote 'India's Contribution to Management' it was the result of the dissatisfaction I experienced when confronted with the prevalent interpretation of India. This was precipitated by my working with a US-based company, A.T. Kearney, in India. I sought to reverse that interpretation with 'India's Contribution to Management' which aimed to capture my understanding of the essence and deeper capacity of synthesis of India.

The next phase was marked by a strong interest in fractals that primarily stemmed from my beginning to see similar patterns in seemingly distinct areas of life. I wrote 'Connecting Inner Power with Global Change: The Fractal Ladder' as a theoretical framework of fractals. The fractals that I envisioned included the added complexities of emotional and thought components. This was followed by 'Redesigning the Stock Market: A Fractal Approach' which was an application of the theoretical fractal framework to the then recent global financial crises of 2008. 'The Flower Chronicles' sought to make the gist of the ubiquitous fractal I had described in the previous two books easily graspable at the visceral level primarily through many practical examples spanning diverse walks of life. 'The Fractal Organization: Creating Enterprises of Tomorrow' was a comprehensive summary of the fractal framework that included the basic theory, the applications, and a practical field guide that had been developed while I was working at the Organizational Development group at Stanford University Medical Center.

The most recent phase has focused on creating a mathematical framework to take the previously developed fractal framework further. The development of such a mathematical framework that seeks to frame innovation in complex adaptive systems was also the focus of my doctoral work. This gave birth to an exciting period and will result in multiple series of books.

The first series, comprising of six books, extended my inquiry into mathematics and complex adaptive systems to an interesting limit culminating in the nature of Light and the Cosmos. The fractal mathematics I propose is at the heart of this series: Cosmology of Light.

The first book, 'A Story of Light: A Simple Exploration of the Creation and Dynamics of this Universe and Others' contains the main ideas in the mathematics, in non-mathematical terms, that are further explored mathematically in the remaining books in this series. The second book 'Oceans of Innovation: The Mathematical Heart of Complex Systems' describes my interpretation of the mathematical foundation of complex systems. The third book, 'Emergence: A Mathematical Journey from the Big Bang to Sustainable Global Civilization' applies the mathematics to several layers of matter and life. The fourth book, 'Quantum Certainty: A Mathematics of Natural and Sustainable Human History' describes a process culminating in space, time, energy, and gravity quantization by which history is made. The fifth book, 'Super-Matter: Functional Richness in an Expanding Universe' describes a process for the creation of super-matter-based on a need for continued functional-richness. A link is made between the resulting quantization of space and an expanding universe. The final book, 'Cosmology of Light: A Mathematical Integration of Matter, Life, History & Civilization, Universe, and Self' proposes an integrated mathematical framework that flows through all things, hence unifying matter, light, civilization, history, universe, and self.

The second series further explores the implications of "one mathematics flowing through all things". The first book in this series 'The Emperor's Quantum Computer: An Alternative Light-Centered Interpretation of Quanta, Superposition, Entanglement and the Computing that Arises from it' describes an alternative narrative for quantum computing to the one commonly expressed today. The second book in the series, 'The Origin and Possibilities of Genetics: A Mathematical Exploration in a Cosmology of Light' explores pre-genetic, genetic, and post-genetic possibilities in a cosmology of light. This book, 'The Second Singularity: A Mathematical Exploration of AI-Based and Other Singularities in a Cosmology of Light' explores the limits of AI-based singularities with respect to light-based singularities. This book, the final in this series explores transhumanism in a cosmology of light.

Pravir's meta-focus these days is on developing a unified theory and mathematics of organization with applications in a range of complex adaptive systems (CAS).

Pravir is the Head of Organizational Sciences at Zappos.com, and in this capacity is leading the creation and incubation of organizational development technologies, in support of establishing a resilient organization that will withstand the test of time.

In the past he has served as a Founding Member of A.T. Kearney India, a top-tier global consulting company; the Managing Director Advisory Services for BSR, a leading global CSR consulting company; as part of the HR leadership teams at Stanford Hospital & Clinics; leading Pricing Operations at Zappos.com; founding a couple of CAS-related initiatives - Deep Order Technologies and Aurosoorya; and experimenting with various forms of AI earlier in his carrier at ZS Associates and GE, amongst other organizations.

He has a Ph.D. in Technology Management with a focus on Mathematics of Innovation in Complex Adaptive Systems from University of Pretoria, an MBA from J.L. Kellogg Graduate School of Management with a focus on Marketing and Organizational Behavior, an MS in Computer Science from University of Florida with a focus on AI, and a BSE in Computer Engineering from Case Western Reserve University.

He is the author of a series of books on fractals and organizations, including 'Redesigning the Stock Market' and 'The Fractal Organization', published by SAGE, a leading global academic publisher. He recently authored a six-book series on 'Cosmology of Light' inspired by his research in CAS, and follow-up four-book series on 'Applications of Cosmology of Light' that explores the implications of "one mathematics" in all things. The books focus on AI, Quantum Computing, Genetics, AI, and Transhumanism, respectively. He has held faculty positions at several institutions of higher learning.

Pravir is a global citizen who has lived, worked, and been educated in many parts of the world.

- Amazon Author Page: https://www.amazon.com/Pravir-Malik/e/B002JVAEZE
- LinkedIn Profile: https://www.linkedin.com/in/pravirmalik/
- Forbes Page & Articles: https://www.forbes.com/sites/forbeshumanresourcescouncil/people/pravirmalik1/#1fa1097c17be
- Forbes Profile: https://profiles.forbes.com/members/hr/profile/Pravir-Malik-Head-Organizational-Sciences-Zappos/44463250-f2ab-434a-b1e2-0a6bdf54d970
- Google Scholar Page: https://scholar.google.com/citations?user=7DWWWZ8AAAAJ&hl=en
- Sage Author Page: https://us.sagepub.com/en-us/nam/author/pravir-malik
- YouTube Page: https://www.youtube.com/user/Aurosoorya
- Twitter: https://twitter.com/PravirMalik
- Research Gate Profile: https://www.researchgate.net/profile/Pravir_Malik
- Medium: https://medium.com/@PravirMalik
- Company website: http://www.deepordertechnologies.com/